Basic Algebraic Geometry 2

Igor R. Shafarevich

Basic Algebraic Geometry 2

Schemes and Complex Manifolds

Third Edition

Igor R. Shafarevich
Algebra Section
Steklov Mathematical Institute
of the Russian Academy of Sciences
Moscow, Russia

Translator
Miles Reid
Mathematics Institute
University of Warwick
Coventry, UK

ISBN 978-3-662-51401-6 ISBN 978-3-642-38010-5 (eBook)
DOI 10.1007/978-3-642-38010-5
Springer Heidelberg New York Dordrecht London

Mathematics Subject Classification (2010): 14-01

Translation of the 3rd Russian edition entitled "Osnovy algebraicheskoj geometrii". MCCME, Moscow 2007, originally published in Russian in one volume

Printed on acid-free paper

Springer is part of Springer Science+Business Media (www.springer.com)

Preface to Books 2–3

Books 2–3 correspond to Chapters V–IX of the first edition. They study schemes and complex manifolds, two notions that generalise in different directions the varieties in projective space studied in Book 1. Introducing them leads also to new results in the theory of projective varieties. For example, it is within the framework of the theory of schemes and abstract varieties that we find the natural proof of the adjunction formula for the genus of a curve, which we have already stated and applied in Section 2.3, Chapter 4. The theory of complex analytic manifolds leads to the study of the topology of projective varieties over the field of complex numbers. For some questions it is only here that the natural and historical logic of the subject can be reasserted; for example, differential forms were constructed in order to be integrated, a process which only makes sense for varieties over the (real or) complex fields.

Changes from the First Edition

As in the Book 1, there are a number of additions to the text, of which the following two are the most important. The first of these is a discussion of the notion of the algebraic variety classifying algebraic or geometric objects of some type. As an example we work out the theory of the Hilbert polynomial and the Hilbert scheme. I am very grateful to V.I. Danilov for a series of recommendations on this subject. In particular the proof of Theorem 6.7 is due to him. The second addition is the definition and basic properties of a Kähler metric, and a description (without proof) of Hodge's theorem.

Prerequisites

Varieties in projective space will provide us with the main supply of examples, and the theoretical apparatus of Book 1 will be used, but by no means all of it. Different sections use different parts, and there is no point in giving exact indications. References to the Appendix are to the Algebraic Appendix at the end of Book 1.

Prerequisites for the reader of Books 2–3 are as follows: for Book 2, the same as for Book 1; for Book 3, the definition of differentiable manifold, the basic theory of analytic functions of a complex variable, and a knowledge of homology, cohomology and differential forms (knowledge of the proofs is not essential); for Chapter 9, familiarity with the notion of fundamental group and the universal cover. References for these topics are given in the text.

Recommendations for Further Reading

For the reader wishing to go further in the study of algebraic geometry, we can recommend the following references.

For the cohomology of algebraic coherent sheaves and their applications: see Hartshorne [37].

An elementary proof of the Riemann–Roch theorem for curves is given in W. Fulton, Algebraic curves. An introduction to algebraic geometry, W.A. Benjamin, Inc., New York–Amsterdam, 1969. This book is available as a free download from http://www.math.lsa.umich.edu/~wfulton/CurveBook.pdf.

For the general case of Riemann–Roch, see A. Borel and J.-P. Serre, Le théorème de Riemann–Roch, Bull. Soc. Math. France **86** (1958) 97–136,

Yu.I. Manin, Lectures on the K-functor in algebraic geometry, Uspehi Mat. Nauk **24**:5 (149) (1969) 3–86, English translation: Russian Math. Surveys **24**:5 (1969) 1–89,

W. Fulton and S. Lang, Riemann–Roch algebra, Grundlehren der mathematischen Wissenschaften **277**, Springer-Verlag, New York, 1985.

Moscow, Russia I.R. Shafarevich

Contents

Book 1: Varieties in Projective Space

Book 2: Schemes and Varieties

Chapter 5
Schemes

In this chapter, we return to the starting point of all our study—the notion of algebraic variety—and attempt to look at it from a more general and invariant point of view. On the one hand, this leads to new ideas and methods that turn out to be exceptionally fertile even for the study of the quasiprojective varieties we have worked with up to now. On the other, we arrive in this way at a generalisation of this notion that vastly extends the range of application of algebraic geometry.

What prompts the desire to reconsider the definition of algebraic variety from scratch? Recalling how affine, projective and quasiprojective varieties were defined, we see that in the final analysis, they are all defined by systems of equations. One and the same variety can of course be given by different equations, and it is precisely the wish to get away from the fortuitous choice of the defining equations and the embedding into an ambient space that leads to the notion of isomorphism of varieties. Put like this, the framework of basic notions of algebraic geometry is reminiscent of the theory of finite field extensions at the time when everything was stated in terms of polynomials: the basic object was an equation and the idea of independence of the fortuitous choice of the equation was discussed in terms of the "Tschirnhaus transformation". In field theory, the invariant treatment of the basic notion considers a finite field extension $k \subset K$, which, although it can be represented in the form $K = k(\theta)$ with $f(\theta) = 0$ (for a separable extension), reflects properties of the equation $f = 0$ invariant under the Tschirnhaus transformation. As another parallel, one can point to the notion of manifold in topology, which was still defined right up to the work of Poincaré as a subset of Euclidean space, before its invariant definition as a particular case of the general notion of topological space.

The nub of this chapter and the next will be the formulation and study of the "abstract" notion of algebraic variety, independent of a concrete realisation. This idea thus plays the role in algebraic geometry of finite extensions in field theory or of the notion of topological space in topology.

The route by which we arrive at such a definition is based on two observations concerning the definition of quasiprojective varieties. In the first place, the basic notions (for example, regular map) are defined for quasiprojective varieties starting

I.R. Shafarevich, *Basic Algebraic Geometry 2*, DOI 10.1007/978-3-642-38010-5_1,

from their covers by affine open sets. Secondly, all the properties of an affine variety X are reflected in the ring $k[X]$, which is associated with it in an invariant way. These arguments suggest that the general notion of algebraic variety should in some sense reduce to that of affine variety; and that in defining affine varieties, one should start from rings of some special type, and define the variety as a geometric object associated with the ring.

It is not hard to carry out this program: in Chapter 1 we studied in detail how properties of an affine variety X are reflected in its coordinate ring $k[X]$, and this allows us to construct a definition of the variety X starting from some ring, which turns out after the event to be $k[X]$. However, proceeding in this way, we can get much more than the invariant definition of an affine algebraic variety. The point is that the coordinate ring of an affine variety is a very special ring: it is an algebra over a field, is finitely generated over it, and has no nilpotent elements. However, as soon as we have worked out a definition of affine variety based on some ring A satisfying these three conditions, the idea arises of replacing A in this definition by a completely arbitrary commutative ring. We thus arrive at a far-reaching generalisation of affine varieties. Since the general definition of algebraic variety reduces to that of an affine variety, it also is the subject of the same degree of generalisation. The general notion which we arrive at in this way is called a scheme.

The notion of scheme embraces a circle of objects incomparably wider than just algebraic varieties. One can point to two reasons why this generalisation has turned out to be exceptionally useful both for "classical" algebraic geometry and for other domains. First of all, the rings appearing in the definition of affine scheme are not now restricted to algebras over a field. For example, this ring may be a ring such as the ring of integers $\mathbb{Z}$, the ring of integers in an algebraic number field, or the polynomial ring $\mathbb{Z}[T]$. Introducing these objects allows us to apply the theory of schemes to number theory, and provides the best currently known paths for using geometric intuition in questions of number theory. Secondly, the rings appearing in the definition of affine scheme may now contain nilpotent elements. Using these schemes allows us, for example, to apply in algebraic geometry the notions of differential geometry related with infinitesimal movements of points or subvarieties $Y \subset X$, even when X and Y are quasiprojective varieties. And we should not forget that, as a particular case of schemes, we get the invariant definition of algebraic varieties which, as we will see, is much more convenient in applications, even when it does not lead to any more general notion.

Since we expect that the reader already has sufficient mastery of the technical material, we drop the usual "from the particular to the general" style of our book. Chapter 5 introduces the general notion of scheme and proves its simplest properties. In Chapter 6 we define "abstract algebraic varieties", which we simply call varieties. After this, we give a number of examples to show how the notions and ideas introduced in this chapter allow us to solve a number of concrete questions that have already occurred repeatedly in the theory of quasiprojective varieties.

1 The Spec of a Ring

1.1 Definition of Spec *A*

We start out on the program sketched in the introduction. We consider a ring A, always assumed to be commutative with 1, but otherwise arbitrary. We attempt to associate with A a geometric object, which, in the case that A is the coordinate ring of an affine variety X, should take us back to X. This object will at first only be defined as a set, but we will subsequently give it a number of other structures, for example a topology, which should justify its claim to be geometric.

The very first definition requires some preliminary explanations. Consider varieties defined over an algebraically closed field. If we want to recover an affine variety X starting from its coordinate ring $k[X]$, it would be most natural to use the relation between subvarieties $Y \subset X$ and their ideals $\mathfrak{a}_Y \subset k[X]$. In particular a point $x \in X$ corresponds to a maximal ideal $\mathfrak{m}_x$, and it is easy to check that $x \mapsto \mathfrak{m}_x \subset k[X]$ establishes a one-to-one correspondence between points $x \in X$ and the maximal ideals of $k[X]$. Hence it would seem natural that the geometric object associated with any ring A should be its set of maximal ideals. This set is called the *maximal spectrum* of A and denoted by m-Spec A. However, in the degree of generality in which we are now considering the problem, the map $A \mapsto$ m-Spec A has certain disadvantages, one of which we now discuss.

It is obviously natural to expect that the map sending A to its geometric set should have the main properties that relate the coordinate ring of an affine algebraic variety with the variety itself. Of these properties, the most important is that homomorphisms of rings correspond to regular maps of varieties. Is there a natural way of associating with a ring homomorphism $f: A \to B$ a map of m-Spec B to m-Spec A? How in general does one send an ideal $\mathfrak{b} \subset B$ to some ideal $\mathfrak{b} \subset A$? There is obviously only one reasonable answer, to take the inverse image $f^{-1}(\mathfrak{b})$. But the trouble is that the inverse image of a maximal ideal is not always maximal. For example, if A is a ring with no zerodivisors that is not a field, and $f: A \hookrightarrow K$ an inclusion of A into a field, then the zero ideal (0) in K is the maximal ideal of K, but its inverse image is the zero ideal (0) in A, which is not maximal.

This trouble does not occur if instead of maximal ideals we consider prime ideals: it is elementary to check that the inverse image of a prime ideal under any ring homomorphism is again prime. In the case that $A = k[X]$ is the coordinate ring of an affine variety X, the set of prime ideals of A has a clear geometric meaning: it is the set of irreducible closed subvarieties of X (points, irreducible curves, irreducible surfaces, and so on). Finally, for a very large class of rings the set of prime ideals is determined by the set of maximal ideals (see Exercise 8). All of this motivates the following definition.

Definition The set of prime ideals of A is called its *prime spectrum* or simply *spectrum* , and denoted by Spec A. Prime ideals are called *points* of Spec A.

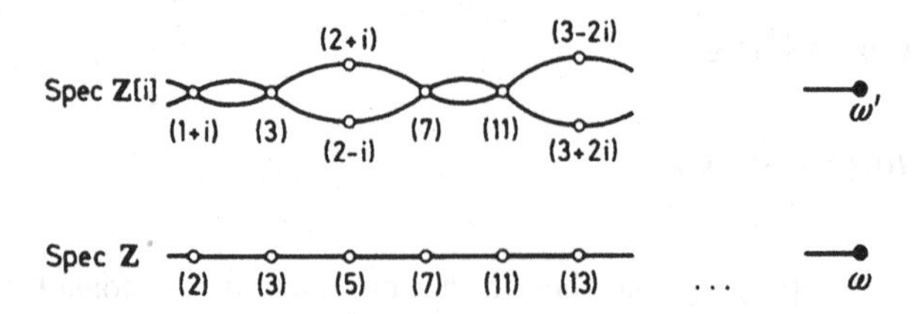

Figure 22 ${}^a\varphi$: Spec($\mathbb{Z}[i]$) $\to$ Spec $\mathbb{Z}$

Since we only consider rings with a 1, the ring itself is not counted as a prime ideal. This is in order that the quotient ring A/P by a prime ideal should always be an integral domain, that is, a subring of a field (with $0 \neq 1$). Every nonzero ring A has at least one maximal ideal. This follows from Zorn's lemma (see for example Atiyah and Macdonald [8, Theorem 1.3]); thus Spec A is always nonempty for $A \neq 0$.

We have already discussed the geometric meaning of Spec A when $A = k[X]$ is the coordinate ring of an affine variety. We consider some other examples.

Example 5.1 Spec $\mathbb{Z}$ consists of the prime ideals (2), (3), (5), (7), (11), ..., and the zero ideal (0).

Example 5.2 Let $\mathcal{O}_x$ be the local ring of a point x of an irreducible algebraic curve. Then Spec $\mathcal{O}_x$ consists of two points, the maximal ideal and the zero ideal.

Consider a ring homomorphism $\varphi\colon A \to B$. In what follows we always consider only homomorphisms that take $1 \in A$ into $1 \in B$. As we remarked above, the inverse image of any prime ideal of B is a prime ideal of A. Sending a prime ideal of B into its inverse image thus defines a map

$$ {}^a\varphi\colon \operatorname{Spec} B \to \operatorname{Spec} A, $$

called the *associated map* of φ.

As a useful exercise, the reader might like to think through the map Spec($\mathbb{C}[T]$)$\to$ Spec($\mathbb{R}[T]$) associated with the inclusion $\mathbb{R}[T] \hookrightarrow \mathbb{C}[T]$.

Example 5.3 We consider the ring $\mathbb{Z}[i]$ with $i^2 = -1$, and try to imagine its prime spectrum Spec($\mathbb{Z}[i]$), using the inclusion map $\varphi\colon \mathbb{Z} \to \mathbb{Z}[i]$. This defines a map

$$ {}^a\varphi\colon \operatorname{Spec}\bigl(\mathbb{Z}[i]\bigr) \to \operatorname{Spec} \mathbb{Z}. $$

We write $\omega = (0) \in \operatorname{Spec} \mathbb{Z}$ and $\omega' = (0) \in \operatorname{Spec}(\mathbb{Z}[i])$ for the points of Spec $\mathbb{Z}$ and Spec($\mathbb{Z}[i]$) corresponding to the zero ideals. Obviously ${}^a\varphi(\omega') = \omega$ and $({}^a\varphi)^{-1}(\{\omega\}) = \{\omega'\}$.

The other points of Spec $\mathbb{Z}$ correspond to the prime numbers. By definition, $({}^a\varphi)^{-1}(\{(p)\})$ is the set of prime ideals of $\mathbb{Z}[i]$ that divide p. As is well known, all such ideals are principal, and there are two of them if $p \equiv 1 \bmod 4$, and only one if $p = 2$ or $p \equiv 3 \bmod 4$. All of this can be pictured as in Figure 22.

We recommend the reader to work out the more complicated example of $\operatorname{Spec}(\mathbb{Z}[T])$, using the inclusion $\mathbb{Z} \hookrightarrow \mathbb{Z}[T]$.

Example 5.4 Recall that a subset $S \subset A$ is a *multiplicative set* if it contains 1 and is closed under multiplication. For every multiplicative set, we can construct a ring of fractions A_S consisting of pairs (a, s) with $a \in A$ and $s \in S$, identified according to the rule

$$(a, s) = (a', s') \iff \exists\, s'' \in S \text{ such that } s''(as' - a's) = 0.$$

Algebraic operations are defined by the rules

$$(a, s) + (a', s') = (as' + a's, ss'),$$
$$(a, s)(a', s') = (aa', ss').$$

The reader will find a more detailed description of this construction in Atiyah and Macdonald [8, Chapter 3]. From now on we write a/s for the pair (a, s). In particular, if S is the set $A \setminus \mathfrak{p}$, where $\mathfrak{p}$ is a prime ideal of A then A_S coincides with the local ring $A_\mathfrak{p}$ of A at a prime ideal (compare Section 1.1, Chapter 2).

There is a map $\varphi\colon A \to A_S$ defined by $a \mapsto (a, 1)$, and hence a map

$${}^a\varphi\colon \operatorname{Spec}(A_S) \to \operatorname{Spec} A.$$

The reader can easily check that ${}^a\varphi$ is an inclusion, and that its image ${}^a\varphi(\operatorname{Spec}(A_S)) = U_S$ is the set of prime ideals of A disjoint from S. The inverse map $\psi\colon U_S \to \operatorname{Spec}(A_S)$ is of the form

$$\psi(\mathfrak{p}) = \mathfrak{p}A_S = \{x/s \mid x \in \mathfrak{p} \text{ and } s \in S\}.$$

In particular, if $f \in A$ and $S = \{f^n \mid n = 0, 1, \dots\}$ then A_S is denoted by A_f.

1.2 *Properties of Points of* Spec A

We can associate with each point $x \in \operatorname{Spec} A$ the field of fractions of the quotient ring by the corresponding prime ideal. This field is called the *residue field* at x and denoted by $k(x)$. Thus we have a homomorphism

$$A \to k(x),$$

whose kernel is the prime ideal we are denoting by x. We write $f(x)$ for the image of $f \in A$ under this homomorphism. If $A = k[X]$ is the coordinate ring of an affine variety X defined over an algebraically closed field k then $k(x) = k$, and for $f \in A$ the element $f(x) \in k(x)$ defined above is the value of f at x. In the general case each element $f \in A$ also defines a "function"

$$x \mapsto f(x) \in k(x)$$

on Spec A, but with the peculiarity that at different points x, it takes values in different sets. For example, when $A = \mathbb{Z}$, we can view any integer as a "function", whose value at (p) is an element of the field $\mathbb{F}_p = \mathbb{Z}/(p)$, and at (0) is an element of the rational number field $\mathbb{Q}$.

We now come up against one of the most serious points at which the "classical" geometric intuition turns out to be inapplicable in our more general situation. The point is that an element $f \in A$ is not always uniquely determined by the corresponding function on Spec A. For example, an element corresponds to the zero function if and only if it is contained in all prime ideals of A. These elements are very simple to characterise.

Proposition *An element $f \in A$ is contained in every prime ideal of A if and only if it is nilpotent* (*that is*, $f^n = 0$ *for some* n).

Proof See Proposition A.10 of Section 6, Appendix,[1] or Atiyah and Macdonald [8, Proposition 1.8]. □

Thus the inapplicability of the "functional" point of view in the general case is related to the presence of nilpotents in the ring. The set of all nilpotent elements of a ring A is an ideal, the *nilradical* of A.

For each point $x \in \operatorname{Spec} A$ there is a local ring $\mathcal{O}_x$, the local ring of A at the prime ideal x. For example, if $A = \mathbb{Z}$ and $x = (p)$ with p a prime number, then $\mathcal{O}_x$ is the ring of rational numbers a/b with denominator b coprime to p; if $x = (0)$ then $\mathcal{O}_x = \mathbb{Q}$.

This invariant of a point of Spec A allows us to extend to our general case a whole series of new geometric notions. For example, the definition of nonsingular points of a variety was related to purely algebraic properties of their local rings (Section 1.3, Chapter 2). This prompts the following definition.

Definition A point $x \in \operatorname{Spec} A$ is *regular* (or *simple*) if the local ring $\mathcal{O}_x$ is Noetherian and is a regular local ring (see Section 2.1, Chapter 2 or Atiyah and Macdonald [8, Theorem 11.22]).

Recall that in general $\operatorname{Spec} A \neq \text{m-}\operatorname{Spec} A$. Suppose that $A = k[X]$ and a point of Spec A corresponds to a prime ideal that is not maximal, that is, to an irreducible subvariety $Y \subset X$ of positive dimension. What is the geometric meaning of regularity of such a point? As the reader can easily check (using Theorem 2.13 of Section 3.2, Chapter 2), in this case, regularity means that Y is not contained in the subvariety of singular points of X.

Let $\mathfrak{m}_x$ be the maximal ideal of the local ring $\mathcal{O}_x$ of a point $x \in \operatorname{Spec} A$. Then obviously

$$\mathcal{O}_x/\mathfrak{m}_x = k(x),$$

[1] Appendix refers to the Algebraic Appendix at the end of Book 1.

and the group $\mathfrak{m}_x/\mathfrak{m}_x^2$ is a vector space over $k(x)$. If $\mathcal{O}_x$ is Noetherian (for example, if A is Noetherian), then this space is finite dimensional. The dual vector space

$$\Theta_x = \mathrm{Hom}_{k(x)}\big(\mathfrak{m}_x/\mathfrak{m}_x^2, k(x)\big)$$

is called the *tangent space* to Spec A at x.

Example 5.5 If A is the ring of integers of an algebraic number field K, for example, $A = \mathbb{Z}$, $K = \mathbb{Q}$, then Spec A consists of maximal ideals together with (0). For $x = (0)$, we have $\mathcal{O}_x = K$ and hence x is regular, with 0-dimensional tangent space. If $x = \mathfrak{p} \neq (0)$ then it is known that $\mathcal{O}_x$ is a principal ideal domain. Hence these points are also regular, with 1-dimensional tangent spaces.

Example 5.6 To find points that are not regular, consider the ring $A = \mathbb{Z}[mi] = \mathbb{Z}[y]/(y^2 + m^2) = \mathbb{Z} + \mathbb{Z}mi$, where $m > 1$ is an integer and $i^2 = -1$. The inclusion $\varphi\colon A \hookrightarrow A' = \mathbb{Z}[i]$ defines a map

$$^a\varphi\colon \mathrm{Spec}\big(A'\big) \to \mathrm{Spec}\, A. \tag{5.1}$$

If we restrict to prime ideals coprime to m then this is a one-to-one correspondence, and it is easy to check that the local rings of corresponding prime ideals are equal. Hence a point $x \in \mathrm{Spec}\, A'$ is not regular only if the prime ideal divides m. Prime ideals of A' dividing m are in one-to-one correspondence with prime factors p of m, and are given by $\mathfrak{p} = (p, mi)$. In this case, $k(x) = \mathbb{F}_p$ is the field of p elements, and $\mathfrak{m}_x/\mathfrak{m}_x^2 = \mathfrak{p}/\mathfrak{p}^2$ is a 2-dimensional $\mathbb{F}_p$-vector space, since $\mathfrak{p}^2 \subset (p)$. Hence $\mathfrak{m}_x$ is not principal, and the local ring $\mathcal{O}_x$ is not regular. Thus all the prime ideals $\mathfrak{p} = (p, mi)$ with $p \mid m$ are singular points of Spec A. The map (5.1) is a resolution of these singularities.

Now that we have defined tangent spaces, it would be natural to proceed to differential forms. The algebraic description of differential forms treated in Section 5.2, Chapter 3, allows us to carry it over to more general rings. In what follows we do not require these constructions, and we will not study them in more detail.

1.3 The Zariski Topology of Spec *A*

The topological notions that we used in connection with algebraic varieties suggest how to put a topology on the set Spec A. For this, we associate with any set $E \subset A$ the subset $V(E) \subset \mathrm{Spec}\, A$ consisting of prime ideals $\mathfrak{p}$ such that $E \subset \mathfrak{p}$. We have the obvious relations

$$V\Big(\bigcup_\alpha E_\alpha\Big) = \bigcap_\alpha V(E_\alpha),$$

and

$$V(I) = V(E') \cup V(E''), \quad \text{where } I = (E') \cap (E'');$$

(that is, I is the intersection of the ideals of A generated by E' and E''). They show that the sets $V(E)$ corresponding to different subsets $E \subset A$ satisfy the axioms for the system of closed sets of a topological space.

Definition The topology on Spec A in which the $V(E)$ are the closed sets is called the *Zariski topology* or *spectral topology*.

In what follows, whenever we refer to Spec A as a topological space, the Zariski topology is always intended. For a homomorphism $\varphi\colon A \to B$ and any set $E \subset A$ we have

$$({}^a\varphi)^{-1}(V(E)) = V(\varphi(E)),$$

from which it follows that the inverse image of a closed set under ${}^a\varphi$ is closed. This shows that ${}^a\varphi$ is a continuous map.

As an example, consider the natural homomorphism $\varphi\colon A \to A/\mathfrak{a}$, where $\mathfrak{a}$ is an ideal of A. Obviously ${}^a\varphi$ is a homeomorphism of $\operatorname{Spec}(A/\mathfrak{a})$ to the closed set $V(\mathfrak{a})$. Any closed subset of Spec A is of the form $V(E) = V(\mathfrak{a})$, where $\mathfrak{a} = (E)$ is the ideal generated by E. Hence every closed subset of Spec A is homeomorphic to Spec of a ring.

We consider another example. For $S \subset A$ a multiplicative set, we let

$$\varphi\colon A \to A_S, \quad U_S = {}^a\varphi(\operatorname{Spec}(A_S)) \quad \text{and} \quad \psi\colon U_S \to \operatorname{Spec}(A_S)$$

be the sets and maps introduced in Example 5.4. We give $U_S \subset \operatorname{Spec} A$ the subspace topology, that is, its closed subsets are of the form $V(E) \cap U_S$. A simple verification shows that not only ${}^a\varphi$, but also ψ is continuous, so that, in other words, $\operatorname{Spec}(A_S)$ is homeomorphic to the subspace $U_S \subset \operatorname{Spec} A$.

Especially important is the special case when $S = \{f^n \mid n = 0, 1, \dots\}$, with $f \in A$ an element that is not nilpotent. Here U_S is the open set $U_S = \operatorname{Spec} A \setminus V(f)$, where $V(f) = V(E)$ with $E = \{f\}$. The open sets of the form $\operatorname{Spec} A \setminus V(f)$ are called *principal open sets*. They are denoted by $D(f)$. It is easy to check that they form a basis for the open sets of the Zariski topology (because every closed set is of the form $V(E) = \bigcap_{f \in E} V(f)$). As in the case of affine varieties, the significance of the principal open sets is that $D(f)$ is homeomorphic to $\operatorname{Spec} A_f$. Using these open sets, one can prove the following important property of Spec A.

Proposition Spec A *is compact.*

Proof We have to prove that given any cover of Spec A by open sets, we can choose a finite subcover. Since principal open sets $D(f)$ form a basis for the open sets of the topology, it is enough to prove this for a cover $\operatorname{Spec} A = \bigcup_\alpha D(f_\alpha)$. This condition means that $\bigcap_\alpha V(f_\alpha) = V(\mathfrak{a}) = \emptyset$, where $\mathfrak{a}$ is the ideal of A generated by all the elements f_α. In other words, there does not exist any prime ideal containing $\mathfrak{a}$; this means that $\mathfrak{a} = A$. But then there exist $f_{\alpha_1}, \dots, f_{\alpha_r}$ and $g_1, \dots, g_r \in A$ such that

$$f_{\alpha_1} g_1 + \cdots + f_{\alpha_r} g_r = 1.$$

From this it follows in turn that $(f_{\alpha_1}, \dots, f_{\alpha_r}) = A$, that is, $\operatorname{Spec} A = D(f_{\alpha_1}) \cup \cdots \cup D(f_{\alpha_r})$. The proposition is proved. □

The Zariski topology is a very "nonclassical" topology; more precisely, it is non-Hausdorff. We have already met this kind of property of affine varieties, for example in Chapter 1: on an irreducible variety, any two nonempty open sets intersect. This property means that the Hausdorff separability axiom is not satisfied: there exist two distinct points, all neighbourhoods of which intersect. But it is "even less Hausdorff" due to the fact that $\operatorname{Spec} A$ includes not just maximal ideals, but all prime ideals: because of this, it contains nonclosed points.

Let us determine what the closure of a point of $\operatorname{Spec} A$ looks like. If our point is the prime ideal $\mathfrak{p} \subset A$ then its closure is $\bigcap_{\{E \supset \mathfrak{p}\}} V(E) = V(\mathfrak{p})$, that is, it consists of all prime ideals $\mathfrak{p}'$ with $\mathfrak{p} \subset \mathfrak{p}'$, and is homeomorphic to $\operatorname{Spec} A/\mathfrak{p}$. In particular, a prime ideal $\mathfrak{p} \subset A$ is a closed point of $\operatorname{Spec} A$ if and only if $\mathfrak{p}$ is a maximal ideal. If A does not have zerodivisors then (0) is prime, and is contained in every prime ideal. Thus its closure is the whole space; (0) is an everywhere dense point.

If a topological space has nonclosed points then there is a certain hierarchy among its points, that we formulate in the following definition: x is a *specialisation* of y if x is contained in the closure of y. An everywhere dense point is called a *generic point* of a space.

When does $\operatorname{Spec} A$ have an everywhere dense point? As we saw in the preceding section, the intersection of all prime ideals $\mathfrak{p} \subset A$ consists of all the nilpotent elements of A, that is, it is the nilradical. If this is a prime ideal then it defines a point of $\operatorname{Spec} A$; but any prime ideal must contain all nilpotent elements, that is, must contain the nilradical. Hence $\operatorname{Spec} A$ has a generic point if and only if its nilradical is prime. The generic point is unique, and is the point defined by the nilradical.

Example Let $\mathcal{O}_x$ be the local ring of a nonsingular point of an algebraic curve (Example 5.2). Then the ideal (0) is the generic point of $\operatorname{Spec} \mathcal{O}_x$, and is open; and the maximal ideal $\mathfrak{m}_x$ a closed point. A picture:

$$\operatorname{Spec} \mathcal{O}_x = \underset{(0)}{\circ} \rightarrow \underset{\mathfrak{m}_x}{\bullet}$$

1.4 Irreducibility, Dimension

The existence of a generic point relates to an important geometric property of X. Namely, a topological space certainly does not have a generic point if it can be written in the form $X = X_1 \cup X_2$, where $X_1, X_2 \subsetneqq X$ are closed sets. A space of this form is *reducible*.

For $\operatorname{Spec} A$, irreducibility is not only a necessary, but also sufficient condition for the existence of a generic point. Indeed, it is enough to prove that if $\operatorname{Spec} A$ is irreducible then the nilradical of A is prime; for, as we said above, it already follows from this that a generic point exists. Suppose that the nilradical N is not prime, and that $fg \in N$, with $f, g \notin N$. Then

$$\operatorname{Spec} A = V(f) \cup V(g) \quad \text{with } V(f), V(g) \neq \operatorname{Spec} A,$$

that is, $\operatorname{Spec} A$ is reducible.

Since every closed set of $\operatorname{Spec} A$ is also homeomorphic to Spec of a ring, the same result carries over to any closed subset. Thus there exists a one-to-one correspondence between points and irreducible closed subsets of $\operatorname{Spec} A$, which sends each point to its closure.

The notion of a reducible space leads us at once to decomposition into irreducible components. If A is a Noetherian ring, then there exists a decomposition

$$\operatorname{Spec} A = X_1 \cup \cdots \cup X_r,$$

where X_i are irreducible closed subsets and $X_i \not\subset X_j$ for $i \neq j$, and this decomposition is unique. The proof of this fact repeats word-for-word the proof of the corresponding assertions for affine varieties (Theorem 1.4 of Section 3.1, Chapter 1), which depended only on $k[X]$ being Noetherian.

Example 5.7 The simplest example of a decomposition of $\operatorname{Spec} A$ into irreducible components is the case of a ring A that is a direct sum of a number of rings having no zerodivisors:

$$A = A_1 \oplus \cdots \oplus A_r.$$

In this case, one checks easily that $\operatorname{Spec} A$ is a disjoint union of irreducible components $\operatorname{Spec}(A_i)$.

Example 5.8 To consider a slightly less trivial example, take the group ring $\mathbb{Z}[\sigma]$ of the cyclic group of order 2:

$$A = \mathbb{Z}[\sigma] = \mathbb{Z} + \mathbb{Z}\sigma, \quad \text{with } \sigma^2 = 1.$$

The nilradical of A equals (0), but this is not a prime ideal, since A has zerodivisors: $(1+\sigma)(1-\sigma) = 0$. Hence

$$\operatorname{Spec} A = X_1 \cup X_2, \quad \text{where } X_1 = V(1+\sigma) \text{ and } X_2 = V(1-\sigma). \tag{5.2}$$

The homomorphisms $\varphi_1, \varphi_2 \colon A \to \mathbb{Z}$ with kernels $(1+\sigma)$ and $(1-\sigma)$ define homeomorphisms

$$^a\varphi_1 \colon \operatorname{Spec} \mathbb{Z} \to V(1+\sigma),$$
$$^a\varphi_2 \colon \operatorname{Spec} \mathbb{Z} \to V(1-\sigma),$$

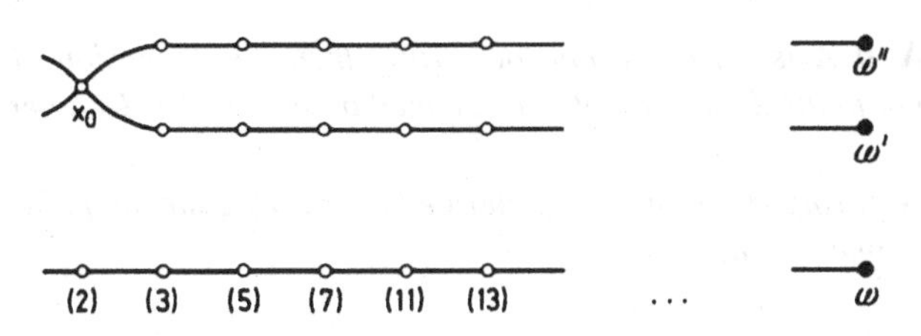

Figure 23 ${}^a\varphi$: $\operatorname{Spec}(\mathbb{Z}[\sigma]/(\sigma^2-1)) \to \operatorname{Spec}\mathbb{Z}$

which show that X_1 and X_2 are irreducible, so that (5.2) is a decomposition of $\operatorname{Spec} A$ into irreducible components.

Let us find the intersection $X_1 \cap X_2$. Obviously

$$X_1 \cap X_2 = V(1+\sigma, 1-\sigma) = V(\mathfrak{a}),$$

where $\mathfrak{a}$ is the ideal $\mathfrak{a} = (1+\sigma, 1-\sigma) = (2, 1-\sigma)$. Since $A/\mathfrak{a} \cong \mathbb{Z}/2$, we see that $\mathfrak{a}$ is a maximal ideal and hence X_1 and X_2 intersect at a unique point $x_0 = X_1 \cap X_2$. It is easy to check that if $x \neq x_0$, for example $x \in X_1$, $x \notin X_2$ then φ_1 establishes an isomorphism of the local rings at the points x and $\varphi_1(x)$. Hence all points $x \neq x_0$ are regular. x_0 is singular, with $\dim \Theta_{x_0} = 2$, and if y_1, y_2 are the points $y_1 = ({}^a\varphi_1^{-1}(x_0))$ and $y_2 = ({}^a\varphi_2^{-1}(x_0))$ then we get that $\mathrm{d}_{y_1}\Theta_{y_1}$ and $\mathrm{d}_{y_2}\Theta_{y_2}$ are two distinct lines in $\Theta_{x_0} \cong \mathbb{F}_2^2$, so that x_0 is a "double point with distinct tangents" of $\operatorname{Spec} A$.

It is convenient to picture $\operatorname{Spec} A$ using the map ${}^a\varphi$: $\operatorname{Spec} A \to \operatorname{Spec}\mathbb{Z}$ where $\mathbb{Z} \hookrightarrow A$ is the natural inclusion (in the same way that we considered $\operatorname{Spec}\mathbb{Z}[i]$ in Example 5.3). We get the picture of Figure 23.

Among the purely topological properties of an affine variety X, that is, the properties that are completely determined by the Zariski topology of $\operatorname{Spec}(k[X])$, we can include the dimension of X. Of course, the definition given in Chapter 1 in terms of the transcendence degree of $k(X)$ uses very specific properties of the ring $k[X]$: it is a k-algebra, it can be embedded in a field, and this field has finite transcendence degree over k. However, Theorem 1.22 and Corollary 1.5 of Section 6.2, Chapter 1 put the definition into a form that can be applied to any topological space.

Definition The *dimension* of a topological space X is the number n such that X has a chain of irreducible closed sets

$$\emptyset \neq X_0 \subsetneqq X_1 \subsetneqq \cdots \subsetneqq X_n,$$

and no such chain with more than n terms.

Of course, not every topological space has finite dimension. This is false in general for $\operatorname{Spec} A$, even if A is Noetherian. Nevertheless there is a series of important types of rings for which the dimension of $\operatorname{Spec} A$ is finite. In this case it is called the *dimension* of A. We run through three of the basic results without proof; for the proofs, see, for example, Atiyah and Macdonald, [8, Chapter 11].

Proposition A *If A is a Noetherian local ring, then the dimension of* Spec A *is finite, and equal to the dimension of A as defined in Section* 2.1, *Chapter* 2.

Proposition B *A ring that is finitely generated over a ring having finite dimension is again finite dimensional.*

Proposition C *If A is Noetherian then*

$$\dim A[T_1, \ldots, T_n] = \dim A + n.$$

Example 5.9 $\mathbb{Z}$ has dimension 1. More generally, the ring of integers of an algebraic number field has dimension 1, so that in it, any prime ideal other than (0) is maximal.

Example 5.10 To give an example of a ring of bigger dimension, consider the case $A = \mathbb{Z}[T]$. Since we expect that the reader has already worked out the structure of $\operatorname{Spec} \mathbb{Z}[T]$ as an exercise in Section 1.1, we assume it known. It is very simple: a maximal ideal is of the form $(p, f(T))$ where p is a prime and $f \in \mathbb{Z}[T]$ a polynomial whose reduction modulo p is irreducible; nonzero prime ideals that are not maximal are principal, and of the form (p) or $(f(T))$, where f is an irreducible polynomial. It follows that the chains of prime ideals of maximal length are as follows:

$$\big(p, f(T)\big) \supset \big(g(t)\big) \supset (0) \quad \text{or} \quad \big(p, f(T)\big) \supset (p) \supset (0).$$

Thus $\dim \mathbb{Z}[T] = 2$, in agreement with Proposition C.

1.5 Exercises to Section 1

1 Let N be the nilradical of a ring A and $\varphi \colon A \to A/N$ the quotient map. Prove that the associated map ${}^a\varphi \colon \operatorname{Spec}(A/N) \to \operatorname{Spec} A$ is a homeomorphism.

2 Prove that a nonzero element $f \in A$ is a zerodivisor if and only if there exists a decomposition $\operatorname{Spec} A = X \cup X'$ where $X \subset \operatorname{Spec} A$ and $X' \subsetneqq \operatorname{Spec} A$ such that $f(x) = 0$ for all $x \in X$. (If f is nilpotent then $X = \operatorname{Spec} A$; if f is not nilpotent, then $X, X' \subsetneqq \operatorname{Spec} A$.)

3 Suppose that $\varphi \colon A \hookrightarrow B$ is an inclusion of rings and B is integral over A. Prove that ${}^a\varphi$ is surjective.

4 Let $\varphi \colon A \to B$ be a ring homomorphism. Does ${}^a\varphi$ always take closed points to closed points? Does this hold under the assumption of Exercise 3?

5 Prove that $\overline{{}^a\varphi(V(E))} = V(\varphi^{-1}(E))$ where $\overline{}$ denotes closure.

6 Suppose that X_1 and X_2 are closed subsets of Spec A and $u_1, u_2 \in A$ satisfy $u_1 + u_2 = 1$, $u_1u_2 = 0$, and that $u_i(x) = 0$ for all $x \in X_i$, for $i = 1, 2$. Prove that then $A = A_1 \oplus A_2$, and that $X_i = {}^a\varphi_i(\operatorname{Spec}(A_i))$, where $\varphi_i \colon A \to A_i$ is the natural homomorphism.

7 Suppose that $\operatorname{Spec} A = X_1 \cup X_2$ is a decomposition into disjoint closed sets. Prove that then $A = A_1 \oplus A_2$ with $X_i = {}^a\varphi_i(\operatorname{Spec}(A_i))$. [Hint: Represent X_i in the form $V(E_i)$, find elements v_1, v_2 such that $v_1 + v_2 = 1$, $v_1v_2 = 0$ and $v_i(x) = 0$ for all $x \in X_i$, for $i = 1, 2$. Using Proposition of Section 1.2, construct functions u_1, u_2 satisfying the conditions of Exercise 6.]

8 Prove that if A is a finitely generated ring over an algebraically closed field then Proposition of Section 1.2 continues to hold if we replace prime ideals by maximal ideals. [Hint: Use the Nullstellensatz.] Deduce that closed points are everywhere dense in any closed subset of Spec A.

9 Let $A = \mathbb{Z}[T]/(F(T))$, where $F(T) \in \mathbb{Z}[T]$, and let p be a prime number such that $F(0) \equiv 0 \bmod p$; suppose that $\mathfrak{p} \in A$ is the maximal ideal of A generated by p and the image of T. Prove that the point $x \in \operatorname{Spec} A$ corresponding to $\mathfrak{p}$ is singular if and only if $F(0) \equiv 0 \bmod p^2$ and $F'(0) \equiv 0 \bmod p$. [Hint: Consider the homomorphism $M/M^2 \to \mathfrak{p}/\mathfrak{p}^2$ where $M = (p, F) \in \mathbb{Z}[T]$.]

10 Prove that a closed subset of $\operatorname{Spec}(\mathbb{Z}[T_1, \dots, T_n])$, each component of which has dimension $n = \dim \operatorname{Spec}(\mathbb{Z}[T_1, \dots, T_n]) - 1$ (that is, codimension 1), is of the form $V(F)$, where $F \in \mathbb{Z}[T_1, \dots, T_n]$.

11 Prove the following universal property of the ring of fractions A_S with respect to a multiplicative system $S \subset A$ (Example 5.4): if $f \colon A \to B$ is a homomorphism such that $f(s)$ is invertible in B for all $s \in S$, then there exists a homomorphism $g \colon A_S \to B$ for which $f = gh$, where $h \colon A \to A_S$ is the natural homomorphism.

2 Sheaves

2.1 Presheaves

The topological space Spec A is just one of the two building blocks of the definition of scheme. The second is the notion of sheaf. In the preceding section, we used the fact that an affine variety X is determined by its ring of regular functions $k[X]$, and then, starting from an arbitrary ring A, we arrived at the corresponding geometric notion, its prime spectrum Spec A. For the definition of the general notion of scheme we also take regular functions on varieties as the starting point. But there may turn out to be too few of these, if we consider functions regular on the whole variety. Therefore it is natural to consider, for any open set $U \subset X$, the ring of regular functions on U. In this way we get, not one ring, but a system of rings, with various connections between them, as we will see. An analogous system is the basis of the definition of scheme. First, however, we need to sort out some definitions and very simple facts relating to this type of object.

Definition Let X be a given topological space. Suppose that with every open set $U \subset X$ we have associated a set $\mathcal{F}(U)$ and with any open sets $U \subset V$ a map

$$\rho_U^V : \mathcal{F}(V) \to \mathcal{F}(U).$$

This system of sets and maps is a *presheaf* if the following conditions hold:

(1) if U is empty, the set $\mathcal{F}(U)$ consists of 1 element;
(2) ρ_U^U is the identity map for any open set U;
(3) for any open sets $U \subset V \subset W$, we have

$$\rho_U^W = \rho_U^V \circ \rho_V^W.$$

A presheaf is sometimes denoted by the single symbol $\mathcal{F}$. If we need to emphasise that the maps ρ_U^V refer to $\mathcal{F}$, we denote them by $\rho_{U,\mathcal{F}}^V$. If all the sets $\mathcal{F}(U)$ are groups, modules over a ring A, or rings, and the maps ρ_U^V are homomorphisms of these structures, then $\mathcal{F}$ is a presheaf of groups, or A-modules, or rings respectively.

A presheaf $\mathcal{F}$ obviously doesn't depend on the choice of the element $\mathcal{F}(\emptyset)$; more precisely, for different choices we get isomorphic presheaves, under a definition of isomorphism that the reader will easily recover. Hence to determine a presheaf, we only need give the sets $\mathcal{F}(U)$ for nonempty sets U. If $\mathcal{F}$ is a presheaf of groups then $\mathcal{F}(\emptyset)$ is the group consisting of one element.

If $\mathcal{F}$ is a presheaf on X and $U \subset X$ an open set, then sending V to $\mathcal{F}(V)$ for open subsets $V \subset U$ obviously defines a presheaf on U. It is called the *restriction* of the presheaf $\mathcal{F}$, and denoted by $\mathcal{F}_{|U}$.

Example 5.11 For a set M, let $\mathcal{F}(U)$ consist of all functions on U with values in M; and for $U \subset V$, let $\rho_U^V : \mathcal{F}(V) \to \mathcal{F}(U)$ be the restriction of functions from V to U. Then properties (1)–(3) are obvious. $\mathcal{F}$ is called the *presheaf of all functions* on X (with values in M).

In order to carry over the intuition of this example to any presheaf, we call the maps ρ_U^V *restriction maps*. There are a number of variations on Example 5.11.

Example 5.12 Let M be a topological space, and let $\mathcal{F}(U)$ consist of all continuous functions on U with values in M, and ρ_U^V the same as in Example 5.11. $\mathcal{F}$ is called the *presheaf of continuous functions* on X.

Example 5.13 Let X be a differentiable manifold, and $\mathcal{F}(U)$ the set of differentiable functions $U \to \mathbb{R}$; once again, ρ_U^V is as in Example 5.11.

Example 5.14 Let X be an irreducible quasiprojective variety, with the topology defined by taking algebraic subvarieties as the closed sets (so that the topological terminology used in Chapter 1 turns into the usual topological notions). For an open set $U \subset X$, $\mathcal{F}(U)$ is the set of rational functions on X that are regular at all points of U; again ρ_U^V is as in Example 5.11. $\mathcal{F}$ is a presheaf of rings. It is called the *presheaf of regular functions*.

2.2 The Structure Presheaf

We proceed to construct the presheaf that will play the principal role in what follows. It is defined on the topological space $X = \operatorname{Spec} A$. The presheaf we define will be called the *structure presheaf* on $\operatorname{Spec} A$, and denoted by $\mathcal{O}$. To clarify the logic of the definition, we go through it first in a more special case.

Suppose first that A has no zerodivisors, and write K for its field of fractions. In this case A is a subfield of K. Now we can copy Example 5.4 exactly. For an open set $U \subset \operatorname{Spec} A$ we denote by $\mathcal{O}(U)$ the set of elements $u \in K$ such that for any point $x \in U$ we have an expression $u = a/b$ with $a, b \in A$ and $b(x) \neq 0$, that is, b is not an element of the prime ideal x. Now $\mathcal{O}(U)$ is obviously a ring. Since all the rings $\mathcal{O}(U)$ are contained in K, we can compare them as subsets of one set. If $U \subset V$ then clearly $\mathcal{O}(V) \subset \mathcal{O}(U)$. We write ρ_U^V for the inclusion $\mathcal{O}(V) \hookrightarrow \mathcal{O}(U)$. A trivial verification shows that we get a presheaf of rings.

Before finishing our consideration of this case, we compute $\mathcal{O}(\operatorname{Spec} A)$. Our arguments repeat the proof of Theorem 1.7 of Section 3.2, Chapter 1. The condition $u \in \mathcal{O}(\operatorname{Spec} A)$ means that for any point $x \in \operatorname{Spec} A$ there exist $a_x, b_x \in A$ such that

$$u = a_x/b_x \quad \text{with} \quad b_x(x) \neq 0. \tag{5.3}$$

Consider the ideal $\mathfrak{a}$ generated by the elements b_x for all $x \in \operatorname{Spec} A$. By condition (5.3), $\mathfrak{a}$ is not contained in any prime ideal of A, and hence $\mathfrak{a} = A$. Thus there exist points $x_1, \dots, x_r \in \operatorname{Spec} A$ and elements $c_1, \dots, c_r \in A$ such that

$$c_1 b_{x_1} + \cdots + c_r b_{x_r} = 1.$$

Taking $x = x_i$ in (5.3), multiplying through by $c_i b_{x_i}$ and adding, we get that

$$u = \sum a_{x_i} c_i \in A.$$

Thus $\mathcal{O}(\operatorname{Spec} A) = A$.

We now proceed to the case of an arbitrary ring A. The final argument suggests that it is natural to set $\mathcal{O}(\operatorname{Spec} A) = A$. But there are some other open sets, for which there are natural candidates for the ring $\mathcal{O}(U)$, namely the principal open sets $D(f)$ for $f \in A$. Indeed, we saw in Section 1.3 that $D(f)$ is homeomorphic to $\operatorname{Spec} A_f$, and hence it is also natural to set

$$\mathcal{O}\big(D(f)\big) = A_f.$$

Thus so far we have defined the presheaf $\mathcal{O}(U)$ on the principal open sets $U = D(f)$. Before defining it on all open sets, we introduce the homomorphisms ρ_U^V, of course only for principal open sets U and V.

We first determine when $D(f) \subset D(g)$. Taking complements, this is equivalent to $V(f) \supset V(g)$, that is, any prime ideal containing g also contains f. In other words, the image $\overline{f}$ of f in the quotient ring $A/(g)$ is contained in any prime ideal of this ring. In Proposition of Section 1.2 we saw that this is equivalent to $\overline{f}$ nilpotent in $A/(g)$, that is, $f^n \in (g)$ for some $n > 0$. Thus $D(f) \subset D(g)$ if and only if

$$f^n = gu \quad \text{for some } n > 0 \text{ and some } u \in A. \tag{5.4}$$

In this case, we can construct the homomorphism

$$\rho_{D(f)}^{D(g)} : A_g \to A_f \quad \text{by} \quad \rho_{D(f)}^{D(g)}(a/g^k) = au^k/f^{nk}.$$

An obvious verification shows that this map does not depend on the expression of an element $t \in A_g$ in the form $t = a/g^k$, and is a homomorphism. The map can be described in a more intrinsic way using the universal property of the ring of fractions A_S, see Exercise 11. In our case, g and its powers are invertible in A_f by (5.4), and the existence of the homomorphism $\rho_{D(f)}^{D(g)}$ follows from this.

Before formulating the definition in its final form, return briefly to the case already considered when A has no zerodivisors. Here we can indicate a method of calculating $\mathcal{O}(U)$ for any open set U in terms of $\mathcal{O}(V)$ where V are various principal open sets. Namely if $\{D(f)\}$ is the set of all principal open sets contained in U then, as one checks easily,

$$\mathcal{O}(U) = \bigcap_{\{U \supset D(f)\}} \mathcal{O}(D(f)).$$

In the general case one would like to take this equality as the definition of $\mathcal{O}(U)$, but this is not possible, since the $\mathcal{O}(D(f))$ are not all contained in a common set. However, they are related to one another by the homomorphisms $\rho_{D(g)}^{D(f)}$ defined whenever $D(g) \subset D(f)$. In this case, the natural generalisation of intersection is the projective limit of sets. We recall the definition.

Let I be a partially ordered set, $\{E_\alpha \mid \alpha \in I\}$ a system of sets indexed by I, and for any $\alpha, \beta \in I$ with $\alpha \le \beta$, let $f_\alpha^\beta : E_\beta \to E_\alpha$ be maps satisfying the conditions (1) f_α^α is the identity map of E_α, and (2) for $\alpha \le \beta \le \gamma$ we have $f_\alpha^\gamma = f_\alpha^\beta \circ f_\beta^\gamma$. Consider the subset of the product $\prod_{\alpha \in I} E_\alpha$ of the sets E_α consisting of elements $x = \{x_\alpha \mid x_\alpha \in E_\alpha\}$ such that $x_\alpha = f_\alpha^\beta(x_\beta)$ for all $\alpha, \beta \in I$ with $\alpha \le \beta$. This subset is called the *projective limit* of the system of sets E_α with respect to the maps f_α^β, and is denote by $\varprojlim E_\alpha$. The maps $\varprojlim E_\alpha \to E_\alpha$ defined by $x \mapsto x_\alpha$ for $x \in \varprojlim E_\alpha$ are called the *natural maps* of the projective limit.

If the E_α are rings, modules or groups, and f_α^β homomorphisms of these structures, then $\varprojlim E_\alpha$ is a structure of the same type. The reader can find a more detailed description of this construction in Atiyah and Macdonald [8, Chapter 10]. Here we should bear in mind that the condition that the partial ordered set I is directed is not essential for the definition of projective limit.

Now we are ready for the final definition:

$$\mathcal{O}(U) = \varprojlim \mathcal{O}\big(D(f)\big),$$

where the projective limit is taken over all $D(f) \subset U$ relative to the system of homomorphisms $\rho_{D(g)}^{D(f)}$ for $D(g) \subset D(f)$ constructed above.

By definition, $\mathcal{O}(U)$ consists of families $\{u_\alpha\}$ with $u_\alpha \in A_{f_\alpha}$, where f_α are all the elements such that $D(f_\alpha) \subset U$, and the u_α are related by

$$u_\alpha = \rho_{D(f_\alpha)}^{D(f_\beta)}(u_\beta) \quad \text{if } D(f_\beta) \supset D(f_\alpha). \tag{5.5}$$

For $U \subset V$ each family $\{v_\alpha\} \in \mathcal{O}(V)$ consisting of $v_\alpha \in A_{f_\alpha}$ with $D(f_\alpha) \subset V$ defines a subfamily $\{v_\beta\}$ consisting of the v_β for those indexes β with $D(f_\beta) \subset U$. Obviously $\{v_\beta\} \in \mathcal{O}(U)$. We set

$$\rho_U^V\big(\{v_\alpha\}\big) = \{v_\beta\}.$$

A trivial verification shows that $\mathcal{O}(U)$ and ρ_U^V define a presheaf of rings on Spec A. This presheaf $\mathcal{O}$ is called the *structure presheaf* of Spec A.

If $U = \operatorname{Spec} A$ then $D(1) = U$, so that 1 is one of the f_α, say f_0. The map

$$\{u_\alpha\} \mapsto u_0$$

defines an isomorphism $\mathcal{O}(\operatorname{Spec} A) \xrightarrow{\sim} A$, as one check easily.

In particular, if $u = \{u_\alpha \mid D(f_\alpha) \subset U\} \in \mathcal{O}(U)$ then by definition $\rho_{D(f)}^U(u) = \{u_\beta \mid D(f_\beta) \subset D(f)\}$. By what we have said above, the map sending $\{u_\beta \mid D(f_\beta) \subset D(f)\}$ to u_α, where $f = f_\alpha$, defines an isomorphism $\mathcal{O}(D(f_\alpha)) \xrightarrow{\sim} A_{f_\alpha}$, under which

$$u_\alpha = \rho_{D(f_\alpha)}^U(u). \tag{5.6}$$

This formula allows us to recover all the u_α from the element $u \in \mathcal{O}(U)$.

2.3 Sheaves

Suppose that a topological space X is a union of open sets U_α. Every function f on X is uniquely determined by its restrictions to the sets U_α; moreover, if on each U_α a function f_α is given such that the restrictions of f_α and of f_β to $U_\alpha \cap U_\beta$ are equal, then there exists a function f on X such that each f_α is the restriction of f to U_α. The same property holds for continuous functions, differentiable functions on a differentiable manifold, and regular functions on a quasiprojective algebraic variety.

This property expresses the local nature of the notion of continuous, differentiable and regular function; it can be formulated for any presheaf, and distinguishes an exceptionally important class of presheaves.

Definition A presheaf $\mathcal{F}$ on a topological space X is a *sheaf* if for any open set $U \subset X$ and any open cover $U = \bigcup U_\alpha$ of U the following two conditions hold:

(1) if $s_1, s_2 \in \mathcal{F}(U)$ and $\rho^U_{U_\alpha}(s_1) = \rho^U_{U_\alpha}(s_2)$ for all U_α then $s_1 = s_2$.

(2) if $s_\alpha \in \mathcal{F}(U_\alpha)$ are such that $\rho^{U_\alpha}_{U_\alpha \cap U_\beta}(s_\alpha) = \rho^{U_\beta}_{U_\alpha \cap U_\beta}(s_\beta)$ for all U_α and U_β, then there exists $s \in \mathcal{F}(U)$ such that $s_\alpha = \rho^U_{U_\alpha}(s)$ for each U_α.

We have already given a series of examples of sheaves before defining the notion. We give the simplest example of a presheaf that is not a sheaf. For X a topological space and M a set, let $\mathcal{F}(U)$ be the set of constant maps $U \to M$ and ρ^V_U the restriction maps. In other words, $\mathcal{F}(U) = M$ for all nonempty open sets $U \subset X$, with ρ^V_U the identity maps, and $\mathcal{F}(\emptyset)$ consists of a single element. Then $\mathcal{F}$ is obviously a presheaf. Suppose that X contains a disconnected open set U represented as a disjoint union of open sets $U = U_1 \cup U_2$ with $U_1 \cap U_2 = \emptyset$. Let $m_1, m_2 \in M$ be two distinct elements and $s_1 = m_1 \in \mathcal{F}(U_1) = M$, $s_2 = m_2 \in \mathcal{F}(U_2) = M$. The condition $\rho^{U_1}_{U_1 \cap U_2}(s_1) = \rho^{U_2}_{U_1 \cap U_2}(s_2)$ holds automatically since $U_1 \cap U_2 = \emptyset$, whereas, because $m_1 \neq m_2$, there does not exist an $s \in \mathcal{F}(U) = M$ such that $\rho^U_{U_1}(s) = s_1$ and $\rho^U_{U_2}(s) = s_2$.

Theorem 5.1 *The structure presheaf $\mathcal{O}$ on* Spec A *is a sheaf; it is denoted by $\mathcal{O}_{\operatorname{Spec} A}$ or $\mathcal{O}_A$, or $\mathcal{O}_X$, where $X = \operatorname{Spec} A$.*

Proof We first check the conditions (1) and (2) in the definition of a sheaf in the case that U and the U_α are principal open sets. First of all, we note that for either of the conditions, it is enough to check it in the case $U = \operatorname{Spec} A$. Indeed, if $U = D(f)$ and $U_\alpha = D(f_\alpha)$ then, as the reader can check easily, conditions (1) and (2) are satisfied for U and U_α if they are satisfied for $\operatorname{Spec}(A_f)$ and the sets $\overline{U}_\alpha = D(\overline{f}_\alpha)$, where $\overline{f}_\alpha$ is the image of f_α under the natural homomorphism $A \to A_f$. We proceed to check conditions (1) and (2), assuming that $U_\alpha = D(f_\alpha)$ and $\operatorname{Spec} A = \bigcup U_\alpha$. □

Proof of (1) Since $\mathcal{O}$ is a presheaf of groups, it is sufficient to prove that if $u \in \mathcal{O}(\operatorname{Spec} A) = A$ and $\rho^{\operatorname{Spec} A}_{U_\alpha}(u) = 0$ for all $U_\alpha = D(F_\alpha)$ then $u = 0$. The condition $\rho^{\operatorname{Spec} A}_{U_\alpha}(u) = 0$ means that

$$f_\alpha^{n_\alpha} u = 0 \quad \text{for all } \alpha \text{ and some } n_\alpha \geq 0. \tag{5.7}$$

Since $D(f_\alpha) = D(f_\alpha^{n_\alpha})$, we have $\bigcup D(f_\alpha^{n_\alpha}) = \operatorname{Spec} A$. We have already seen that this implies an identity

$$f_{\alpha_1}^{n_1} g_1 + \cdots + f_{\alpha_r}^{n_r} g_r = 1 \quad \text{for some } g_1, \ldots, g_r \in A.$$

Multiplying (5.7) for $\alpha = \alpha_1, \dots, \alpha_r$ by $g_1, \dots, g_r$ and adding, we get that $u = 0$. □

Proof of (2) Since Spec A is compact, we can restrict to the case of a finite cover. Indeed, the reader can easily check that if the assertion holds for a subcover then it also holds for the whole cover.

Suppose that $\operatorname{Spec} A = D(f_1) \cup \cdots \cup D(f_r)$ and $u_i \in A_{f_i}$ with $u_i = v_i/f_i^n$; we can take all the n the same in view of the finiteness of the cover. Note first that $D(f) \cap D(g) = D(fg)$, by an obvious verification. By definition,

$$\rho^{D(f_i)}_{D(f_i f_j)}(u_i) = \frac{v_i f_j^n}{(f_i f_j)^n},$$

and by assumption

$$(f_i f_j)^m (v_i f_j^n - v_j f_i^n) = 0.$$

Setting $v_j f_j^m = w_j$ and $m + n = l$, we get that

$$u_i = \frac{w_i}{f_i^l} \quad \text{and} \quad w_i f_j^l = w_j f_i^l. \tag{5.8}$$

As in the proof of (1), we see that

$$\sum f_i^l g_i = 1.$$

We set $u = \sum w_j g_j$. By the assumptions,

$$f_i^l u = \sum_j w_j g_j f_i^l = \sum_j w_i g_j f_j^l = w_i.$$

Hence $\rho^{\operatorname{Spec} A}_{D(f_i)} = w_i/f_i^l = u_i$, as required. This proves (1) and (2) for basic open sets.

Verifying (1) and (2) for any open sets is now a formal consequence of what we have already proved. In terms of general nonsense, our situation is described as follows: on a topological space X we are given some basis $\mathcal{V} = \{V_\alpha\}$ for the open sets of the topology that is closed under intersections. Suppose that a presheaf of groups $\mathcal{F}$ on X satisfies the following two conditions:

(a) for all open sets $U \subset X$ we have

$$\mathcal{F}(U) = \varprojlim \mathcal{F}(V_\alpha),$$

where the limit is taken over all $V_\alpha \in \mathcal{V}$ such that $V_\alpha \subset U$, under the homomorphisms $\rho^{V_\alpha}_{V_\beta}$; and

(b) the $\rho^U_{V_\alpha}$ coincide with the natural homomorphisms of the projective limit.

It follows from these conditions and the definition of $\varprojlim$ that the patching conditions (1) and (2) in the definition of sheaf hold for U and $V_\alpha \in \mathcal{V}$. The structure presheaf $\mathcal{O}$ satisfies both of these properties: the first by definition, and the second by the equality (5.6). We prove that if in addition conditions (1) and (2) are satisfied for open sets $V_\alpha \in \mathcal{V}$, then $\mathcal{F}$ is a sheaf. □

Proof of (1) Suppose that $U = \bigcup_\xi U_\xi$ and $U_\xi = \bigcup_\lambda V_{\xi,\lambda}$ with $V_{\xi,\lambda} \in \mathcal{V}$. If $\rho^U_{U_\xi}(u) = 0$ for all U_ξ then $\rho^U_{V_{\xi,\lambda}}(u) = 0$. Introducing new indexes $(\xi, \lambda) = \gamma$ we get $U = \bigcup V_\gamma$ and $\rho^U_{V_\gamma}(u) = 0$ for all γ. To prove that $u = 0$ it is enough by (b) to check that $\rho^U_{V_\alpha}(u) = 0$ for all $V_\alpha \subset U$. This follows at once by considering the restriction maps corresponding to the open sets in the following diagram:

$$\begin{array}{ccc} V_\alpha \cap V_\gamma & \subset & V_\gamma \\ \cap & & \cap \\ V_\alpha & \subset & U. \end{array}$$

Indeed,

$$(\rho^{V_\alpha}_{V_\alpha \cap V_\gamma} \circ \rho^U_{V_\alpha})(u) = \rho^U_{V_\alpha \cap V_\gamma}(u) = (\rho^{V_\gamma}_{V_\alpha \cap V_\gamma} \circ \rho^U_{V_\gamma})(u) = 0$$

for all V_γ, and hence $\rho^U_{V_\alpha}(u) = 0$, since $V_\alpha = \bigcup (V_\alpha \cap V_\gamma)$, and condition (1) holds for the sets V_α by assumption. □

Proof of (2) Let $u_\xi \in \mathcal{F}(U_\xi)$ be given, satisfying

$$\rho^{U_{\xi_1}}_{U_{\xi_1} \cap U_{\xi_2}}(u_{\xi_1}) = \rho^{U_{\xi_2}}_{U_{\xi_1} \cap U_{\xi_2}}(u_{\xi_2}) \qquad \text{for all } \xi_1, \xi_2,$$

where $U = \bigcup U_\xi$. Suppose that $U_\xi = \bigcup_\lambda V_{\xi,\lambda}$ with $V_{\xi,\lambda} \in \mathcal{V}$. Setting $v_{\xi,\lambda} = \rho^{U_\xi}_{V_{\xi,\lambda}}(u_\xi)$ and choosing new indexes $\gamma = (\xi, \lambda)$, we verify that

$$\rho^{V_{\gamma_1}}_{V_{\gamma_1} \cap V_{\gamma_2}}(v_{\gamma_1}) = \rho^{V_{\gamma_2}}_{V_{\gamma_1} \cap V_{\gamma_2}}(v_{\gamma_2}). \tag{5.9}$$

This follows at once by considering the restriction maps ρ corresponding to the open sets in the following diagram:

$$\begin{array}{ccccc} U_{\xi_1} & \supset & U_{\xi_1} \cap U_{\xi_2} & \subset & U_{\xi_2} \\ \cup & & \cup & & \cup \\ V_{\gamma_1} & \supset & V_{\gamma_1} \cap V_{\gamma_2} & \subset & V_{\gamma_2} \end{array}$$

where $\gamma_1 = (\xi_1, \lambda_1)$ and $\gamma_2 = (\xi_2, \lambda_2)$; indeed, the left-hand side of (5.9) is equal to

$$\rho^{U_{\xi_1}}_{V_{\gamma_1} \cap V_{\gamma_2}}(u_{\xi_1}) = (\rho^{U_{\xi_1} \cap U_{\xi_2}}_{V_{\gamma_1} \cap V_{\gamma_2}} \circ \rho^{U_{\xi_1}}_{U_{\xi_1} \cap U_{\xi_2}})(u_{\xi_1}),$$

and the right-hand side is obviously equal to the same thing.

By (5.9), for any $V_\alpha \in \mathcal{V}$ with $V_\alpha \subset U$ the elements $\rho^{V_\gamma}_{V_\alpha \cap V_\gamma}(v_\gamma)$ satisfy the analogous relation, and hence, by assumption there exists an element $v_\alpha \in \mathcal{F}(V_\alpha)$ such that $\rho^{V_\alpha}_{V_\alpha \cap V_\gamma}(v_\alpha) = \rho^{V_\gamma}_{V_\alpha \cap V_\gamma}(v_\gamma)$. An obvious verification shows that these elements determine an element $u \in \varprojlim \mathcal{F}(V_\alpha) = \mathcal{F}(U)$. This satisfies $\rho^U_{V_\alpha}(u) = v_\alpha$. Hence the elements $u'_\xi = \rho^U_{U_\xi}(u)$ satisfy $\rho^{U_\xi}_{V_\tau}(u'_\xi) = \rho^{U_\xi}_{V_\tau}(u_\xi)$ for all $V_\tau \subset U_\xi$ with $V_\tau \in \mathcal{V}$, and hence $u'_\xi = u_\xi$. The theorem is proved. □

2.4 Stalks of a Sheaf

We return to the analysis of the general notion of sheaf and presheaf. Consider first a subsheaf $\mathcal{F}$ for which all the sets $\mathcal{F}(U)$ are subsets of a common set, and the restriction maps are inclusions $\rho^U_V : \mathcal{F}(V) \subset \mathcal{F}(U)$. This holds, for example, for the sheaf $\mathcal{O}$ on $\operatorname{Spec} A$, where A is an integral domain, since then all the $\mathcal{O}(U) \subset K$ are subrings of the field of fractions K of A. Then we can consider the union $\mathcal{F}_x = \bigcup \mathcal{F}(U)$ of the sets $\mathcal{F}(U)$ taken over all the open sets U containing a given point x. For the sheaf of regular functions on a quasiprojective variety, $\mathcal{F}_x$ is the local ring $\mathcal{O}_x$ of x (see Exercise 1 of Section 1.6, Chapter 2).

In the general case, the sets $\mathcal{F}(U)$ are not all subsets of some ambient set, but are related by homomorphisms ρ^V_U; this allows us to replace the union $\bigcup \mathcal{F}(U)$ by the inductive limit $\varinjlim \mathcal{F}(U)$. This definition is analogous to that of projective limit, and can be found in Atiyah and Macdonald [8, Ex. 14–19, Chapter 2]. If $\mathcal{F}$ is the sheaf of continuous functions, the stalk $\mathcal{F}_x$ consists of the germs of functions continuous in some neighbourhood of x, that is, the result of identifying functions that are equal in some neighbourhood of x.

Definition The *stalk* $\mathcal{F}_x$ of a presheaf at a point $x \in X$ is the inductive limit of the sets $\mathcal{F}(U)$ taken over all open sets $U \ni x$ with respect to the system of maps ρ^V_U for $U \subset V$.

By definition, an element of $\mathcal{F}_x$ is an element of $\mathcal{F}(U)$ for U some neighbourhood U of x, with elements $u \in \mathcal{F}(U)$ and $v \in \mathcal{F}(V)$ identified if there exists a neighbourhood $x \in W \subset U \cap V$ such that $\rho^U_W(u) = \rho^V_W(v)$.

Example Applying this definition to the case of the structure sheaf $\mathcal{O}$ on $\operatorname{Spec} A$ for a ring A, we see that the stalk $\mathcal{O}_x$ at a point $x \in \operatorname{Spec} A$ corresponding to a prime ideal $\mathfrak{p}$ is just the local ring $A_\mathfrak{p}$ of A at $\mathfrak{p}$. Indeed, the principal open sets $D(f)$ with $f \notin \mathfrak{p}$ provide arbitrarily small neighbourhoods of x, and $\mathcal{O}(D(f)) = A_f$; therefore $\mathcal{O}_x = \varinjlim A_f$, where the limit is taken over the multiplicative system $f \in A \setminus \mathfrak{p}$, and it is easy to see that this is equal to $A_\mathfrak{p}$.

In the general case, for any open set $U \ni x$, there is a natural homomorphism

$$\rho^U_x : \mathcal{F}(U) \to \mathcal{F}_x.$$

If $\mathcal{F}$ is a sheaf and $\rho_x^U(u_1) = \rho_x^U(u_2)$ for all points $x \in U$ then $u_1 = u_2$. Indeed, by definition, this means that any point $x \in U$ has a neighbourhood $x \in W \subset U$ such that $\rho_W^U(u_1) = \rho_W^U(u_2)$. From the definition of sheaf it follows that $u_1 = u_2$.

Thus for a sheaf $\mathcal{F}$, the elements of $\mathcal{F}(U)$ can be specified as families of germs $\{u_x \in \mathcal{F}_x\}_{x\in U}$. Of course, we do not get all families of germs in this way. The following is an obvious necessary condition:

$$\text{for all } x \in U, \text{ there exists a neighbourhood } x \in W \subset U \text{ and}$$
$$\text{an element } w \in \mathcal{F}(W) \text{ such that } u_y = \rho_y^W(w) \text{ for all } y \in W.$$

The reader can easily check that conversely, any family satisfying this condition corresponds to some element $u \in \mathcal{F}(U)$.

This holds, of course, only if $\mathcal{F}$ is a sheaf. But if $\mathcal{F}$ is an arbitrary presheaf, then we can still consider the set $\mathcal{F}'(U)$ of all families of germs $\{u_x \in \mathcal{F}_x\}_{x\in U}$ satisfying the above condition. For $U \subset V$ the map

$$\rho_U^V : \{v_x \in \mathcal{F}_x\}_{x\in V} \to \{v_y \in \mathcal{F}_y\}_{y\in U}$$

makes $\mathcal{F}'(U)$ into a presheaf. It is easy to check that in this way we get a sheaf. $\mathcal{F}'$ is called the *sheafication* or *associated sheaf* of $\mathcal{F}$. It is the sheaf $\mathcal{F}'$ "closest" to $\mathcal{F}$. For example, if $\mathcal{F}$ is the presheaf of constant M-valued functions on X, that is, $\mathcal{F}(U) = M$ for all U, then $\mathcal{F}'$ is the sheaf of *locally* constant M-valued functions on X, that is, $\mathcal{F}'(U)$ is the set of functions on U that are constant on each connected component of U.

2.5 Exercises to Section 2

1 Let X be a discrete topological space and $\mathcal{F}(U)$ the set of all maps $f : U \to M$ such that $f(U)$ is finite; for $U \subset V$, ρ_U^V is restriction. Is $\mathcal{F}$ a presheaf? A sheaf?

2 Let X be a nonsingular quasiprojective variety with the Zariski topology (introduced in Example 5.14). For an open set $U \subset X$ we set $\mathcal{F}(U) = \Omega^r[U]$ (Section 5.1, Chapter 3) and for $U \subset V$, ρ_U^V is restriction of differential r-forms. Is $\mathcal{F}$ a presheaf? A sheaf?

3 Let A be a ring and $\mathfrak{a} \subset A$ an ideal. For any element $f \in A$, not a nilpotent, write $\mathfrak{a}_f$ for the ideal of A_f generated by the images of elements of $\mathfrak{a}$ under the homomorphism $A \to A_f$. Construct, by analogy with Section 2.2, a presheaf $\mathcal{F}$ such that $\mathcal{F}(D(f)) = \mathfrak{a}_f$, and prove that $\mathcal{F}$ is a sheaf. Start with the simpler version when A is an integral domain.

4 Let X be a topological space, M an Abelian group, and $\mathcal{F}(U)$ the quotient group of all locally constant functions on U with values in M by the constant functions; for $U \subset V$, ρ_U^V is restriction. Prove that $\mathcal{F}$ is a presheaf and determine its sheafication $\mathcal{F}'$.

5 Prove that the structure sheaf $\mathcal{O}$ on $\operatorname{Spec} A$ can be defined as follows: for $\mathcal{O}(U)$ we take families of elements $\{u_x \in \mathcal{O}_x\}_{x\in U}$, where $\mathcal{O}_x = A_x$ is the local ring of A at the prime ideal $x \in \operatorname{Spec} A$, satisfying the following condition: for every $y \in U$ there exists a principal open set $y \in D(f) \subset U$ and an element $u \in A_f$ such that all the u_x for $x \in D(f)$ are images of u under the natural homomorphisms $A_f \to \mathcal{O}_x$. If $U \subset V$ are open sets, the restriction $\rho_U^V(v)$ is obtained by choosing from the family $v = \{v_x \in \mathcal{O}_x\}_{x\in V}$ the elements v_x with $x \in U$.

6 Let A be a 1-dimensional local ring, $\xi \in \operatorname{Spec} A$ a generic point. Prove that ξ is an open point and find $\mathcal{O}(\xi)$.

7 Let A be the local ring of the origin in $\mathbb{A}^2$. Find $\mathcal{O}(U)$ where $U = (\operatorname{Spec} A) \setminus x$, and x is the closed point.

3 Schemes

3.1 Definition of a Scheme

Definition 5.1 A *ringed space* is a pair $X, \mathcal{O}$ consisting of a topological space X and a sheaf of rings $\mathcal{O}$. The sheaf $\mathcal{O}$ is sometimes denoted by $\mathcal{O}_X$, and is called the *structure sheaf* of X.

A word of caution on the definition of maps of ringed spaces is in order. The point is that, as discussed in Section 2.3, Chapter 1, any map $\varphi\colon X \to Y$ of sets induces a map of functions (with values in a third set K): a function $f\colon Y \to K$ pulls back to the function $\varphi^*(f)\colon X \to K$ given by

$$\varphi^*(f)(x) = f\bigl(\varphi(x)\bigr) \quad \text{for } x \in X. \tag{5.10}$$

But in connection with Spec of a ring, we meet ringed spaces for which, although the elements of $\mathcal{O}(U)$ can be interpreted as "functions" (see Section 1.2), these functions are in the first place not determined by their values, and secondly, the values on the left- and right-hand sides of (5.10) are elements of different sets. In other words, in this case, the set-theoretic map will not determine the pullback of functions. Thus in the definition of the analogue of map for ringed spaces, we also demand that the "pullback of functions" is specified, requiring only a certain natural compatibility. In view of this, we introduce a new term for the analogue of map of ringed spaces, a *morphism*.

Definition 5.2 A *morphism of ringed spaces* $\varphi\colon (X, \mathcal{O}_X) \to (Y, \mathcal{O}_Y)$ is a continuous map $\varphi\colon X \to Y$ and a collection of homomorphisms $\psi_U\colon \mathcal{O}_Y(U) \to \mathcal{O}_X(\varphi^{-1}(U))$ for any open sets $U \subset Y$. We require that the diagram

$$\begin{array}{ccc} \mathcal{O}_X(\varphi^{-1}(V)) & \xrightarrow{\rho^{\varphi^{-1}V}_{\varphi^{-1}U}} & \mathcal{O}_X(\varphi^{-1}(U)) \\ \psi_V \downarrow & & \downarrow \psi_U \\ \mathcal{O}_Y(V) & \xrightarrow[\rho^V_U]{} & \mathcal{O}_Y(U) \end{array}$$

is commutative for any open sets $U \subset V$ of Y.

Example 5.15 Any topological space X is a ringed space if we take $\mathcal{O}_X$ to be the sheaf of continuous functions. Any continuous map $\varphi\colon X \to Y$ defines a morphism if we set $\psi_U(f) = \varphi^*(f)$ for $f \in \mathcal{O}_Y(U)$.

Example 5.16 Any differentiable manifold is a ringed space if we take $\mathcal{O}_X$ to be the sheaf of differentiable functions. Any differentiable map defines a morphism in the same way as in Example 5.15.

Example 5.17 Any ring A defines a ringed space $\operatorname{Spec} A, \mathcal{O}_A$ where $\mathcal{O}_A$ is the structure sheaf constructed in Section 2.2. From now on, we denote this ringed space by $\operatorname{Spec} A$. We show that a homomorphism $\lambda\colon A \to B$ defines a morphism $\varphi\colon \operatorname{Spec} B \to \operatorname{Spec} A$. We first set $\varphi = {}^a\lambda$. For $U = D(f) \subset \operatorname{Spec} A$ we have $\varphi^{-1}(U) = D(\lambda(f))$. Sending $a/f^n \mapsto \lambda(a)/\lambda(f)^n$ defines a homomorphism ψ_U of the ring $A_f = \mathcal{O}_A(U)$ to $B_{\lambda(f)} = \mathcal{O}_B(\varphi^{-1}(U))$. The reader can easily verify that these homomorphisms extend to homomorphisms $\psi : \mathcal{O}_A(U) \to \mathcal{O}_B(\varphi^{-1}(U))$ for every open set $U \subset \operatorname{Spec} A$, and define a morphism φ of ringed spaces.

It is not true, of course, that any morphism $\varphi\colon \operatorname{Spec} A \to \operatorname{Spec} B$ of ringed spaces is of the form ${}^a\lambda$ (as incorrectly stated in the first edition of this book!). The point is that some relic of the relation (5.10) remains in our general situation: although the left- and right-hand sides of (5.10) take values in different sets, so that equality between them does not make sense, we can nevertheless ask whether they are both zero. For $U \subset \operatorname{Spec} A$, let $\psi_U : \mathcal{O}_A(U) \to \mathcal{O}_B(\varphi^{-1}(U))$ be the homomorphism in the definition of morphism of ringed space. Let $x \in \varphi^{-1}(U)$ and $a \in \mathcal{O}_A(U)$. We can now compare the two properties

$$a\big(\varphi(x)\big) = 0 \quad \text{and} \quad \big(\psi_U(a)\big)(x) = 0.$$

(Recall that the left-hand side is an element of $k(\varphi(x))$, and the right-hand side of $k(x)$, see the discussion at the start of Section 1.2.) The second of these implies the first. Indeed, if $a(\varphi(x)) \neq 0$ then there exists an open set $\varphi(x) \in V \subset U$ in which a is invertible, that is, $aa_1 = 1$ for some $a_1 \in \mathcal{O}(V)$. Hence $(\psi_V(a))(x) \neq 0$, which, together with the commutative diagram in the definition of morphism, contradicts $(\psi_U(a))(x) = 0$. But for an arbitrary map of ringed spaces $\operatorname{Spec} B \to \operatorname{Spec} A$, the first equality does not imply the second (see Exercise 11), while this is true tautologically for morphisms of the form ${}^a\lambda$, as we have just seen.

Definition 5.3 A morphism of ringed spaces $\varphi\colon \operatorname{Spec} B \to \operatorname{Spec} A$ is *local* if for any $U \subset \operatorname{Spec} A$, any $x \in \operatorname{Spec} B$ with $\varphi(x) \in U$, and any $a \in \mathcal{O}_A(U)$, we have

$$a\bigl(\varphi(x)\bigr) = 0 \quad \Longrightarrow \quad \bigl(\psi_U(a)\bigr)(x) = 0.$$

It follows from what we said above that, for a local morphism of ringed spaces, the two conditions $a(\varphi(x)) = 0$ and $(\psi_U(a))(x) = 0$ are equivalent. The same condition can be expressed as follows. At the level of affines, if $U = \operatorname{Spec} B$, then $\psi_U\colon \mathcal{O}_Y(U) = A \to \mathcal{O}_X(\varphi^{-1}(U)) = B$ maps the prime ideal $\mathfrak{p}_x \in \operatorname{Spec} A$ into the prime ideal $\mathfrak{p}_{\varphi(x)} \in \operatorname{Spec} B$. At the level of stalks of the structure sheaf, ψ induces maps $\mathcal{O}_{Y,\varphi(x)} = B_{\mathfrak{p}_{\varphi(x)}} \to \mathcal{O}_{X,x} = A_{\mathfrak{p}_x}$, which must take the maximal ideals to one another, that is, $\psi(\mathfrak{m}_{\varphi(x)}) \subset \mathfrak{m}_x$. The latter is called a *local homomorphism* of local rings.

Theorem 5.2 *Every local morphism $\varphi\colon \operatorname{Spec} B \to \operatorname{Spec} A$ can be expressed uniquely in the form $\varphi = {}^a\lambda$, where $\lambda\colon A \to B$ is a homomorphism.*

Proof There is, of course, only one candidate for λ, namely ψ_U, where $U = \operatorname{Spec} A$. We must prove that $\varphi = {}^a\lambda$. First we need to check this equality on the set $\operatorname{Spec} B$. This follows at once from the fact that φ is local. Indeed, the equalities $a(\varphi(x)) = 0$ and $(\psi_U(a))(x) = 0$ are equivalent for $x \in \operatorname{Spec} B$ and $U = \operatorname{Spec} A$; that is, $\varphi(x) = \psi_U^{-1}(x) = {}^a\lambda(x)$. The equality of the two homomorphisms ψ_U for φ and for ${}^a\lambda$ holds for $U = \operatorname{Spec} A$ by definition, and it then follows for any U from the commutativity of the diagram in the definition of morphism of ringed space. The theorem is proved. □

From now on, we often denote a ringed space $X, \mathcal{O}_X$ by the single letter X, and a morphism $X \to Y$, which is defined by a map φ and homomorphisms ψ_U, by the single letter φ.

A simple verification shows that composing morphisms $\varphi\colon X \to Y$ and $\varphi'\colon Y \to Z$ of ringed spaces (that is, composing both the φ and the ψ_U) gives a morphism $\varphi' \circ \varphi\colon X \to Z$. A morphism that has an inverse is called an *isomorphism of ringed spaces*. If $X, \mathcal{O}_X$ is a ringed space and $U \subset X$ an open subset, then restricting the sheaf $\mathcal{O}_X$ to U defines a ringed space $U, \mathcal{O}_{X|U}$. In this sense we will in what follows often consider an open subset $U \subset X$ as a ringed space. We make two comments on Examples 5.15–5.17 above.

Remark 5.1 Whereas in Examples 5.15–5.16 a morphism was uniquely determined by the map $\varphi\colon X \to Y$ on sets, because the corresponding homomorphisms ψ_U where given by pullback of functions, this is not the case in Example 5.17. For example, if A has a nonzero nilradical N, $B = A/N$ and $\lambda\colon A \to B$ is the natural quotient map, then as sets, $\operatorname{Spec} A = \operatorname{Spec} B$, and $\varphi = {}^a\lambda$ is the identity map, whereas even on $U = \operatorname{Spec} B$ the map $\psi_U = \lambda$ is not an isomorphism. Thus a morphism of ringed spaces cannot be reduced to the map of the corresponding topological spaces.

Remark 5.2 The notion of ringed space provides a convenient principle for the classification of geometric objects. Consider, for example, differentiable manifolds. They can be defined as ringed spaces, namely, as those for which every point has a neighbourhood U such that the ringed space $U, \mathcal{O}_{|U}$ is isomorphic to $\overline{U}, \overline{\mathcal{O}}$, where $\overline{U}$ is a domain in n-dimensional Euclidean space, and $\overline{\mathcal{O}}$ is the sheaf of differentiable functions on it. This is precisely the definition used in de Rham [25], except that he does not use the terminology of sheaves.

The general idea of this method for defining geometric objects is as follows: we impose a restriction on the local structure of a ringed space, that is, we fix in advance a class of ringed spaces, and require that every point has a neighbourhood isomorphic as a ringed space to one of these.

The last remark leads us to the basic definition.

Definition 5.4 A *scheme* is a ringed space $X, \mathcal{O}_X$ for which every point has a neighbourhood U such that the ringed space $U, \mathcal{O}_{X|U}$ is isomorphic to $\operatorname{Spec} A$, where A is some ring.

A neighbourhood U of x for which $U, \mathcal{O}_{X|U}$ is isomorphic to $\operatorname{Spec} A$ is called an *affine neighbourhood* of x. The residue field $k(x)$ and the tangent space Θ_x (compare Section 1.2) are independent of the choice of affine neighbourhood. In exactly the same way, the stalk $\mathcal{O}_x$ of the structure sheaf $\mathcal{O}$ does not depend on whether we consider x as a point of X or of its neighbourhood U. Hence $\mathcal{O}_x$ is a local ring and if $\mathfrak{m}_x$ is its maximal ideal then $\mathcal{O}_x/\mathfrak{m}_x \cong k(x)$.

A *morphism of schemes* $f: X \to Y$ is defined as a local morphism of the corresponding ringed spaces; that is, $f: X \to Y$ is a morphism of ringed spaces such that for every point $x \in X$, every affine neighbourhood $U \subset X$ of x, and every affine set $V \subset Y$ with $f(U) \subset V$, the morphism of affine schemes $f: U \to V$ is local.

For a morphism of schemes $f: X \to Y$ and any point $x \in X$ there exist affine neighbourhoods $U \subset X$ of x and $V \subset Y$ of $f(x)$ such that $f(U) \subset V$. Since the morphism f is local, $f: U \to V$ is of the form $f = {}^a\lambda$, with $\lambda: A \to B$ a homomorphism, where $U = \operatorname{Spec} B$, $V = \operatorname{Spec} A$. Hence it defines an inclusion of fields $\psi_x: k(f(x)) \to k(x)$. For $a \in \mathcal{O}_Y(V) = A$, we then have a relation analogous to (5.10)

$$\psi_x\big(a\big(f(x)\big)\big) = \lambda(a)(x).$$

If X is a scheme and A a ring then a morphism $X \to \operatorname{Spec} A$ defines a homomorphism $A \to \mathcal{O}_X(U)$ for any open set $U \subset X$, that is, it makes $\mathcal{O}_X$ into a sheaf of A-algebras. It is not hard to prove that, conversely, if $\mathcal{O}_X$ is a sheaf of A-algebras, then this determines a canonical morphism $X \to \operatorname{Spec} A$. A scheme X having a morphism $X \to \operatorname{Spec} A$ is called a scheme *over* A or an A-scheme. A morphism of A-schemes is defined by requiring that the diagram

$$\begin{array}{ccc} Y & \xrightarrow{\varphi} & X \\ & \searrow \quad \swarrow & \\ & \operatorname{Spec} A & \end{array}$$

is commutative; this is equivalent to saying that all the ψ_U are A-algebra homomorphisms. The case that appears most frequently is when $A = k$, that is, the case of schemes over a field k or k-schemes.

Since any ring is an algebra over the ring of integers $\mathbb{Z}$, every scheme is a scheme over $\mathbb{Z}$. In this sense, schemes over A are a generalisation of schemes.

Here are the two simplest examples of schemes.

Example 5.18 Example 5.17 of a ringed space shows that $\operatorname{Spec} A$ is a scheme for any ring A. Schemes of this type are called *affine schemes*. Ring homomorphisms $\lambda\colon A \to B$ and morphisms of schemes $\operatorname{Spec} B \to \operatorname{Spec} A$ are in one-to-one correspondence; the correspondence is given by $\varphi = {}^a\lambda$.

Example 5.19 We explain how the notion of quasiprojective variety fits into the framework of schemes. We start from the case of an affine variety X over an algebraically closed field k. The scheme $\operatorname{Spec}(k[X])$ defined in Example 5.18 is not equal to X even as a set: the points of $\operatorname{Spec}(k[X])$ are all the prime ideals of $k[X]$, which correspond in turn to all the irreducible subvarieties of X, not just its points. Notwithstanding this, the variety X and the scheme $\operatorname{Spec}(k[X])$ are related to one another in a very natural way: the set of points of X is contained in $\operatorname{Spec}(k[X])$ as a topological space, and the regular maps $X \to Y$ of affine varieties and the morphisms of schemes $\operatorname{Spec}(k[X]) \to \operatorname{Spec}(k[Y])$ are the same thing: both correspond to algebra homomorphisms $k[Y] \to k[X]$. Thus we have here an isomorphism of categories.

We now consider an arbitrary quasiprojective variety X over k, and associate with X in a similar way a k-scheme $\widetilde{X}$. As the set $\widetilde{X}$ we take the set of irreducible subvarieties of X. Let $U \subset X$ be an open subset and $\widetilde{U}$ the set of irreducible subvarieties of U. Sending an irreducible subvariety $Z \subset U$ to its closure $\overline{Z} \subset X$ defines an embedding $\widetilde{U} \hookrightarrow \widetilde{X}$. The subsets $\widetilde{U} \subset \widetilde{X}$ define a topology on $\widetilde{X}$. Finally, we define a sheaf $\mathcal{O}_{\widetilde{X}}$ by the condition $\mathcal{O}_{\widetilde{X}}(\widetilde{U}) = k[U]$, with the natural restriction maps. We leave to the reader the task of checking that in this way we make $\widetilde{X}$ into a k-scheme.

A regular map $f\colon X \to Y$ defines a map of sets $\widetilde{f}\colon \widetilde{X} \to \widetilde{Y}$ in which an irreducible variety $Z \subset X$ corresponds to $\widetilde{f}(Z)$, the closure of $f(Z)$ in Y. Finally, for $U \subset Y$ we define $\widetilde{\psi}_U\colon \mathcal{O}_{\widetilde{Y}}(\widetilde{U}) \to \mathcal{O}_{\widetilde{X}}(\widetilde{f}^{-1}(\widetilde{U}))$ to be the pullback $f^*\colon k[U] \to k[f^{-1}(U)]$. The reader will easily verify that $\widetilde{f}\colon \widetilde{X} \to \widetilde{Y}$ is a morphism of schemes, and that $f \mapsto \widetilde{f}$ defines a one-to-one correspondence between regular maps $X \to Y$ and morphisms $\widetilde{X} \to \widetilde{Y}$. We again have an isomorphism of categories.

In what follows we will often no longer distinguish between a quasiprojective variety and the corresponding scheme.

Example 5.20 We are now in a position to clear up the question of describing the set of algebras or of multiplication laws by the equations that express associativity

((1.28) of Section 4.1, Chapter 1). Multiplication laws correspond to the points of an affine *subscheme* of $\mathbb{A}^{n^3}$, whose ideal is generated by the left-hand side of the associativity relations. Example 2.5 of Volume 1 shows that the tangent space to a closed point of this scheme coincides with the space of cocycles.

In conclusion we make some obvious remarks concerning the definition of scheme. The structure sheaf $\mathcal{O}$ of a scheme X has an important property: its stalk $\mathcal{O}_x$ at any point $x \in X$ is a local ring. Indeed, the stalk $\mathcal{F}_x$ of any sheaf $\mathcal{F}$ on a space X is not changed if we pass from X to an open subset U containing x. For the structure sheaf on an affine scheme $\operatorname{Spec} A$ we have already seen that $\mathcal{O}_x$ is the local ring of A at the prime ideal $x \in \operatorname{Spec} A$. Because of this, local properties of affine schemes, such as regularity of a point, tangent space, and so on, carry over automatically to arbitrary schemes.

The properties of irreducibility and dimension, formulated in Section 1.4 in terms of the topology of a scheme apply also to arbitrary schemes.

Finally, certain notions introduced earlier for quasiprojective varieties carry over at once to schemes. A *rational map* of a scheme X to a scheme Y is an equivalence class of morphisms $\varphi\colon U \to Y$ where U is an open dense subset of X, and two morphisms $\varphi\colon U \to Y$ and $\psi\colon V \to Y$ are equal if they coincide on $U \cap V$. Schemes X and Y are *birational* or *birationally equivalent* if they have isomorphic dense open subsets (compare Proposition of Section 4.3, Chapter 1).

3.2 Glueing Schemes

By definition any scheme is covered by open sets isomorphic to affine schemes, which we call simply affine open sets. Can we recover X from such a cover $X = \bigcup X_\alpha$? We consider this question in somewhat greater generality, without presupposing the open sets U_α to be affine.

We note first that any open set $U \subset X$ is a scheme. This follows from that fact that each point has an affine neighbourhood V, and the open sets $D(f) \subset V$ form a basis for the open sets. If $X = \bigcup U_\alpha$ is an open cover then the schemes U_α are not independent: U_α and U_β have an isomorphic open set $U_\alpha \cap U_\beta$. Hence we start from the following data: a system of schemes U_α with $\alpha \in I$, in each U_α a system of open subsets $U_{\alpha\beta} \subset U_\alpha$ for $\alpha, \beta \in I$, with $U_{\alpha\alpha} = U_\alpha$, and a system of isomorphisms of schemes $\varphi_{\alpha\beta}\colon U_{\alpha\beta} \to U_{\beta\alpha}$. We determine under what conditions it is possible to construct a scheme X, an open cover $X = \bigcup V_\alpha$ and a system of isomorphisms $\psi_\alpha\colon U_\alpha \to V_\alpha$ such that ψ_α restricted to $U_{\alpha\beta}$ defines an isomorphism $U_{\alpha\beta} \xrightarrow{\sim} V_\alpha \cap V_\beta$, and $\psi_\beta \circ \varphi_{\alpha\beta} \circ \psi_\alpha^{-1}$ is the identity map of $V_\alpha \cap V_\beta$. If X exists, we say that it is obtained by glueing the U_α.

It is easy to check that, in order for glueing to be possible, the following conditions must hold:

$$\varphi_{\alpha\alpha} = \mathrm{id} \quad \text{for } \alpha \in I \quad \text{and} \quad \varphi_{\alpha\beta} \circ \varphi_{\beta\alpha} = \mathrm{id} \quad \text{for } \alpha, \beta \in I. \tag{5.11}$$

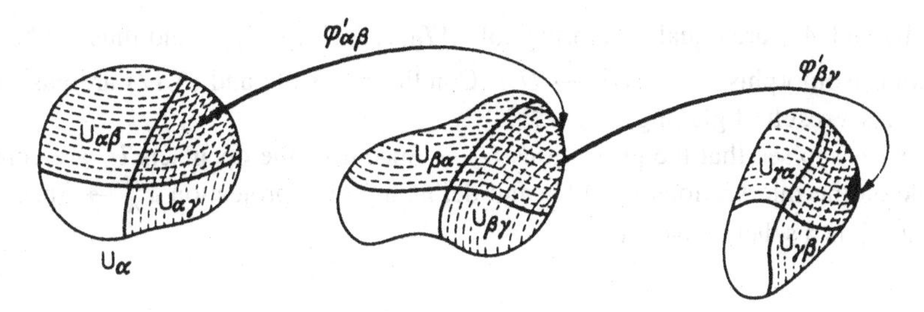

Figure 24 The glueing conditions

The restriction $\varphi'_{\alpha\beta}$ of $\varphi_{\alpha\beta}$ to $U_{\alpha\beta} \cap U_{\alpha\gamma}$ is an isomorphism of $U_{\alpha\beta} \cap U_{\alpha\gamma}$ with $U_{\beta\alpha} \cap U_{\beta\gamma}$, and these isomorphisms are related by

$$\varphi'_{\alpha\gamma} = \varphi'_{\beta\gamma} \circ \varphi'_{\alpha\beta} \quad \text{for } \alpha, \beta, \gamma \in I. \tag{5.12}$$

The morphisms and schemes appearing in conditions (5.11) and (5.12) are illustrated in Figure 24.

Let us prove that if (5.11) and (5.12) hold then glueing is possible. First we determine X as a set. For this we set T to be the disjoint union of all the U_α, and introduce the equivalence relation $x \sim y$ if $x \in U_{\alpha\beta}$, $y \in U_{\beta\alpha}$ and $y = \varphi_{\alpha\beta}(x)$. Conditions (5.11) and (5.12) guarantee that $\sim$ is an equivalence relation. We write $X = T/\sim$ for the quotient set and $p\colon T \to X$ for the quotient map.

Introduce the quotient topology on X, setting $U \subset X$ to be open if $p^{-1}(U) \subset T$ is open (the topology of T is defined by the open sets $\bigcup W_\alpha$ with $W_\alpha \subset U_\alpha$ open sets). It is easy to see that p establishes a homeomorphism ψ_α of the sets U_α with open subsets $V_\alpha \subset X$, and that $X = \bigcup V_\alpha$.

Finally, we define a sheaf $\mathcal{O}_X$ on X as follows. For W contained in some V_α, we make some choice of $V_\alpha \supset W$, and set $\mathcal{O}_X(W) = \mathcal{O}_{V_\alpha}(\psi_\alpha^{-1}(W))$. The choice of a different $V_\beta \supset W$ replaces $\mathcal{O}_X(W)$ by an isomorphic ring. The restriction homomorphisms $\rho^W_{W'}$ for $W' \subset W \subset V_\alpha$ are defined in the obvious way. Thus the presheaf $\mathcal{O}_X$ is not defined on all open subsets of X, but the W on which it is defined form a basis for the open sets of the topology. The situation is as for the definition of the structure sheaf of Spec A. In the same way as there, we can extend the definition of $\mathcal{O}_X(U)$ to all open sets $U \subset X$ as the projective limit $\varprojlim \mathcal{O}_X(W)$, taken over open sets $W \subset U$ where $\mathcal{O}_X(W)$ is already defined. There remains the standard verification of a large number of properties (that $\mathcal{O}_X$ is a sheaf, that X is a scheme, and so on), that we omit.

Example We define a scheme $\mathbb{P}^N_A$, called *projective N-space* over A. For this, consider the polynomial ring $A[T_0, \dots, T_N]$ in $N+1$ independent variables $T_0, \dots, T_N$. In the ring of fractions $A[T_0, \dots, T_N]_{(T_0 \cdots T_N)}$, consider the subrings $A_i = A[T_0/T_i, \dots, T_N/T_i]$. We set $U_i = \operatorname{Spec} A_i$ and $U_{ij} = D(T_j/T_i) \subset U_i$. By definition $U_{ij} = \operatorname{Spec} A_{ij}$, where $A_{ij} = (A_i)_{(T_j/T_i)}$ consists of elements $F(T_0, \dots, T_N)/(T_i^p T_j^q)$ where F is a form of degree $p+q$. It follows from this that A_{ij} and A_{ji} are equal as subrings of $A[T_0, \dots, T_N]_{(T_0 \cdots T_N)}$, and thus we have a natural isomorphism $\varphi_{ij} : U_{ij} \xrightarrow{\sim} U_{ji}$. Conditions (5.11) and (5.12) are easy to check. As a result of glueing we get $\mathbb{P}^N_A$.

It is easy to see that the projections $U_i \to \operatorname{Spec} A$ of the open sets U_i are compatible on the intersections $U_i \cap U_j$ and define a global projection $\mathbb{P}^N_A \to \operatorname{Spec} A$, so that $\mathbb{P}^N_A$ is a scheme over A.

3.3 Closed Subschemes

If $\lambda\colon A \to B$ is a surjective ring homomorphism then the associated map ${}^a\lambda\colon \operatorname{Spec} B \to \operatorname{Spec} A$ defines a homeomorphism of $\operatorname{Spec} B$ and a closed subset $V(\mathfrak{a}) \subset \operatorname{Spec} A$, where $\mathfrak{a} = \ker \lambda$. In this case, we say that $\operatorname{Spec} B$ is a *closed subscheme* of $\operatorname{Spec} A$, and ${}^a\lambda$ a *closed embedding* or *closed immersion.* We now generalise these notions to arbitrary schemes.

Definition 5.5 A morphism of schemes $\varphi\colon Y \to X$ is a *closed embedding* if every point $x \in X$ has an affine neighbourhood U such that $\varphi^{-1}(U) \subset Y$ is an affine subscheme and the homomorphism $\psi_U : \mathcal{O}_X(U) \to \mathcal{O}_Y(\varphi^{-1}(U))$ is surjective. In this case we say that Y is a *closed subscheme* of X.

Since closed is a local property, in this case $\varphi(Y)$ is a closed subspace of X. To show that the example we started from is covered by this definition, we prove the following assertion.

Proposition *If* $X = \operatorname{Spec} A$ *is an affine scheme and* $\varphi\colon Y \to X$ *a closed embedding then* Y *is also affine,* $Y = \operatorname{Spec} B$*, and* $\varphi = {}^a\lambda$*, where* $\lambda\colon A \to B$ *is a surjective ring homomorphism.*

Proof By definition of a closed embedding, we can find a cover $X = \bigcup U_i$, where $U_i = D(f_i)$ are principal open sets with $f_i \in A$ such that $\varphi^{-1}(U_i) = \operatorname{Spec}(A_i)$ and the $\psi_i\colon A_{f_i} \to A_i$ are surjective. Set $\mathfrak{a}_i = \ker \psi_i \subset A_{f_i}$, so that $\rho^X_{U_i} = {}^a\lambda_i$, and finally set $\mathfrak{a} = \bigcap \lambda_i^{-1}(\mathfrak{a}_i)$. The morphism φ makes Y into a scheme over A, that is, the sheaf $\mathcal{O}_Y$ has an A-module structure. But since $\mathfrak{a} \subset \lambda_i^{-1}(\mathfrak{a}_i)$, under the action of A on $\mathcal{O}_Y(\varphi^{-1}(U_i))$, the ideal $\mathfrak{a}$ acts trivially. In other words, Y is a scheme over $A/\mathfrak{a}$. This means that there is a commutative diagram

$$\begin{array}{ccc} Y & \xrightarrow{\varphi} & X \\ & {}_u\searrow \quad \nearrow_v & \\ & \operatorname{Spec}(A/\mathfrak{a}) & \end{array}$$

with v a closed embedding.

The proposition will be proved if we check that u is an isomorphism. Locally, u is given (in open sets $\varphi^{-1}(U_i)$ and $v^{-1}(U_i)$) by

$$u_i\colon (A/\mathfrak{a})_{\overline{f_i}} \to A_{f_i}/\mathfrak{a}_i,$$

where $\overline{f_i}$ is the image of f_i in $A/\mathfrak{a}$. It is enough to prove that all the u_i are isomorphisms.

The surjectivity of the u_i follows at once from the fact that $\mathfrak{a} \subset \lambda_i^{-1}(\mathfrak{a}_i)$. The proof that they are injective uses the following remark. The ring $\mathcal{O}_Y(\varphi^{-1}(U_i \cap U_j))$ can be described in two ways:

$$\mathcal{O}_Y\big(\varphi^{-1}(U_i \cap U_j)\big) = (A_i)_{\psi_i(\lambda_i f_j)} = (A_j)_{\psi_j(\lambda_j f_i)}. \tag{5.13}$$

Consider the localisation $\lambda_j^i\colon A_{f_i} \to (A_{f_i})_{\lambda_i(f_j)} = A_{(f_i f_j)}$. It follows at once from (5.13) that

$$\lambda_j^i(\mathfrak{a}_i) = \lambda_i^j(\mathfrak{a}_j), \tag{5.14}$$

where $\lambda_j^i(\mathfrak{a}_i)$, say, is the ideal of $A_{(f_i f_j)}$ generated by elements $\lambda_j^i(\alpha)$ with $\alpha \in \mathfrak{a}_i$.

Suppose that $a \in A$ defines an element of the kernel of u_i. Then $\lambda_i(a) \in \mathfrak{a}_i$. By (5.14) it follows that

$$\lambda_j^i\big(\lambda_i(a)\big) \in \lambda_i^j(\mathfrak{a}_j).$$

The left-hand side is the image of a under the localisation $A \to A_{(f_i f_j)}$ and is hence equal to $\lambda_i^j(\lambda_j(a))$, and elements of the right-hand side are of the form $\lambda_i^j(a_j)/\lambda_j(f_i)^k$. Thus

$$\lambda_i^j\big(\lambda_j(f_i)^k \lambda_j(a) - a_j\big) = 0.$$

Hence

$$\lambda_j(f_i)^{k+l}\lambda_j(a) = \lambda_j(f_i)^l a_j \in \mathfrak{a}_j$$

for some l. We see that

$$\lambda_j\big(f_i^{k+l} a\big) \in \mathfrak{a}_j, \tag{5.15}$$

and moreover, k and l can be chosen the same for all j. The relation (5.15), which is now proved for all j, shows that $f_i^{k+l} a \in \mathfrak{a}$, that is, $(\overline{f_i})^{k+l}\overline{a} = 0$, where $\overline{a}$ is the image of a in $A/\mathfrak{a}$. But this means that $\overline{a}$ defines the zero element of $(A/\mathfrak{a})_{\overline{f_i}} = 0$. The proposition is proved. □

Definition 5.6 A scheme isomorphic to a closed subscheme of $\mathbb{P}_A^N$ (see Example of Section 3.2) is a *projective scheme* over Spec A (or over A).

By definition, a closed subscheme $X \subset \mathbb{P}_A^N$ can be obtained by glueing $N+1$ affine schemes $V_i = U_i \cap X$ for $i = 0, \dots, N$, where $V_i = X \cap \mathbb{A}_i^N$, and the structure sheaf of V_i is the restriction $\mathcal{O}_{X|V_i}$. Then $V_i = \operatorname{Spec} C_i$, where $C_i = A_i/\mathfrak{a}_i$ with

$A_i = A[T_0/T_i, \dots, T_N/T_i]$ and $\mathfrak{a}_i$ an ideal of A_i. But, as in the case of projective varieties (see Section 4.1, Chapter 1), projective schemes can also be defined by homogeneous ideals. For this, we set $\Gamma = A[T_0, \dots, T_N]$. If $\Gamma^{(r)}$ is the submodule of forms of degree r in Γ then $\Gamma = \bigoplus \Gamma^{(r)}$. We write $\mathfrak{a}^{(r)}$ for the module of forms $F \in \Gamma^{(r)}$ such that $F/T_i^r \in \mathfrak{a}_i$ for $i = 0, \dots, N$, and set $\mathfrak{a}_X = \bigoplus \mathfrak{a}^{(r)}$. Obviously $\mathfrak{a}_X$ is a homogeneous ideal of Γ, called the ideal of the projective scheme $X \subset \mathbb{P}^n_A$. It follows from the definition that for $F \in \Gamma$,

$$T_i F \in \mathfrak{a}_X \quad \text{for } i = 0, \dots, N \quad \Longrightarrow \quad F \in \mathfrak{a}_X. \tag{5.16}$$

Conversely, any homogeneous ideal $\mathfrak{a} \subsetneqq \Gamma$ satisfying (5.16) defines a closed subscheme $X \subset \mathbb{P}^N_A$ for which $\mathfrak{a}_X = \mathfrak{a}$. We need only take $\mathfrak{a}_i$ to be the ideal $\mathfrak{a}_i \subset A_i$ consisting of elements of the form F/T_i^r with $F \in \mathfrak{a}^{(r)}$. It is not hard to check that the closed affine subschemes $V_i \subset U_i$ defined by $\mathfrak{a}_i$ for $i = 0, \dots, N$ glue together into a global closed subscheme $X \subset \mathbb{P}^N_A$ with the required properties. We omit the elementary but boring verification of these assertions.

All the arguments in the proof of Theorem 1.11 of Section 5.2, Chapter 1 go through on replacing k by an arbitrary ring A.

Already the familiar quasiprojective varieties contain many more closed subschemes than closed subvarieties. For example, a closed subscheme of the affine line $X = \operatorname{Spec} k[T]$ is of the form $\operatorname{Spec} k[T]/(F)$, where $F(T)$ is an arbitrary polynomial, whereas closed subvarieties correspond only to the set of roots of these polynomials, taking no account of their multiplicities.

If $\varphi\colon X \to Y$ is a morphism of schemes and $Y' \subset Y$ a closed subscheme of Y then we can define its pullback or scheme-theoretic inverse image $\varphi^{-1}(Y')$, which will be a closed subscheme of X. We restrict ourselves for the time being to the case that X and Y are affine schemes, with $X = \operatorname{Spec} A$, $Y = \operatorname{Spec} B$ and $\varphi = {}^a\lambda$ for $\lambda\colon B \to A$ a homomorphism; the general case will be treated in Section 4.1 below. Then the closed embedding $Y' \hookrightarrow Y$ is defined by a quotient homomorphism $B \to B/\mathfrak{b}$. If $\lambda(\mathfrak{b})A = A$ then the set-theoretic inverse image $\varphi^{-1}(Y')$ is empty. Otherwise, the scheme $X' = \operatorname{Spec}(A/(\lambda(\mathfrak{b})A))$ is obviously a closed subscheme of X. It is called the *scheme-theoretic inverse image* of $Y' \subset Y$. Its topological space is indeed the inverse image of $Y' \subset Y$.

For example, if X and Y are both isomorphic to the affine line $\mathbb{A}^1_k$ over an algebraically closed field k of characteristic $\neq 2$, and φ is the map given by $\varphi(x) = x^2$, then $\varphi^{-1}(y)$ for $y \neq 0$ consists of two connected components isomorphic to $\operatorname{Spec} k$ (that is, two "ordinary" points), but for $y = 0$ it is the subscheme $\varphi^{-1}(y) = \operatorname{Spec}(k[T]/(T^2))$. This example show that schemes for which the rings $\mathcal{O}(U)$ have nilpotent elements can arise in classical situations. We have already seen that it is natural to define the inverse image of a codimension 1 subvariety as a divisor, that is, as a sum of codimension 1 subvarieties with multiplicities. In simple cases, these multiplicities turn out to be sufficient to specify the scheme-theoretic structure sheaf on these subvarieties. In the general case, this is a palliative: it is clear that the inverse image under a morphism between two objects should be an object of the same type, in the present case a scheme. But then we very quickly arrive at schemes with nilpotents.

An even more extreme case is the example where $X = Y = \mathbb{A}^1$ is the affine line over an algebraically closed field k of characteristic p and $\varphi(x) = x^p$. This map is a one-to-one correspondence, but is not an isomorphism. Applying our notion of scheme-theoretic inverse image, we get that $\varphi^{-1}(y) \cong \operatorname{Spec} k[T]/(T^p)$ for every closed point $y \in Y$, that is, the inverse image of every point contains nilpotent elements in its structure sheaf. It is interesting that in this case X and Y are algebraic groups with respect to addition, and φ is a homomorphism of algebraic groups. Thus it is natural to expect that $\varphi^{-1}(0)$ will again be a "group" of some new type. We will see in the next section that this is indeed the case.

3.4 Reduced Schemes and Nilpotents

We say that a scheme X is *reduced* if the rings $\mathcal{O}_X(U)$ have no nilpotent elements. We can associate with every scheme X a closed reduced subscheme $X' \subset X$ which coincides with X as a topological space: for an open set $U \subset X$ the ring $\mathcal{O}_{X'}(U)$ is defined as the quotient of $\mathcal{O}_X(U)$ by its nilradical (that is, by the ideal formed by all nilpotent elements). This scheme is denoted by X_{red}.

Example 5.21 Let X be a scheme over an algebraically closed field k. Any morphism of k-schemes $\varphi\colon \operatorname{Spec} k \to X$ takes the closed point $o \in \operatorname{Spec} k$ to a closed point $x = \varphi(o) \in X$, with $k(x) = k$. Conversely, any point $x \in X$ with these properties obviously defines a morphism $\operatorname{Spec} k \to X$: it is enough to specify the morphism $\operatorname{Spec} k \to U$, where U is an affine neighbourhood of x; now if $U = \operatorname{Spec} A$, and $\mathfrak{m}_x$ is the maximal ideal of x then the homomorphism $A \to k$ with kernel $\mathfrak{m}_x$ determines φ. If X is of finite type over k (see Section 3.5) then the condition $k(x) = k$ is automatic for all closed points; this is a consequence of the Nullstellensatz. Then the one-to-one correspondence

$$\{\text{morphisms } \operatorname{Spec} k \to X\} \xrightarrow{\sim} X_{\max},$$

where $X_{\max}$ is the set of closed points of a scheme X of finite type over k, obviously commutes with morphisms $X \to X'$, that it, is a functor.

Example 5.22 After $\operatorname{Spec} k$, the next simplest scheme is $\operatorname{Spec} D$, where $D = k[\varepsilon]/\varepsilon^2 = k + k\varepsilon$ is the so-called algebra of *dual numbers*. We describe the morphisms of $\operatorname{Spec} D$ to a scheme X of finite type over k. Since D has a unique maximal ideal (ε), $\operatorname{Spec} D$ has a unique closed point, which we denote by $\overline{o}$. The homomorphism $D \to k$ with kernel (ε) defines a canonical embedding $i\colon \operatorname{Spec} k \hookrightarrow \operatorname{Spec} D$ with $\overline{o} = i(o)$. Any morphism $\varphi\colon \operatorname{Spec} D \to X$ determines the composite morphism $\varphi \circ i\colon \operatorname{Spec} k \to X$, which, as we have seen, determines a closed point $x \in X$. Obviously $x = \varphi(\overline{o})$. Let U be some affine neighbourhood of x and $U = \operatorname{Spec} A$. If we write $\mathcal{M}_x(\operatorname{Spec} D, X)$ for the set of morphisms of schemes $\operatorname{Spec} D \to X$ such that $\varphi(\overline{o}) = x$, then $\mathcal{M}_x(\operatorname{Spec} D, X) = \mathcal{M}_x(\operatorname{Spec} D, U)$. The latter equals the set of homomorphisms $f\colon A \to D$ such that $f(\mathfrak{m}_x) \subset (\varepsilon)$. Since, as a vector space, $A = k + \mathfrak{m}_x$, such a homomorphism is determined by its effect on $\mathfrak{m}_x$, where it defines a linear form $\mathfrak{m}_x \to (\varepsilon) \cong k$. Since $\varepsilon^2 = 0$, we must have $f(\mathfrak{m}_x^2) = 0$, that is, f is a linear function on $\mathfrak{m}_x/\mathfrak{m}_x^2$. Conversely, any such function f, extended to be 0 on $\mathfrak{m}_x^2$, determines a homomorphism $A \to D$ taking $\mathfrak{m}_x$ into (ε), that is, an element of $\mathcal{M}_x(\operatorname{Spec} D, X)$. A linear function on the space $\mathfrak{m}_x/\mathfrak{m}_x^2$ is an element of the tangent space at x to X. Thus we have verified the following result.

Proposition *If X is a k-scheme and $x \in X$, the set $\mathcal{M}_x(\operatorname{Spec} D, X)$ is in one-to-one correspondence with the tangent space $\Theta_{X,x}$ to X at x. It is easy to check that the correspondence $\mathcal{M}_x(\operatorname{Spec} D, X) \xrightarrow{\sim} \Theta_{X,x}$ commutes with morphisms $f\colon X \to X'$ and their differentials $\mathrm{d}_x\colon \Theta_{X,x} \to \Theta_{X',f(x)}$.*

The one-to-one correspondence between $\mathcal{M}_x(\operatorname{Spec} D, X)$ and Θ_x allows us to give a new description of Θ_x not just as a set, but as a vector space. For this, for any element $\lambda \in k$, consider the homomorphism of the algebra D given by $\mu_\lambda\colon \varepsilon \mapsto \varepsilon\lambda$. It defines a morphism ${}^a\mu_\lambda\colon \operatorname{Spec} D \to \operatorname{Spec} D$, and for $\varphi \in \mathcal{M}_x(\operatorname{Spec} D, X)$, taking the composite $\varphi \mapsto \varphi \circ {}^a\mu_\lambda$ defines an operation of scalar product of elements of $\mathcal{M}_x(\operatorname{Spec} D, X)$ by λ, which is compatible with multiplication in Θ_x, as one sees easily. To describe the addition of vectors, we need to consider the algebra $D' = k[\varepsilon_1, \varepsilon_2]/(\varepsilon_1, \varepsilon_2)^2$. This is obviously a local ring with $\mathfrak{m} = (\varepsilon_1, \varepsilon_2)$ and $\mathfrak{m}/\mathfrak{m}^2 = k\varepsilon_1 + k\varepsilon_2$. There are inclusions $i_1, i_2\colon D \to D'$ given by $i_1(\varepsilon) = \varepsilon_1$ and $i_2(\varepsilon) = \varepsilon_2$, and a projection $\pi\colon D' \to D$ given by $\pi(\varepsilon_1) = \pi(\varepsilon_2) = \varepsilon$. Two morphisms $\varphi_1, \varphi_2\colon \operatorname{Spec} D \to X$ define homomorphisms $f_1, f_2\colon A \to D$. From them we get a homomorphism $(i_1 f_1, i_2 f_2)\colon A \to D'$ and $\pi(i_1 f_1, i_2 f_2)\colon A \to D$ that defines the sum of φ_1 and φ_2.

The proposition can be given the following geometric interpretation. Consider an affine neighbourhood U of x and the closed subscheme $T_x = \operatorname{Spec} k[U]/\mathfrak{m}_x^2 \subset U$, where $\mathfrak{m}_x$ is the maximal ideal of x in $k[U]$. The homomorphism $k[U] \to k[U]/\mathfrak{m}_x^2$ defines the closed embedding $T_x \to U$. It is easy to see that T_x is also a closed subscheme of X, and does not depend on the choice of the neighbourhood U. The arguments given above prove that any morphism $\varphi\colon \operatorname{Spec} D \to X$ is of the form $\varphi = j \circ \psi$, where ψ is a morphism $\operatorname{Spec} D \to T_x$ and $j\colon T_x \hookrightarrow X$ the closed embedding. Thus morphisms $\operatorname{Spec} D \to X$ that take $\operatorname{Spec} k$ to x are in one-to-one correspondence with morphisms $\operatorname{Spec} D \to T_x$. The subscheme T_x is quite big: it has the same tangent space at x as X itself. But it is also sufficiently small that a morphism $\varphi\colon \operatorname{Spec} D \to T_x$ is uniquely determined by its differential $\mathrm{d}_o\varphi$, where $o \in \operatorname{Spec} k \subset \operatorname{Spec} D$. This is the geometric interpretation of our computations. They justify the term *first order infinitesimal neighbourhood* of x in X for the subscheme T_x. We can define in a similar way the nth order infinitesimal neighbourhood of x in X.

3.5 Finiteness Conditions

We now treat two properties of schemes having the nature of "finite dimensionality" conditions.

Definition A scheme X is *Noetherian* if it has a finite cover by affine sets

$$X = \bigcup U_i \quad \text{with } U_i = \operatorname{Spec} A_i, \tag{5.17}$$

such that the A_i are Noetherian rings.

A scheme X over a ring B (see Section 3.1) is *of finite type* over B if X has a finite covering (5.17) such that the A_i are algebras of finite type over B.

A scheme of finite type over a Noetherian ring is obviously Noetherian. For each of the notions just introduced, we now prove an assertion having the same format in each case.

Proposition 5.1 *If the affine scheme* Spec A *is Noetherian then* A *is a Noetherian ring.*

Proof By assumption there exists a cover (5.17) such that the A_i are Noetherian rings. Let $\mathfrak{a}_1 \subset \mathfrak{a}_2 \subset \cdots$ be a chain of ideals of A. As we showed in Section 2.2, $A = \mathcal{O}(X)$, where $\mathcal{O}$ is the structure sheaf of X. Consider the ideals $\mathfrak{a}_n^{(i)} = \rho_{U_i}^X(\mathfrak{a}_n)A_i \subset A_i$. Since the A_i are finitely many Noetherian rings, there exists N such that

$$\mathfrak{a}_{n+1}^{(i)} = \mathfrak{a}_n^{(i)} \quad \text{for all } i \text{ and all } n \geq N. \tag{5.18}$$

We prove that then $\mathfrak{a}_{n+1} = \mathfrak{a}_n$ for $n \geq N$. Indeed, since the U_i form a finite cover of X, it follows from (5.18) that

$$\rho_x^X(\mathfrak{a}_{n+1})\mathcal{O}_x = \rho_x^X(\mathfrak{a}_n)\mathcal{O}_x \quad \text{for all } x \in X \text{ and all } n \geq N.$$

It remains to repeat the argument of Section 2.2. If $u \in \mathfrak{a}_{n+1}$ then

$$u = a_x/b_x \quad \text{with } a_x \in \mathfrak{a}_n \text{ and } b_x \in A \text{ with } b_x(x) \neq 0.$$

There exist points $x_1, \ldots, x_r$ and elements $c_1, \ldots, c_r \in A$ such that $c_1 b_{x_1} + \cdots + c_r b_{x_r} = 1$. Then

$$u = \sum a_{x_i} c_i \in \mathfrak{a}_n,$$

that is, $\mathfrak{a}_n = \mathfrak{a}_{n+1}$. The proposition is proved. □

Proposition 5.2 *If an affine scheme* Spec A *is of finite type over a ring* B *then* A *is an algebra of finite type over* B.

Proof By assumption there exist a cover (5.17) such that the algebras A_i are of finite type over B. Since each Spec A_i is compact, it has a finite cover by principal open sets $D(f)$ with $f \in A$. The corresponding algebras $(A_i)_f = A_f$ are of finite type over A. Thus we can assume from the start that $U_i = D(f_i)$ in (5.17). Suppose that the generators of A_i over B are of the form $x_{ij}/f_j^{n_{ij}}$. On the other hand, since $\bigcup D(f_i) = \operatorname{Spec} A$, there exist elements $g_i \in A$ such that

$$\sum f_i g_i = 1. \tag{5.19}$$

Let us write $A' \subset A$ for the algebra generated over B by all the x_{ij}, f_i and g_i, and prove that $A' = A$.

Let $x \in A$. By assumption $x \in A_{f_i}$ for all f_i. This means that there exists an integer n such that $f_i^n x$ belongs to the subalgebra of A generated by the elements x_{ij} and f_i (we can assume that n does not depend on i by choosing it sufficiently large). Then in particular,

$$f_i^n x \in A' \quad \text{for all } f_i. \tag{5.20}$$

Raising (5.19) to a sufficiently high power, we get a relation $\sum f_i^n g_i^{(n)} = 1$, where the $g_i^{(n)}$ belong to the subalgebra of A generated over B by the f_j and g_j. In particular, $g_i^{(n)} \in A'$. Multiplying relations (5.20) by $g_i^{(n)}$ and adding gives $x \in A'$. The proposition is proved. □

3.6 Exercises to Section 3

1 Let $X, \mathcal{O}_X$ be a ringed space and G a group of automorphisms of $X, \mathcal{O}_X$. Define the set $Y = X/G$ to be the quotient set of X by G, and let $p\colon X \to Y$ be the quotient map. Give Y the quotient topology, in which a set $U \subset Y$ is open if and only if $p^{-1}(U) \subset X$ is open. Finally, define $\mathcal{O}_Y$ by the condition $\mathcal{O}_Y(U) = \mathcal{O}_X(p^{-1}(U))^G$. Here A^G denotes the set of G-invariant elements of a ring A—you have to check that G is in a natural way a group of automorphisms of the ring $\mathcal{O}_X(p^{-1}(U))$.

Prove that $Y, \mathcal{O}_Y$ is a ringed space. It is called the *quotient ringed space* of X by G, and denoted by X/G.

2 Let k be an infinite field, $\mathbb{A}^2$ the affine plane over k, $X = \mathbb{A}^2 \setminus (0, 0)$, and suppose that G consists of the automorphisms $(x, y) \mapsto (\alpha x, \alpha y)$ for all $\alpha \in k$ and $\alpha \neq 0$. In the notation of Exercise 1, prove that the ringed space Y equals the projective line $\mathbb{P}^1$ over k.

3 Let X be as in Exercise 2, but G consists of automorphisms $(x, y) \mapsto (\alpha x, \alpha^{-1} y)$ with $\alpha \in k$ and $\alpha \neq 0$. Prove that Y is a scheme. Prove that if $X = \mathbb{A}^2$ and G is as above, then Y is not a scheme.

4 Study the scheme-theoretic inverse image of a point $x \in \operatorname{Spec} \mathbb{Z}$ under the morphism ${}^a\varphi$ of Example 5.3.

5 Study the scheme-theoretic inverse images of points under the morphism $f\colon X \to Y$ projecting the circle $x^2 + y^2 = 1$ onto the x-axis, $f(x, y) = x$, where all varieties are defined over $\mathbb{R}$. In other words $X = \operatorname{Spec} \mathbb{R}[T_1, T_2]/(T_1^2 + T_2^2 - 1)$ and $Y = \operatorname{Spec} \mathbb{R}[T_1]$.

6 In Example 5.19, prove that the points of the variety X are just the closed points of the scheme $\widetilde{X}$.

7 Let Γ be a graded ring: $\Gamma = \bigoplus_{n \geq 0} \Gamma_n$ with $\Gamma_n \cdot \Gamma_m \subset \Gamma_{n+m}$. An ideal $\mathfrak{a} \subset \Gamma$ is *graded* or *homogeneous* if $\mathfrak{a} = \bigoplus_{n \geq 0} (\mathfrak{a} \cap \Gamma_n)$. Write $\operatorname{Proj} \Gamma$ for the *homogeneous prime spectrum* of Γ, that is, the set of homogeneous prime ideals $\mathfrak{p} \subset \Gamma$ not containing the ideal $\bigoplus_{n>0} \Gamma_n$, and introduce in this set the topology induced by the inclusion $\operatorname{Proj} \Gamma \subset \operatorname{Spec} \Gamma$. For a homogeneous element $f \in \Gamma_m$ with $m > 0$, we write $\Gamma_{(f)}$ for the subring of the ring of fractions Γ_f consisting of ratios g/f^k with $g \in \Gamma_{mk}$ for $k \geq 0$. Set $D_+(f) = D(f) \cap \operatorname{Proj} \Gamma$, which we call a *principal open set*. Let ψ_f be the composite map $D_+(f) \to D(f) \to \operatorname{Spec}(\Gamma_f) \to \operatorname{Spec}(\Gamma_{(f)})$. Prove that the structure sheaves on the affine schemes $\operatorname{Spec}(\Gamma_{(f)})$, for homogeneous f, carry over under ψ_f to $\operatorname{Proj} \Gamma$ to define a global sheaf $\mathcal{O}$, and that the ringed space $\operatorname{Proj} \Gamma, \mathcal{O}$ is a scheme. This scheme is also denoted by $\operatorname{Proj} \Gamma$.

8 In the notation of Exercise 7, prove that if Γ is a graded algebra over a ring A, that is, $A \cdot \Gamma_n \subset \Gamma_n$, then $\operatorname{Proj} \Gamma$ has a natural structure of scheme over A.

9 In the notation of Exercise 7, suppose that $\Gamma = A[T_0, \dots, T_n]$ with the usual grading by the degree. Prove that the scheme $\operatorname{Proj} \Gamma$ is isomorphic to $\mathbb{P}^n_A$ (Example of Section 3.2).

10 Let Y be an affine n-dimensional variety over a field k, $y \in Y$ a nonsingular point and $\mathfrak{m}_y \subset k[Y]$ the corresponding maximal ideal. In the notation of Exercise 7, set $\Gamma = \bigoplus_{n \geq 0} \mathfrak{m}_y^n$, (with $\mathfrak{m}_y^0 = k[Y]$). Prove that $\operatorname{Proj} \Gamma = \widetilde{X}$, where X is the variety obtained by blowing up Y with centre y (Sections 4.1–4.3, Chapter 2), and that the morphism $\widetilde{\sigma}\colon \operatorname{Proj} \Gamma \to \operatorname{Spec}(k[Y])$ given by the natural $k[Y]$-algebra structure on Γ (see Exercise 8) corresponds to the blowup up $\sigma\colon X \to Y$.

11 Let $\mathcal{O}$ be the local ring of a nonsingular point of an algebraic curve, η the generic and ζ the closed points of $\operatorname{Spec} \mathcal{O}$. Write K for the field of fractions of $\mathcal{O}$ and ξ for the point of $\operatorname{Spec} K$. Define a morphism of ringed spaces $\operatorname{Spec} K \to \operatorname{Spec} \mathcal{O}$ by setting $\varphi(\xi) = \zeta$ and $\psi_U\colon \mathcal{O} \hookrightarrow K$ the natural inclusion for $U = \operatorname{Spec} \mathcal{O}$, and $\psi_U = 0$ if $U = \{\eta\}$. Prove that φ is a morphism of ringed spaces, but is not of the form ${}^a\lambda$ for any ring homomorphism $\lambda\colon \mathcal{O} \to K$.

12 Let $X = \operatorname{Spec} B$ and $Y = \operatorname{Spec} A$, and suppose that $\varphi\colon X \to Y$ is a morphism of ringed spaces; for $x \in X$, write $y = \varphi(x)$. By considering all possible neighbourhoods U of y, prove that the homomorphisms $\psi_U\colon \mathcal{O}_Y(U) \to \mathcal{O}_X(\varphi^{-1}(U))$ define a homomorphism $\psi_x\colon \mathcal{O}_{Y,y} \to \mathcal{O}_{X,x}$. Prove that φ satisfies the condition for a local morphism of ringed space at x (see Definition 5.3) if and only if $\psi_x(\mathfrak{m}_{Y,y}) \subset \mathfrak{m}_{X,x}$, where $\mathfrak{m}_{X,x} \subset \mathcal{O}_{X,x}$ and $\mathfrak{m}_{Y,y} \subset \mathcal{O}_{Y,y}$ are the maximal ideals.

4 Products of Schemes

4.1 Definition of Product

It is quite hopeless to try to define the product of schemes X and Y in terms of the set of pairs (x, y) with $x \in X$ and $y \in Y$. Indeed, for $X = Y = \mathbb{A}^1$, we have $X \times Y = \mathbb{A}^2$. The points of the scheme $X \times Y$ thus correspond to irreducible subvarieties of the plane $\mathbb{A}^2$. Therefore these include all irreducible plane curves, which, however, cannot be expressed in the form of pairs (x, y). Because of this, we start by trying to establish the properties that we want products of scheme to satisfy, postponing the question of the existence of a scheme with these properties until later. We used a similar process in Section 5.1, Chapter 1 to arrive at the definition of products of quasiprojective varieties.

We consider schemes over an arbitrary ring A. By definition (see Section 3.1), this means a scheme X together with a morphism $X \to \operatorname{Spec} A$. We consider a still more general situation, a morphism of two arbitrary schemes $X \to S$. Such an object is called a scheme over S or an S-scheme. It is obvious how to define a morphism between two S-schemes $\varphi\colon X \to S$ and $\psi\colon Y \to S$: this is a morphism $f\colon X \to Y$ for which $\varphi = \psi \circ f$.

If $\varphi\colon X \to S$ and $\psi\colon Y \to S$ are two schemes over S, then their product over S (which we denote by $X \times_S Y$) should obviously have projections to the factors, that is, two morphisms of S-schemes $p_X\colon X \times_S Y \to X$ and $p_Y\colon X \times_S Y \to Y$ fitting in a commutative diagram

$$\begin{array}{ccccc} & & X \times_S Y & & \\ & {}^{p_X}\swarrow & & \searrow^{p_Y} & \\ Y & & & & X \\ & {}^{\varphi}\searrow & & \swarrow^{\psi} & \\ & & S & & \end{array}$$

Moreover, it is natural to require that the product is universal. This means that for any scheme Z, and morphisms $u\colon Z \to X$ and $v\colon Z \to Y$ for which the diagram

$$\begin{array}{ccccc} & & Z & & \\ & \swarrow & & \searrow & \\ Y & & & & X \\ & {}^{\varphi}\searrow & & \swarrow^{\psi} & \\ & & S & & \end{array}$$

is commutative, there should exist a morphism $h\colon Z \to X \times_S Y$ such that $p_X \circ h = u$, $p_Y \circ h = v$, and the morphism h with these properties should be unique; it is be denoted by (u, v).

If a scheme $X \times_S Y$ satisfying these properties exists, then it is obviously unique up to isomorphism. It is called the *product* of X and Y over S. Sometimes, instead of schemes over S, we speak simply of morphisms $\varphi\colon X \to S$, and then $X \times_S Y$ is called the *fibre product* of φ and ψ.

The definition we have just given is that of product of two objects in a category. In the present case, we consider the category of schemes over S. In the category of sets the fibre product of two maps $\varphi\colon X \to S$ and $\psi\colon Y \to S$ exists and is equal to the subset $Z \subset X \times Y$ consisting of pairs (x, y) with $x \in X$ and $y \in Y$ such that $\varphi(x) = \psi(y)$. The same holds in the category of quasiprojective varieties over an algebraically closed field k. We have already seen the definition of fibre product in this case in Theorem 4.13 of Section 4.3, Chapter 4.

The product of two schemes over a scheme S exists. The proof of this assertion is essentially elementary, but rather lengthy. It can be found in Hartshorne [37, Theorem 3.3, Chapter II]. We confine ourselves to a few remarks that will allow the reader to recover the proof for him or herself.

If X, Y and S are affine schemes with $X = \operatorname{Spec} A$, $Y = \operatorname{Spec} B$ and $S = \operatorname{Spec} C$ then the S-scheme structures of X and Y define C-algebra structures on A and B. In this case, the scheme $Z = \operatorname{Spec}(A \otimes_C B)$ is the product of X and Y over S, if we give it the projections $p_X = {}^a f\colon Z \to X$ and $p_Y = {}^a g\colon Z \to Y$ corresponding to the homomorphisms $f\colon A \to A \otimes_C B$ given by $f(a) = a \otimes 1$ and $g\colon B \to A \otimes_C B$ given by $g(b) = 1 \otimes b$. This assertion is a simple corollary of the definition of tensor product (see Atiyah and Macdonald [8, Proposition 2.12]).

In the general case, we must consider affine covers $S = \bigcup W_\alpha$, $X = \bigcup U_{\alpha\beta}$ and $Y = \bigcup V_{\alpha\gamma}$ by affine open sets such that $\varphi(U_{\alpha\beta}), \psi(V_{\alpha\gamma}) \subset W_\alpha$. Then $\varphi\colon U_{\alpha\beta} \to W_\alpha$ and $\psi\colon V_{\alpha\gamma} \to W_\alpha$ are affine schemes over W_α and by what we have seen, the products $U_{\alpha\beta} \times_{W_\alpha} V_{\alpha\gamma}$ exist. It is not hard to check that these schemes satisfy conditions (5.11) and (5.12) (for suitable choices of open subsets and isomorphisms that one can easily specify), so that they glue together into a global scheme. After this one has to define the projections of this scheme to X and Y and to verify the universality condition.

It follows easily from the definition of product that it is associative, that is, $(X \times_S Y) \times_S Z = X \times_S (Y \times_S Z)$. If $S = \operatorname{Spec} A$ is an affine scheme then $X \times_{\operatorname{Spec} A} Y$ is denoted by $X \times_A Y$, and if $Y = \operatorname{Spec} B$ then by $X \otimes_A B$. An arbitrary scheme can be viewed as a scheme over $\mathbb{Z}$. Hence for any two schemes X and Y their product $X \times_{\mathbb{Z}} Y$ is defined. It is called simply the product of X and Y, and denoted by $X \times Y$.

As a first application of the notion of product we treat the definition of scheme-theoretic inverse image of a closed subscheme; in Section 3.3, this definition was given only for affine schemes. If Y is a closed subscheme of X with $j\colon Y \hookrightarrow X$ the closed embedding, and $\varphi\colon X' \to Y$ any morphism, then by definition the scheme $Y' = Y \times_X X'$ has a morphism $j'\colon Y \times_X X' \to X'$. It is not hard to see that j' is a closed embedding, so that Y' is a closed subscheme of X'. It is called the *scheme-theoretic inverse image* of Y under φ. It is easy to check that for affine schemes this definition coincides with that given earlier.

The advantage of the new definition is that it is also applicable in some other situations. Suppose for example that x is a point of a scheme X, not necessarily closed. Set $T = \operatorname{Spec} k(x)$ and define a morphism $T \to X$ by setting $\varphi(T) = x$ and $\psi_U(\mathcal{O}(U)) = 0$ if the open set U does not contain x. If $x \in U = \operatorname{Spec} A$ then x is a prime ideal of A and we define ψ_U as the natural homomorphism $A \to k(x)$ into the field of fractions of A/x. The homomorphisms ψ_U extends automatically to all open sets $U \subset X$, defining a morphism $\varphi\colon T \to X$.

If $\varphi: X' \to X$ is another morphism then the scheme $X' \times_X T$ is called the scheme-theoretic inverse image of x, or the *fibre* of φ over x. It has a morphism $X' \times_X T \to T$, that is, it is a scheme over $k(x)$, and is denoted by $\varphi^{-1}(x)$. In connection with this terminology, any morphism of schemes $\varphi: X' \to X$ is sometimes viewed as a family of schemes $\varphi^{-1}(x)$ parametrised by X. Thus *families* of schemes and *morphisms* of schemes are synonyms. These definitions are analogous to the definition of the fibre of a regular map and algebraic family of varieties, but are more precise, since fibres may turn out to be nonreduced schemes, as we have already seen at the end of Section 3.3 in the case of morphisms of affine schemes.

4.2 Group Schemes

The notion of direct product allows us to carry over to schemes the definition of algebraic groups. For this we need only reformulate the definition of algebraic group given in Section 4.1, Chapter 3 in such a way that it only involves morphisms, and not points.

Let $\varphi: X \to S$ be a scheme over S. A group law is defined by a morphism

$$\mu: X \times_S X \to X.$$

The operation of taking the inverse of an element is replaced by a morphism

$$i: X \to X.$$

The role of the identity element is played by a morphism

$$\varepsilon: S \to X$$

such that $\varphi \circ \varepsilon = \mathrm{id}_S$, where id_S denote the identity morphism of S. We have already seen repeatedly that for a scheme X over a field k, say, a morphism $\operatorname{Spec} k \to X$ defines a point of X. The neutral property of the identity element is expressed by

$$\mu \circ (\varepsilon \circ \varphi, \mathrm{id}_X) = \mu \circ (\mathrm{id}_X, \varepsilon \circ \varphi) = \mathrm{id}_X, \tag{5.21}$$

where id_X is the identity map of X. The property of the inverse is expressed by

$$\mu \circ (i, \mathrm{id}_X) = \mu \circ (\mathrm{id}_X, i) = \varepsilon \circ \varphi. \tag{5.22}$$

It remains to write out the associativity condition. For this note that, by associativity of the product of schemes, we have two morphisms (μ, id_X) and $(\mathrm{id}_X, \mu): X \times_S X \times_S X \to X \times_S X$. Our associativity requirement is then

$$\mu \circ (\mu, \mathrm{id}_X) = \mu \circ (\mathrm{id}_X, \mu). \tag{5.23}$$

If conditions (5.21)–(5.23) are satisfied, the scheme X over S with the morphisms μ, i and ε is called a *group scheme* over S. We leave the reader to formulate the definitions of homomorphism and isomorphism of group schemes.

Here is a typical example illustrating that it is reasonable to extend the notion of algebraic group to that of group scheme. Let $X = \mathbb{G}_a$ be the scheme $\mathbb{A}^1$ over an algebraically closed field k of characteristic p, with the group law defined by $\mu(x, y) = x + y$. This is an algebraic group, and has already appeared in Example 3.10 of Section 4.1, Chapter 3. Consider the morphism $f\colon X \to X$ given by $f(x) = x^p$, which is a homomorphism of algebraic groups because $\operatorname{char} k = p$. As a map of point sets, f is a one-to-one correspondence, and as a map of groups an isomorphism, but as a regular map of algebraic varieties it is not an isomorphism. This is a serious departure from the familiar situation of group theory.

At the end of Section 3.3 we saw that as a morphism of schemes, the scheme-theoretic inverse image $f^{-1}(x)$ of any point $x \in X$ is a nontrivial scheme, that is, is not $\operatorname{Spec} k$. It is natural to try to make $f^{-1}(0)$ into a group scheme. For this, write $Z = f^{-1}(0)$, and $j\colon Z \hookrightarrow X$ for the closed embedding. Consider the morphism

$$\mu \circ (j, j)\colon Z \times_k Z \to X.$$

As an exercise we propose that the reader proves that there exists a unique morphism

$$\mu'\colon Z \times_k Z \to Z$$

such that $\mu \circ (j, j) = j \circ \mu'$, and that μ' makes Z into a group scheme.

One can show that Z is the kernel of the homomorphism f in the sense of category theory. The category of commutative algebraic groups over a field k of characteristic p is not an Abelian category, whereas extending it to the category of commutative group schemes over k makes it into an Abelian category.

4.3 Separatedness

Finally, we treat what is arguably the most important application of the notion of products, the question of separatedness of schemes.

The image $\Delta(X)$ of the morphism $\Delta = (\mathrm{id}, \mathrm{id})\colon X \to (X \times_S X)$ is called the *diagonal*. A scheme X over S is *separated* if its diagonal is closed. A scheme X is separated if it is separated over $\operatorname{Spec} \mathbb{Z}$.

The same condition characterises Hausdorff spaces among topological spaces (see any book on point set topology, for example, Bourbaki [16, I.8.1]). In the case of schemes, the meaning of the requirement is somewhat different. In any case, the topological space associated with a scheme is almost never Hausdorff. To get the feel for the meaning of the separated condition, we give an example of a nonseparated scheme.

Let $U_1, U_2 \cong \mathbb{A}^1_k$ be two copies of the affine line over k and $U_{12} \subset U_1$, $U_{21} \subset U_2$ the open sets obtained in some fixed choice of coordinate T_1 on U_1 and T_2 on U_2 by deleting the origin 0. The map φ that sends a point $u \in U_{12}$ into the point $u' \in U_{21}$ with the same coordinate is obviously an isomorphism $U_{12} \xrightarrow{\sim} U_{21}$. The conditions (5.11) and (5.12) that are required in order to be able to glue U_1 and U_2 along U_{12}

and U_{21} are obviously satisfied. As a result of this glueing, we get a scheme X over k called the *affine line with doubled-up origin* or the *bug-eyed* affine line. In fact X contains 2 points 0_1 and 0_2 obtained from the origin 0 in U_1 and U_2. We prove that this scheme is not separated over k.

The closed points of $X \times_k X$ are of the form (x_1, x_2) where x_1, x_2 are points of X (see Exercise 1), and the diagonal map Δ is given by $\Delta(x) = (x, x)$. Since X is by construction covered by two affine open sets V_1 and V_2 isomorphic to U_1 and U_2, the product $X \times X$ is covered by four affine open sets $V_1 \times V_1$, $V_1 \times V_2$, $V_2 \times V_1$, and $V_2 \times V_2$.

Consider for example the set $V_1 \times V_2$. It is isomorphic to $\mathbb{A}^1 \times \mathbb{A}^1$, and its intersection with $\Delta(X)$ consists of points (x, x) with $x \in V_1 \cap V_2 = U_{12}$. Already from this one sees that $\Delta(X)$ is not closed, since already its intersection with $V_1 \times V_2$ is not closed. To complete the picture, we can compute the closure of $\Delta(X)$. The closure of $\Delta(X) \cap (V_1 \times V_2)$ in $V_1 \times V_2$ is obviously obtained by adding the point $(0_1, 0_2)$. Considering all four open sets $V_i \times V_j$ (for $1 \le i, j \le 2$) in the same way, we discover that the closure of $\Delta(X)$ is obtained by adding the two points $(0_1, 0_2)$ and $(0_2, 0_1)$. It follows that the closure of $\Delta(X)$ is isomorphic to the line $\mathbb{A}^1$ in which the points o has split into four points, $(0_1, 0_1)$, $(0_2, 0_2)$, $(0_1, 0_2)$ and $(0_2, 0_1)$, of which the first two are in $\Delta(X)$, and the last two not.

To grasp more clearly the way in which the nonseparated nature of a scheme affects its properties, we work out the example X just constructed in a little more detail. The fields $k(V_1)$ and $k(V_2)$ are isomorphic and define a field, naturally called the function field of X. The local ring $\mathcal{O}_x$ of a point $x \in X$ is a subring of $k(X)$. What are $\mathcal{O}_{0_1}$ and $\mathcal{O}_{0_2}$?

Obviously $\mathcal{O}_{0_1}$ is equal to the local ring of the point $0_1 \in V_1$. Since the isomorphism between U_{12} and U_{21} extends to the identity isomorphism between U_1 and U_2, functions in $\mathcal{O}_{0_1}$ correspond under this to functions of $\mathcal{O}_{0_2}$, which means simply that $\mathcal{O}_{0_1} = \mathcal{O}_{0_2}$. Thus two distinct points have one and the same local ring. Moreover, any function in this local ring takes the same values at 0_1 and 0_2; in other words, the two points cannot be distinguished by means of rational functions. It can be shown that nonseparatedness is quite generally associated with this type of phenomenon.

We now proceed to a general analysis of the notion of separatedness.

Proposition 5.3 *An affine scheme X over a ring B is separated, and $\Delta: X \to X \times_B X$ is a closed embedding.*

Proof Let $X = \operatorname{Spec} A$, where A is a B-algebra. Since by definition $X \times_B X = \operatorname{Spec}(A \otimes_B A)$, the morphism $\Delta: X \to X \times_B X$ is associated with the homomorphism $\lambda: A \otimes_B A \to A$. By definition, λ is determined by

$$\lambda \circ u = \mathrm{id}, \qquad \lambda \circ v = \mathrm{id}, \tag{5.24}$$

where $u, v: A \to A \otimes_B A$ are the homomorphisms given by

$$u(a) = a \otimes 1, \qquad v(a) = 1 \otimes a.$$

It follows at once from this that $\lambda(a \otimes b) = ab$. From this, or in fact already from (5.24), it follows that λ is surjective, and this means that Δ is a closed embedding. The proposition is proved. □

Since every scheme is covered by affine sets, which are separated, non separated should be related to some properties of the glueing of affine schemes. This is confirmed by the following result, in which we only consider the case when X is an S-scheme over an affine scheme $S = \operatorname{Spec} B$.

Proposition 5.4 *Let* $X = \bigcup U_\alpha$ *be an affine cover that satisfies the conditions*: (1) *all the sets* $U_\alpha \cap U_\beta$ *are affine, and* (2) *the ring* $\mathcal{O}_X(U_\alpha \cap U_\beta)$ *is generated by its subrings* $\rho^{U_\alpha}_{U_\alpha \cap U_\beta}(\mathcal{O}_X(U_\alpha))$ *and* $\rho^{U_\beta}_{U_\alpha \cap U_\beta}(\mathcal{O}_X(U_\beta))$ *for each* α, β. *Then* X *is separated over* B.

Proof Let $u, v\colon X \times_B X \to X$ be the standard maps of the product. Then

$$\Delta^{-1}\bigl(u^{-1}(U_\alpha) \cap v^{-1}(U_\beta)\bigr) = \Delta^{-1}\bigl(u^{-1}(U_\alpha)\bigr) \cap \Delta^{-1}\bigl(v^{-1}(U_\beta)\bigr) = U_\alpha \cap U_\beta. \quad (5.25)$$

On the other hand, it follows easily from the definition of the product that for any open sets $U, V \subset X$ the open set $u^{-1}(U) \cap v^{-1}(V) \subset X \times X$ is isomorphic to the product $U \times V$. Together with (5.25), this shows that to prove that X is separated, it is enough that the restriction of Δ

$$\Delta_{\alpha\beta} = \Delta_{|U_\alpha \cap U_\beta}\colon U_\alpha \cap U_\beta \to U_\alpha \times_B U_\beta$$

has closed image. But $U_\alpha \cap U_\beta$ is affine by assumption (5.24), say $U_\alpha \cap U_\beta = \operatorname{Spec} C_{\alpha\beta}$, and by condition (5.25), the corresponding ring homomorphism $A_\alpha \otimes_B A_\beta \to C_{\alpha\beta}$ is surjective, where $U_\alpha = \operatorname{Spec} A_\alpha$. This means that $\Delta_{\alpha\beta}$ is a closed embedding. The proposition is proved. □

It is not hard to prove that the converse also holds. We verify one implication here, which is useful, although completely obvious.

Proposition 5.5 *In a separated scheme, the intersection of two affine open sets is affine.*

Proof Indeed,

$$U \cap V = \Delta^{-1}(U \times V).$$

If U and V are affine then so is $U \times V$, and if X is separated then Δ is a closed embedding, and hence $U \cap V$ is a closed subscheme of an affine scheme. This is affine by Proposition of Section 3.3. □

We draw attention to one interesting feature of the criterion stated in Proposition 5.4: it is independent of the morphism $X \to S$. Thus the property that a scheme

X over an affine scheme S is separated over S does not depend on the choice of S or of the morphism $X \to S$ (note here that the base S is affine, hence itself separated). It could be stated, for example, in terms of X over $\mathbb{Z}$.

An important application of Proposition 5.4 is verifying that projective space $\mathbb{P}^n_A$ is separated over any ring A. In this case $\mathbb{P}^n_A = \bigcup_{i=0}^n U_i$, with $U_i = \operatorname{Spec} A[T_0/T_i, \dots, T_n/T_i]$. Since $U_i \cap U_j = (\operatorname{Spec} A[T_0/T_i, \dots, T_n/T_i])_{(T_j/T_i)}$, it is obviously an affine set. $\mathcal{O}_{\mathbb{P}^n_A}(U_i \cap U_j)$ consists of rational functions $F(T_0, \dots, T_n)/T_i^p T_j^q$, where $F \in A[T_0, \dots, T_n]$ is a form of degree $p+q$. Its subrings $\rho^{U_i}_{U_i \cap U_j}(\mathcal{O}_{\mathbb{P}^n_A}(U_i))$ and $\rho^{U_j}_{U_i \cap U_j}(\mathcal{O}_{\mathbb{P}^n_A}(U_j))$ consist of elements F/T_i^p and G/T_j^q where F and G are forms of degree p and q. They obviously generate $\mathcal{O}_{\mathbb{P}^n_A}(U_i \cap U_j)$.

It is easy to check that an open subset or closed subscheme of a separated scheme is again separated. It follows from this that projective and quasiprojective schemes are separated.

We draw attention to some properties of quasiprojective varieties related to separatedness. We have made use especially often of the fact that regular maps are uniquely determined by their restriction to any dense open subset. The analogous property of schemes is closely related to the separated property. Namely, if X is separated, then for any scheme Y and morphisms $f, g\colon Y \to X$, the set $Z \subset Y$ consisting of $y \in Y$ with $f(y) = g(y)$ is closed. Indeed, we have a morphism $(f, g)\colon Y \to X \times X$, and Z is the inverse image of the diagonal under this morphism.

This shows that rational maps are only a natural generalisation of morphisms for separated schemes. If X is not separated then two different morphisms $Y \to X$ may define the same rational map.

Another property that appears frequently is the closed graph of a regular map. If $f\colon Y \to X$ is a morphism of schemes then its graph is the image of the morphism $(\mathrm{id}, f)\colon Y \to Y \times X$. It is the inverse image of the diagonal of $X \times X$ under the map $f \times \mathrm{id}\colon Y \times X \to X \times X$ defined by $f \times \mathrm{id} = (f \circ p_Y, p_X)$ where $p_Y\colon Y \times X \to Y$ and $p_X\colon Y \times X \to X$ are the natural projections. Thus the graph is closed provided that X is separated.

4.4 Exercises to Section 4

1 Let X and Y be schemes over an algebraically closed field k. Prove that the correspondence $u \mapsto (p_X(u), p_Y(u))$ establishes a one-to-one correspondence between closed points of $X \times_k Y$ and pairs (x, y), where $x \in X$ and $y \in Y$ are closed points.

2 Find all points of the scheme $\operatorname{Spec} \mathbb{C} \times_{\mathbb{R}} \operatorname{Spec} \mathbb{C}$, where $\mathbb{R}$ and $\mathbb{C}$ are the real and complex number fields.

3 Let X be an affine group scheme over an affine scheme $S = \operatorname{Spec} B$ with $X = \operatorname{Spec} A$, where A is a B-algebra. Prove that the group law defines a homomorphism

$\mu\colon A \to A \otimes_B A$, the identity morphism a homomorphism $\varepsilon\colon A \to B$, and the inverse an automorphism $i\colon A \to A$. State conditions (5.21)–(5.23) in terms of these homomorphisms.

4 Write $\mathbb{G}_a = \mathbb{A}^1_k$ for the additive group discussed in Section 4.2, where $chak = p$. Prove that the kernel Z of the homomorphism $\mathbb{G}_a \to \mathbb{G}_a$ given by $x \mapsto x^p$ constructed in Section 4.2 is an affine group scheme, with $Z = \operatorname{Spec} A$, where $A = k[T]/(T^p)$. Compute in this case all the homomorphisms introduced in Exercise 3.

5 As in Exercise 4, write $\mathbb{G}_m$ for the multiplicative group $\mathbb{A}^1 \setminus 0$ (with the group law $(a,b) \mapsto ab$), and consider the analogous homomorphism $f\colon \mathbb{G}_m \to \mathbb{G}_m$ given by $x \mapsto x^p$; compute its kernel Z', the scheme theoretic inverse image $f^{-1}(1)$. Prove that Z (in Exercise 4) and Z' are not isomorphic as group schemes.

6 Let k be a field of characteristic 2. Prove that up to isomorphism the scheme $X = \operatorname{Spec} k[T]/T^2$ has only two structures of group scheme, namely the group schemes Z and Z' of Exercises 4–5.

7 Prove that the nonseparated scheme of Section 4.3 is isomorphic to that of Exercise 3 of Section 3.6.

8 Prove that a scheme of the form $\operatorname{Proj} \Gamma$ as in Exercise 8 of Section 3.6 is always separated.

Chapter 6
Varieties

1 Definitions and Examples

1.1 Definitions

In this chapter we consider the schemes most closely related to projective varieties; they will be called algebraic varieties. This is exactly what we arrive at on attempting to give an intrinsic definition of algebraic variety.

Definition A *variety* over an algebraically closed field k is a reduced separated scheme of finite type over k. A *morphism* of varieties is a morphism of schemes over k. A variety X that is an affine scheme is called an *affine variety*.

We saw in Example 5.19 that every quasiprojective variety defines a scheme. This scheme is a variety, that we will also call quasiprojective.

By definition, any variety X has a finite cover $X = \bigcup U_i$, where the U_i are affine varieties. It follows from this that X is finite dimensional. If X is irreducible then all the U_i are dense in X and $\dim X = \dim U_i$. Moreover, they are all birational, since $U_i \cap U_j$ is open and dense in both U_i and U_j. Hence the function fields $k(U_i)$ are isomorphic; these fields can be identified. The resulting field is called the *function field* of X and denoted $k(X)$. The dimension of X equals the transcendence degree of $k(X)$.

A closed point of a variety X that is contained in an affine open set U is also a closed point of U, and is a point of the corresponding affine variety with coordinates in k. There are sufficiently many such points on X.

Proposition *Closed points are dense in every closed subset of* X.

Proof We note first that in an affine variety (and even in an affine scheme), every nonempty closed subset contains a closed point. Indeed, a nonempty closed subset Z of $\operatorname{Spec} A$ is of the form $\operatorname{Spec} B$, where B is a quotient ring of A. Since every ring has a maximal ideal, Z has a closed point.

I.R. Shafarevich, *Basic Algebraic Geometry* 2, DOI 10.1007/978-3-642-38010-5_2,

If X is an arbitrary variety, $Z \subset X$ a closed subset and $z \in Z$, then it is enough to prove that $Z \cap U$ contains a closed point for any neighbourhood U of z. We can restrict to affine neighbourhoods U, since these form a basis of all open sets. For affine U, by what we have just said, $Z \cap U$ has a closed point.

But there is a trap here for the unwary—a point may be closed in U, but not in X. This actually happens, for example, in the case of the subset $U = \operatorname{Spec} \mathcal{O} \setminus \{x\}$ where $\mathcal{O}$ is a local ring of a closed point x of a curve. Fortunately, everything turns out to be all right in the case of a variety: if $z \in X$ is a closed point of some neighbourhood U of z then it is also closed in X. This follows from the fact that the closed points x of a variety are characterised by $k(x) = k$. Indeed, a point x is closed in X if and only if it is closed in all affine open sets containing it, and for affine varieties the condition $k(x) = k$ obviously characterises closed points. The field $k(x)$ depends only on the local ring of x, and hence does not change on passing from X to an open subset $U \ni x$. The proposition is proved. □

Since a variety is a reduced scheme, an element $f \in \mathcal{O}_X(U)$ is uniquely determined by its values $f(x) \in k(x)$ at all $x \in U$. By the proposition, it is determined by its values at closed points. Moreover $k = k(x)$, so that an element $f \in \mathcal{O}_X(U)$ can be interpreted as a k-valued function on the set of closed points of U.

If $\varphi\colon X \to Y$ is a morphism of varieties, $x \in X$ and $y = \varphi(x)$, then the homomorphism of local rings $\varphi^*\colon \mathcal{O}_y \to \mathcal{O}_x$ induces an inclusion of residue fields $k(y) \hookrightarrow k(x)$. If x is a closed point then $k(x) = k$, and hence also $k(y) = k$, that is, y is also closed. Therefore the image of a closed point is again closed. Thus interpreting elements $f \in \mathcal{O}_Y(U)$ as functions on closed points, the homomorphism $\psi_U\colon \mathcal{O}_Y(U) \to \mathcal{O}_X(\varphi^{-1}(U))$ is determined by $\psi_U(f)(x) = f(\varphi(x))$. In other words, by specifying the map $\varphi\colon X \to Y$, or even its restriction to the set of closed points, we determine the morphism itself.

A variety X has of course any number of nonreduced closed subschemes. But any closed subset $Z \subset X$ can be made into a reduced scheme, or as we will say from now on, into a *closed subvariety*. If X is an affine variety, $X = \operatorname{Spec} A$ and $Z = V(\mathfrak{a})$ then we set $Z = \operatorname{Spec} A/\mathfrak{a}'$ where $\mathfrak{a}'$ is the radical of $\mathfrak{a}$, the ideal of elements $a \in A$ such that $a^r \in \mathfrak{a}$ for some r. The general case is obtained by glueing.

All this shows how close varieties are to quasiprojective varieties. Indeed, all the local notions and properties treated in Chapter 2 carry over word-for-word for algebraic varieties: nonsingular points, the theorem that the set of singular points is closed, properties of normal varieties. The same is true for properties of divisors and differential forms.

The only properties not carrying over in an obvious way to algebraic varieties are those related to the property of being projective. We now explain what condition replaces projective for the case of arbitrary varieties. Projectivity is of course very far from being an "abstract" property. But we have at our disposal one assertion, Theorem 1.11 of Section 5.2, Chapter 1, which is an intrinsic property that is characteristic of projective varieties. We take this as a definition.

Definition A variety X is *complete* if for any variety Y, the projection morphism $p\colon X \times Y \to Y$ takes closed sets to closed sets.

The main properties of projective varieties, for example, the fact that the image of a morphism is closed, and the fact that there are no everywhere regular functions except the constants (that is, $\mathcal{O}_X(X) = k$), were deduced in Section 5.2, Chapter 1 from Theorem 1.11, and therefore hold for complete varieties. Note that the proof that the image of a morphism is closed used the fact that a morphism has a closed graph. As we saw in Section 4.3, Chapter 5, this follows from the separated assumption on a variety.

Of all the properties of projective varieties proved in Chapters 1–4, there is only one that used projectivity directly, rather than via an appeal to Theorem 1.11 of Section 5.2, Chapter 1. This is the extremely important result Theorem 2.10 of Section 3.1, Chapter 2. Here we prove a generalisation of it to arbitrary complete varieties.

Theorem 6.1 *If X is a nonsingular irreducible variety and $\varphi\colon X \to Y$ a rational map to a complete variety Y, the locus of indeterminacy of φ has codimension ≥ 2.*

Proof Let $V \subset X$ be the set of points at which φ is defined, $\Gamma_\varphi \subset V \times Y$ the graph of the morphism $\varphi\colon V \to Y$ and Z its closure in $X \times Y$. The image of Z under the projection $p\colon X \times Y \to X$ is closed, since Y is complete. Since $p(Z) \supset V$, it follows that $p(Z) = X$. The restriction $p\colon Z \to p(Z)$ is a birational morphism, since it is an isomorphism of Γ_φ and V. The theorem thus follows from the next result. □

Lemma *If $p\colon Z \to X$ is a surjective birational morphism with X a nonsingular variety then the set of points of indeterminacy of the inverse rational map p^{-1} has codimension ≥ 2.*

Indeed, $\varphi = q \circ p^{-1}$, where q is the restriction to Z of the projection $p\colon X \times Y \to X$. Therefore φ is defined wherever p^{-1} is. This proves Theorem 6.1.

Proof of the Lemma Suppose that there exists a codimension 1 subvariety $T \subset X$ such that p^{-1} is not defined at any point of T. Replacing Z, X and T by affine open subsets, we can assume that they are affine and $T \subset p(Z) \subset X$. Let $Z \subset \mathbb{A}^m$, and write $u_1, \dots, u_m$ for the coordinates of $\mathbb{A}^m$ as elements of $\mathcal{O}_Z(Z)$. Consider a point $t \in T$ and represent the rational functions $(p^{-1})^*(u_i)$ in the form

$$(p^{-1})^*(u_i) = g_i/h,$$

where $g_1, \dots, g_m, h \in \mathcal{O}_t$ and have no common factor. Then

$$h(p^{-1})^*(u_i) = g_i, \quad \text{so that} \quad p^*(h)u_i = p^*(g_i).$$

Hence $g_i(\tau) = 0$ for every point $\tau \in T$ at which $h(\tau) = 0$, and this contradicts the assumption that $g_1, \dots, g_m, h \in \mathcal{O}_t$ have no common factor. The lemma is proved. □

The complete varieties just introduced turn out to have properties so close to those of quasiprojective varieties that the question arises as to whether the two notions might not coincide. We will see a little later in Section 2.3 that this is not the case; there exist varieties that cannot be embedded in any projective space. However, what is much more important is that the intrinsic, invariant nature of the notion of variety makes it into a much more flexible tool. Many constructions can be performed very simply and naturally within the framework of this notion. It may sometimes turn out a posteori that we have not actually left the framework of projective or quasiprojective varieties, but this is often already of secondary importance. In Sections 1.2–1.4 we give some important examples of this kind of constructions.

A very simple example is provided by the definition of the product of varieties. The definition in the framework of varieties is extremely simple: the arguments of Section 4.1, Chapter 5 simplify substantially if we use the fact that the set of closed points of the variety $X \times Y$ is the set of pairs of the form (x, y), where $x \in X$ and $y \in Y$ are closed points (see Exercises 1 and 2). But we spent quite a lot of effort on this definition in Section 5, Chapter 1, since there we needed to be sure that the product of quasiprojective varieties was again a quasiprojective variety.

Another example that we now consider is the notion of normalisation of a variety. Let X be an irreducible variety, K a finite field extension of the function field $k(X)$. We show that there exist a normal irreducible variety X_K^ν and a morphism $\nu_K \colon X_K^\nu \to X$ with the properties that $k(X_K^\nu) = K$ and the induced map $\nu_K^* \colon k(X) \to k(X_K^\nu) = K$ is the given field extension. Such a variety is unique: for any two normalisations X_K^ν and $\widetilde{X_K^\nu}$ there exists an isomorphism $f \colon X_K^\nu \to \widetilde{X_K^\nu}$ such that the diagram

$$\begin{array}{ccc} X_K^\nu & \xrightarrow{f} & \widetilde{X_K^\nu} \\ & {\scriptstyle \nu_K} \searrow \quad \swarrow {\scriptstyle \widetilde{\nu_K}} & \\ & X & \end{array}$$

is commutative. X_K^ν is called the *normalisation* of X in K. The uniqueness of the normalisation X_K^ν is proved exactly as in Section 5.2, Chapter 2, where we considered the case $K = k(X)$. To prove the existence, consider an affine cover $X = \bigcup U_i$. The integral closure A_i^ν of $k[U_i]$ in K is a finitely generated algebra over k, as we saw in Sect 5.2, Chapter 2. Hence the normalisation $U_{i,K}^\nu \to U_i$ in K of the affine variety U_i exists and is affine. From the uniqueness of normalisation it follows that $\nu_{i,K}^{-1}(U_i \cap U_j)$ and $\nu_{j,K}^{-1}(U_j \cap U_i)$ are isomorphic. This allows us to glue the $U_{i,K}^\nu$ together into a single scheme X_K^ν, which is obviously a reduced irreducible scheme of finite type over k.

We prove that X_K^ν is separated (Section 4.3, Chapter 5). It is enough to prove that the diagonal in $X_K^\nu \times X_K^\nu$ is closed, and for this it is enough to show that it is closed in the neighbourhood of any point $\xi \in X_K^\nu \times X_K^\nu$. Suppose that the morphism $\nu \times \nu \colon X_K^\nu \times X_K^\nu \to X \times X$ takes ξ into $\eta \in X \times X$, and let U' be an affine neighbourhood of η such that $(\nu \times \nu)^{-1}(U') = V'$ is affine. The existence of U' follows from the existence of the normalisation in the affine case. Since X is a separated scheme, if we write $\Delta \subset X \times X$ for the diagonal then the scheme $U = \Delta \cap U'$ is closed in U', and hence is affine. It follows that the scheme $(\nu \times \nu)^{-1}(U)$ is affine, and hence also its irreducible component V containing ξ. Write $\delta^\nu \colon X_K^\nu \to X_K^\nu \times X_K^\nu$ and $\delta \colon X \to X \times X$ for the diagonal morphisms, and set $W = (\delta^\nu)^{-1}(V) = \nu^{-1}(U)$. We obtain the commutative diagram

$$\begin{array}{ccc} W & \xrightarrow{\delta^\nu} & V \\ & {}_{\delta\circ\nu}\searrow \quad \swarrow_{\nu\times\nu} & \\ & U & \end{array}$$

in which δ^ν corresponds to a finite regular map of affine varieties. This holds a fortiori for the morphism $\delta \circ \nu$ (because a finite module over a ring is a fortiori finite over a bigger ring). Applying Theorem 1.13 of Section 5.3, Chapter 1, we get that $\delta^\nu(W) = V$, which means that the diagonal is closed in the neighbourhood V' of ξ.

Thus the scheme X_K^ν is an irreducible variety, and a trivial verification shows that it is the required normalisation.

We see that in the framework of arbitrary varieties, the construction of the normalisation is quite trivial. It remains to consider the question of whether the normalisation of a quasiprojective variety is again quasiprojective. This is true, but we do not give the proof, which is based, naturally enough, on purely projective considerations; it can be found, for example, in Lang [55, Proposition 4 of Section 4, Chapter V] or Hartshorne [37, Ex. 5.7 of Chapter III]. In the case of curves, we can repeat the proofs of Theorems 2.22–2.23 of Section 5.3, Chapter 2. These results imply that the normalisation of any curve is quasiprojective (in the case $K = k(X)$), and that of a complete curve is projective. In particular, it follows from this that a nonsingular curve is quasiprojective. In fact this is true for any curve, but the proof is more complicated, and we omit it here.

1.2 Vector Bundles

The idea of a vector bundle is one of the most important constructions of algebraic varieties, and is typically "abstract" or "nonprojective" in nature. We recall that the general notion of *fibration* is nothing other than a morphism of varieties $p: X \to S$, that is, a variety over S. We are interested in fibrations whose fibres are vector spaces. In formulating this notion we must bear in mind that an n-dimensional vector space over a field k has a natural structure of algebraic variety isomorphic to $\mathbb{A}^n$.

Definition A *family of vector spaces* over X is a fibration $p: E \to X$ such that each fibre $E_x = p^{-1}(x)$ for $x \in X$ is a vector space over $k(x)$, and the structure of algebraic variety of E_x as a vector space coincides with that of $E_x \subset E$ as the inverse image of x under p.

A *morphism* of a family of vector space $p: E \to X$ into another family $q: F \to X$ is a morphism $f: E \to F$ for which the diagram

$$\begin{array}{ccc} E & \xrightarrow{f} & F \\ & {}_{p}\searrow \quad \swarrow_{q} & \\ & X & \end{array}$$

commutes (so that in particular f maps E_x to F_x for all $x \in X$), and the map $f_x: E_x \to F_x$ is linear over $k(x)$. It's obvious how to define an isomorphism of families.

The simplest example of a family is the direct product $E = X \times V$, where V is a vector space over k, and p the first projection of $X \times V \to X$. A family of this type, or isomorphic to it, is said to be *trivial*.

Example 6.1 Let V and W be two vector spaces of dimension m and n. We determine the general form of a morphism $f\colon X \times V \to X \times W$ between two trivial families. We let $e_1, \dots, e_m$ and $u_1, \dots, u_n$ be bases of V and W, and write $\xi_1, \dots, \xi_m$ and $\eta_1, \dots, \eta_n$ for the corresponding coordinates. The projections $p\colon X \times V \to V$ and $q\colon X \times W \to W$ define elements $x_i = p^*(\xi_i) \in \mathcal{O}_{X\times V}(X \times V)$ and $y_j = q^*(\eta_i) \in \mathcal{O}_{X\times W}(X \times W)$. Obviously, closed points $\alpha \in X \times V$ and $\beta \in X \times W$ are uniquely determined by the values of $x_i(\alpha)$ and $y_j(\beta) \in k$. Therefore the morphism f is uniquely determined by specifying the elements $f^*(y_j) \in \mathcal{O}_{X\times V}(X \times V)$.

The composite of the isomorphism $X \to X \times e_i$ and the embedding $X \times e_i \to X \times V$ defines a morphism $\varphi_i\colon X \to X \times V$. Set $a_{ij} = \varphi_i^*(f^*(y_j)) \in \mathcal{O}_X(X)$. Then

$$f^*(y_j) = \sum a_{ij} x_i. \tag{6.1}$$

Indeed, it is enough to check this equality at all closed points $\alpha \in X \times V$, and there it follows at once from the definition of morphism of family of vector spaces (because $f_x\colon E_x \to F_x$ is linear).

Conversely, any matrix (a_{ij}) with $a_{ij} \in \mathcal{O}_X(X)$ defines a morphism $f\colon X \times V \to X \times W$ by means of formula (6.1). Obviously we get an isomorphism if and only if $m = n$ and the determinant $\det|a_{ij}|$ is an invertible element of $\mathcal{O}_X(X)$.

If $p\colon E \to X$ is a family of vector spaces and $U \subset X$ any open set, the fibration $p^{-1}(U) \to U$ is a family of vector spaces over U. It is called the *restriction* of E to U and denoted $E_{|U}$.

Definition A family of vector spaces $p\colon E \to X$ is a *vector bundle* if every point $x \in X$ has a neighbourhood U such that the restriction $E_{|U}$ is trivial.

The dimension of the fibre E_x of a vector bundle is obviously a locally constant function on X, and, in particular, is constant if X is connected. In this case the number $\dim E_x$ is called the *rank* of E, and denoted by $\operatorname{rank} E$.

Example 6.2 Let V be an $(n+1)$-dimensional vector space and $\mathbb{P}^n$ the vector space of lines $l \subset V$ through 0. Write l_x for the line corresponding to a point $x \in \mathbb{P}^n$. Consider the subset $E \subset \mathbb{P}^n \times V$ of pairs (x, v) such that $x \in \mathbb{P}^n$ and $v \in V$ are closed points, with $v \in l_x$. Obviously E is the set of closed points of some quasiprojective subvariety of $\mathbb{P}^n \times V$, which we continue to denote by E. The projection $\mathbb{P}^n \times V \to \mathbb{P}^n$ defines a morphism $p\colon E \to \mathbb{P}^n$. We prove that $p\colon E \to \mathbb{P}^n$ is a vector bundle. In V, we introduce a coordinate system $(x_0, \dots, x_n)$. The restriction of E to the open set U_α given by $x_\alpha \neq 0$ consists of points

$$\xi = (t_1, \ldots, t_n; y_0, \ldots, y_n) \quad \text{such that} \quad y_i = t_i y_\alpha,$$

where $t_i = x_i/x_\alpha$, and the map $\xi \mapsto ((t_1, \ldots, t_n), y_\alpha)$ defines an isomorphism of $E_{|U_\alpha}$ with $U_\alpha \times k$.

Therefore E is a vector bundle of rank 1. The projection $\mathbb{P}^n \times V \to V$ defines a morphism $q\colon E \to V$. The reader can easily check that q coincides with the blowup of the origin $0 = (0, \ldots, 0) \in V$, and $q^{-1}(0) = \mathbb{P}^n \times 0$.

Consider a vector bundle $p\colon E \to X$ and a morphism $f\colon X' \to X$. The fibre product $E' = E \times_X X'$ over X has a morphism $p'\colon E' \to X'$. This morphism defines a vector bundle. Indeed, if $E_{|U} \cong U \times V$ with $U \subset X$ then writing $U' = f^{-1}(U)$, we get $E'_{|U'} = E \times_U U' \cong U' \times V$. This vector bundle is called the pullback of E, and denoted by $f^*(E)$. Obviously rank $f^*(E) = \operatorname{rank} E$.

Example 6.3 Let X be a projective variety and $f\colon X \hookrightarrow \mathbb{P}^1$ a closed embedding to projective space. Let $p\colon E \to \mathbb{P}^n$ be the vector bundle of Example 6.2. Then $f^*(E)$ is a vector bundle over X of rank 1. It depends in general on the embedding f, and is a very important invariant of f.

Example 6.4 Let $X = \operatorname{Grass}(r, n)$ be the Grassmannian of r-dimensional vector subspaces of an n-dimensional vector space with basis $e_1, \ldots, e_n$ (Example 1.24 of Section 4.1, Chapter 1). Consider in $X \times V$ the subvariety E consisting of points (x, v) such that $v \in L_x$, where we write L_x for the r-dimensional vector subspace corresponding to $x \in \operatorname{Grass}(r, n)$. Obviously the projection $p\colon X \times V \to X$ gives E the structure of a family of vector spaces. Let us prove that it is locally trivial. Consider the open subset $U_{k_1 \ldots k_r} \subset \operatorname{Grass}(r, n)$ defined by $p_{k_1 \ldots k_r} \neq 0$; then for $x \in U_{k_1 \ldots k_r}$, the vector subspace $L_x = p^{-1}(x)$ has a basis

$$\Big\{ e_i - \sum_{j \neq k_1 \ldots k_r} a_{ij} e_j \Big\} \quad \text{where } a_{ij} = \frac{p_{k_1 \ldots \widehat{k_i} j \ldots k_r}}{p_{k_1 \ldots k_r}}.$$

This determines an isomorphism $p^{-1}(U_{k_1 \ldots k_r}) \to U_{k_1 \ldots k_r} \times L$, where $L = k^r$.

Since a vector bundle is locally trivial, it can be obtained by glueing together trivial bundles over a number of open sets. This leads to an effective method of constructing vector bundles.

Let $X = \bigcup U_\alpha$ be a cover such that the bundle $p\colon E \to X$ is trivial on each U_α. For each U_α, we fix an isomorphism

$$\varphi_\alpha\colon p^{-1}(U_\alpha) \xrightarrow{\sim} U_\alpha \times V.$$

Over the intersection $U_\alpha \cap U_\beta$ we have two isomorphisms of $p^{-1}(U_\alpha \cap U_\beta)$ with $(U_\alpha \cap U_\beta) \times V$, namely $\varphi_{\alpha|p^{-1}(U_\alpha \cap U_\beta)}$ and $\varphi_{\beta|p^{-1}(U_\alpha \cap U_\beta)}$. Hence $\varphi_\beta \circ \varphi_\alpha^{-1}$ defines an automorphism of the trivial vector bundle $(U_\alpha \cap U_\beta) \times V$ over $U_\alpha \cap U_\beta$.

We now use the result of Example 6.1. We choose a basis of V, and write the automorphism $\varphi_\beta \circ \varphi_\alpha^{-1}$ as an $n \times n$ matrix $C_{\alpha\beta} = (a_{ij})_{\alpha\beta}$ with entries in the ring $\mathcal{O}_X(U_\alpha \cap U_\beta)$. These matrixes obviously satisfy the glueing conditions

$$\begin{aligned} C_{\alpha\alpha} &= \text{id}, \quad \text{and} \\ C_{\alpha\gamma} &= C_{\alpha\beta} C_{\beta\gamma} \quad \text{on } U_\alpha \cap U_\beta \cap U_\gamma. \end{aligned} \tag{6.2}$$

Conversely, specifying matrixes $C_{\alpha\beta}$ with entries in $\mathcal{O}_X(U_\alpha \cap U_\beta)$ defines a vector bundle, provided the $C_{\alpha\beta}$ satisfy (6.2).

The matrixes $C_{\alpha\beta}$ are called *transition matrixes* of the vector bundle. For example, if $\mathcal{L}$ is the rank 1 vector bundle over $\mathbb{P}^n$ introduced in Example 6.2, the maps φ_α are of the form $\varphi_\alpha(x, y) = (x, y_\alpha)$, so that the transition matrix $C_{\alpha\beta}$ is the 1×1 matrix $x_\beta x_\alpha^{-1}$.

It is easy to determine how the matrixes $C_{\alpha\beta}$ depend on the choice of the isomorphisms φ_α. Any other isomorphism φ'_α is of the form $\varphi'_\alpha = f_\alpha \varphi_\alpha$ where f_α is an automorphism of the trivial bundle $U_\alpha \times V$. By Example 6.1 again, f_α can be expressed as a matrix B_α with entries in $\mathcal{O}_X(U_\alpha)$ having an inverse matrix of the same form. We thus arrive at new matrixes

$$C'_{\alpha\beta} = B_\beta C_{\alpha\beta} B_\alpha^{-1}.$$

Conversely making any such change of the matrixes $C_{\alpha\beta}$ leads to an isomorphic vector bundle.

1.3 Vector Bundles and Sheaves

A vector bundle is a generalisation of a vector space. We now introduce the analogue of a point of a vector space.

Definition A *section* of a vector bundle $p \colon E \to X$ is a morphism $s \colon X \to E$ such that $p \circ s = 1$ on X.

In particular $s(x) = 0_x$ (the zero vector in E_x) is a section, called the *zero section* of E. The set of sections of a vector bundle E is written $\mathcal{L}(E)$.

Example 6.5 A section f of the trivial rank 1 bundle $X \times k$ is simply a morphism of X to $\mathbb{A}^1$, that is, an element $f \in \mathcal{O}_X(X)$. Thus $\mathcal{L}(X \times k) = \mathcal{O}_X(X)$. In particular $\mathcal{L}(\mathbb{P}^n \times k) = k$; similarly, $\mathcal{L}(\mathbb{P}^n \times V) = V$.

Consider the vector bundle E of Example 6.2. Every section $s \colon \mathbb{P}^n \to E$ determines, in particular, a section $s \colon \mathbb{P}^n \to \mathbb{P}^n \times V$, and hence by Corollary 1.2 of Section 5.2, Chapter 1 is of the form $s(x) = (x, v)$ for some fixed $v \in V$. But since $s(x) \in E$, it follows that $v \in l_x$ for every $x \in \mathbb{P}^n$, and hence $v = 0$. Thus $\mathcal{L}(E) = 0$. This proves in particular that E is not a trivial bundle.

In terms of transition functions, a section s is given by sending each set U_α to a vector $s_\alpha = (f_{\alpha,1}, \dots, f_{\alpha,n})$ with $f_{\alpha,i} \in \mathcal{O}_X(U_\alpha)$, such that $s_\beta = C_{\alpha\beta} s_\alpha$ over $U_\alpha \cap U_\beta$.

It is easy to check from the definition of vector bundle that if s_1 and s_2 are sections of E then there exists a section $s_1 + s_2$ such that

$$(s_1 + s_2)(x) = s_1(x) + s_2(x)$$

for any point $x \in X$. The sum on the right-hand side is meaningful, since $s_1(x)$ and $s_2(x) \in E_x$, and E_x is a vector space. In a similar way the equality

$$(fs)(x) = f(x)s(x)$$

determines a multiplication of a section s by an element $f \in \mathcal{O}_X(X)$.

Thus the set $\mathcal{L}(E)$ is a module over the ring $\mathcal{O}_X(X)$. We associate with any open set $U \subset X$ the set $\mathcal{L}(E, U)$ of sections of the bundle E restricted to U. An obvious check shows that we obtain a sheaf. We denote it by $\mathcal{L}_E$; it is a sheaf of Abelian groups, but has an extra structure, which we now define in a general form.

Definition Let X be a topological space, and suppose given on X a sheaf of rings $\mathcal{G}$, a sheaf of Abelian groups $\mathcal{F}$, and in addition, for each $U \subset X$, a $\mathcal{G}(U)$-module structure on $\mathcal{F}(U)$. In this situation we say that $\mathcal{F}$ is *sheaf of $\mathcal{G}$-modules* if the multiplication map $\mathcal{F}(U) \otimes \mathcal{G}(U) \to \mathcal{F}(U)$ is compatible with the restriction homomorphisms ρ_U^V; that is, the diagram

$$\begin{array}{ccc} \mathcal{F}(V) \otimes \mathcal{G}(V) & \longrightarrow & \mathcal{F}(V) \\ {\scriptstyle \rho^V_{U,\mathcal{F}} \otimes \rho^V_{U,\mathcal{G}}} \downarrow & & \downarrow {\scriptstyle \rho^V_{U,\mathcal{F}}} \\ \mathcal{F}(U) \otimes \mathcal{G}(U) & \longrightarrow & sF(U) \end{array}$$

is commutative for each $U \subset V$. Under these circumstances, each stalk $\mathcal{F}_x$ of $\mathcal{F}$ is a module over the stalk $\mathcal{G}_x$ of $\mathcal{G}$.

A *homomorphism* $\mathcal{F} \to \mathcal{F}'$ of sheaves of $\mathcal{G}$-modules is a system of homomorphisms $\varphi_U : \mathcal{F}(U) \to \mathcal{F}'(U)$ of $\mathcal{G}(U)$-modules such that the diagram

$$\begin{array}{ccc} \mathcal{F}(V) & \xrightarrow{\varphi_V} & \mathcal{F}'(V) \\ {\scriptstyle \rho^V_{U,\mathcal{F}}} \downarrow & & \downarrow {\scriptstyle \rho^V_{U,\mathcal{F}'}} \\ \mathcal{F}(U) & \xrightarrow[\varphi_U]{} & \mathcal{F}'(U) \end{array}$$

is commutative for all $U \subset V$.

Obviously the sheaf $\mathcal{L}_E$ corresponding to a vector bundle is a sheaf of modules over the structure sheaf $\mathcal{O}_X$.

Every operation on modules that can be defined intrinsically can be carried over to sheaves of modules. In particular, for any modules over a ring A, the operations

$$M \oplus M', \quad M \otimes_A M', \quad M^* = \mathrm{Hom}(M, A), \quad \bigwedge\nolimits_A^p M$$

are defined. Applying these to the modules $\mathcal{F}(U)$ and $\mathcal{F}'(U)$ over the ring $\mathcal{G}(U)$, we arrive at sheaves $\mathcal{F} \oplus \mathcal{F}'$, $\mathcal{F} \otimes_{\mathcal{G}} \mathcal{F}'$, $\mathcal{F}^*$ and $\bigwedge_{\mathcal{G}}^p \mathcal{F}$, that we call the *direct sum, tensor product, dual sheaf and exterior power.*

The sheaf of a trivial bundle of rank n is determined by $\mathcal{L}_E(U) = \mathcal{O}_X(U)^n$; that is, $\mathcal{L}_E$ is the direct sum of n copies of $\mathcal{O}_X$. This sheaf is called the *free sheaf* of rank n. Let $\mathcal{F}$ be a sheaf of $\mathcal{O}_X$-modules. If every point has a neighbourhood U such that $\mathcal{F}_{|U}$ is free and of finite rank then we say that $\mathcal{F}$ is a *locally free sheaf* of finite rank. If $\mathcal{F}$ is a locally free sheaf then obviously every stalk $\mathcal{F}_x$ is a free $\mathcal{O}_x$-module. The sheaf $\mathcal{L}_E$ corresponding to any vector bundle E is locally free of finite rank, since E is locally isomorphic to a trivial bundle.

Theorem 6.2 *The correspondence $E \mapsto \mathcal{L}_E$ establishes a one-to-one correspondence between vector bundles and locally free sheaves of finite rank (here objects of either type are considered up to isomorphism).*

Proof We show how to recover a vector bundle from a locally free sheaf $\mathcal{F}$. We can obviously assume that X is connected. Suppose that $X = \bigcup U_\alpha$ is a cover such that $\mathcal{F}_{|U_\alpha}$ is a free sheaf, and let $\varphi_\alpha : \mathcal{F}_{|U_\alpha} \xrightarrow{\sim} \mathcal{O}_{U_\alpha}^{n_\alpha}$ be the corresponding isomorphism. Then

$$\varphi_\beta \circ \varphi_\alpha^{-1} : \mathcal{O}_{U_\alpha \cap U_\beta}^{n_\alpha} \to \mathcal{O}_{U_\alpha \cap U_\beta}^{n_\beta} \tag{6.3}$$

is an isomorphism of sheaves of modules. Since X is connected, it follows that all the numbers n_α are equal; set $n_\alpha = n$. Any endomorphism of the sheaf of modules $\mathcal{O}_U^n$ is given by a matrix $C = (c_{ij})$ with $c_{ij} \in \mathcal{O}_X(U)$. Thus the isomorphism (6.3) defines a matrix $C_{\alpha\beta}$ and obviously these matrixes satisfy the relations (6.2). Hence they define some vector bundle E. A trivial verification, which we omit, shows that $\mathcal{L}_E = \mathcal{F}$. The theorem is proved. □

One checks easily that the correspondence $E \mapsto \mathcal{L}_E$ between vector bundles and locally free sheaves allows us to associate a homomorphism of sheaves of $\mathcal{O}_X$-modules to any homomorphism of bundles. In other words, the correspondence is an equivalence of the two categories.

We should point out that the fibre of a vector bundle and the stalk of the corresponding sheaf are entirely different objects. For example, if $E = X \times k$ then $\mathcal{L}_E = \mathcal{O}_X$, so that $E_x = k$, whereas $(\mathcal{L}_E)_x = \mathcal{O}_x$. In the general case the fibre E_x can be recovered from the stalk $(\mathcal{L}_E)_x$ using the relation

$$E_x = (\mathcal{L}_E)_x / \mathfrak{m}_x (\mathcal{L}_E)_x, \tag{6.4}$$

where $\mathfrak{m}_x$ is the maximal ideal of $\mathcal{O}_x$. It is enough to verify this locally; then we can write $E = U \times k^n$ and $\mathcal{L}_E = \mathcal{O}_U^n$, and (6.4) is obvious.

Whereas the notion of vector bundle was introduced here in a set-theoretic way, that of a locally free sheaf is adapted for the more general situation, and is meaningful for arbitrary schemes. It gives a natural analogue of the language of vector bundles. Moreover, the description in terms of transition matrixes also carries over: the matrixes $C_{\alpha\beta}$ must have entries belonging to the ring $\mathcal{O}_X(U_\alpha \cap U_\beta)$, and their determinants must be invertible elements of this ring.

We can also define a vector bundle over an arbitrary scheme X as a scheme locally isomorphic to $U \times \mathbb{A}^n$, with $U_\alpha \times \mathbb{A}^n$ and $U_\beta \times \mathbb{A}^n$ glued together by transition matrixes $C_{\alpha\beta}$. Then the operations that determine the vector space structure in the fibres are defined invariantly (because the matrixes $C_{\alpha\beta}$ define linear maps). The sheaf of sections $\mathcal{L}_E$ of a vector bundle E is defined just as before, and the correspondence $E \mapsto \mathcal{L}_E$ is described by Theorem 6.2.

But even for varieties, Theorem 6.2 is convenient because it gives a method of constructing new vector bundles.

Example 6.6 Let E and F be vector bundles, and $\mathcal{L}_E, \mathcal{L}_F$ the corresponding locally free sheaves. It is obvious that the sheaves $\mathcal{L}_E \oplus \mathcal{L}_F$, $\mathcal{L}_E \otimes \mathcal{L}_F$, $\mathcal{L}_E^*$, $\bigwedge_{\mathcal{O}}^p \mathcal{L}_E$ are all locally free. The corresponding vector bundles are denoted by $E \oplus F$, $E \otimes F$, E^*, $\bigwedge^p E$. In case $p = \operatorname{rank} E$ we write $\bigwedge^p E = \det E$; this is a rank 1 vector bundle, called the *determinant line bundle* of E.

If $X = \bigcup U_\alpha$ is a cover in which E and F are defined by transition matrixes $C_{\alpha\beta}$ and $D_{\alpha\beta}$ then in the same cover, $E \oplus F$, $E \otimes F$, E^*, $\bigwedge^p E$ are defined by the transition matrixes

$$\begin{pmatrix} C_{\alpha\beta} & 0 \\ 0 & D_{\alpha\beta} \end{pmatrix}, \quad C_{\alpha\beta} \otimes D_{\alpha\beta}, \quad \left({}^tC_{\alpha\beta}\right)^{-1}, \quad \bigwedge^p C_{\alpha\beta} \tag{6.5}$$

where ${}^tC$ denotes the transpose matrix. Setting $p = \operatorname{rank} E$, we see that the bundle $\det E$ is defined by the 1×1 matrixes $\det C_{\alpha\beta}$.

Corollary *For any bundle E, the dual bundle E^* has* $\det E^* = (\det E)^{-1}$.

It follows from (6.4) that where these operations are performed on vector bundles, the corresponding operations on vector spaces are performed on each fibre.

Example 6.7 Let X be a nonsingular variety. Taking an open set U to the group $\Omega^p[U]$ of differential p-forms regular on U defines in an obvious way a sheaf of $\mathcal{O}_X$-modules. It is called the *sheaf of differential p-forms*.

Theorem 3.18 of Section 5.3, Chapter 3 asserts that this sheaf is locally free. Hence by Theorem 6.2 it defines a vector bundle, denoted by Ω^p. In particular, Ω^1 is called the *cotangent bundle*.

The stalk of the sheaf $\mathcal{F}$ at a point $x \in X$ is of the form $\mathcal{F}_x = \mathcal{O}_x dt_1 + \cdots + \mathcal{O}_x dt_n$, where $t_1, \ldots, t_n$ are local parameters at x, and the sum is a direct sum. The homomorphism $\mathcal{F}_x \to \mathcal{F}_x / \mathfrak{m}_x \mathcal{F}_x$ can be written in the form

$$u_1 dt_1 + \cdots + u_n dt_n \mapsto u_1(x) dt_1 + \cdots + u_n(x) dt_n,$$

and hence by (6.4) it follows that

$$\Omega_x^1 \cong \mathcal{F}_x / \mathfrak{m}_x \mathcal{F}_x \cong \mathfrak{m}_x / \mathfrak{m}_x^2. \tag{6.6}$$

Obviously $\bigwedge^p \Omega^1 = \Omega^p$ and $\det \Omega^1 = \Omega^n$, where $n = \dim X$.

Example 6.8 The vector bundle dual to the cotangent bundle is called the *tangent bundle*, and is denoted by Θ. By (6.6), for any point $x \in X$ we have

$$\Theta_x = \left(\mathfrak{m}_x/\mathfrak{m}_x^2\right)^*,$$

that is, it is the tangent space at x. By Remark of Section 5.2, Chapter 3 it follows that for an affine subset $U \subset X$ with $U = \operatorname{Spec} A$ we have $\mathcal{O}_X(U) = \operatorname{Der}_k(A, A)$.

The final general question we want to discuss in connection with vector bundles is the notion of subbundle and quotient bundle.

Definition A morphism of vector bundles $\varphi\colon F \to E$ which is a closed embedding of varieties is an *embedding of vector bundles*. In this case the image $\varphi(F)$ is called a *subbundle* of E.

Proposition *A subbundle $F \subset E$ of a vector bundle is locally a direct summand.*

Proof The assertion means that for any point $x \in X$ there exists a neighbourhood U of x and a subbundle G of the restriction $E_{|U}$ such that

$$E_{|U} = F_{|U} \oplus G. \tag{6.7}$$

By Theorem 6.2, this equality is equivalent to the same equality for sheaves of modules, or simply for modules over $\mathcal{O}_X(U)$. As always, the local assertion can be reformulated in terms of local rings, but for this we must first translate the assumption that $\varphi\colon F \to E$ is a closed embedding in terms of the sheaves $\mathcal{L}_E$ and $\mathcal{L}_F$. Obviously, in this case, for any closed point $x \in X$ the homomorphism $\varphi_x\colon F_x \to E_x$ is an embedding. This means that if $\mathcal{L}_{F|U} = \mathcal{O}^r$ and $\mathcal{L}_{E|U} = \mathcal{O}^n$, and $\varphi\colon \mathcal{O}^r \to \mathcal{O}^n$ is the sheaf homomorphism corresponding to the homomorphism of vector bundles, then a free basis $e_1, \dots, e_r$ of $\mathcal{O}^r$ goes into a system of elements $\varphi(e_1), \dots, \varphi(e_r) \in \mathcal{O}^n$ that are linearly independent at each point. Thus we must show that if $\mathcal{O}$ is a local ring with maximal ideal $\mathfrak{m}$, and $\varphi\colon \mathcal{O}^r \to \mathcal{O}^n$ a homomorphism, and $e_1, \dots, e_r$ a free basis of $\mathcal{O}^r$ such that $\varphi(e_1), \dots, \varphi(e_r) \in \mathcal{O}^n$ are linearly independent modulo $\mathfrak{m}\mathcal{O}^n$ then φ is an embedding and $\mathcal{O}^n$ is a direct sum of $\varphi(\mathcal{O}^r)$ and a submodule isomorphic to $\mathcal{O}^{n-r}$. Indeed, set $\overline{e}_i = \varphi(e_i)$. Since $\dim(\mathcal{O}^n/\mathfrak{m}\mathcal{O}^n) = n$, the images of the elements $\overline{e}_i$ can be lifted to a basis of $\mathcal{O}^n/\mathfrak{m}\mathcal{O}^n$. Then by Nakayama's lemma (Proposition A.11 of Appendix to Volume 1) the elements $\overline{e}_1, \dots, \overline{e}_r$ can be extended to a system of generators $\overline{e}_1, \dots, \overline{e}_n$ of $\mathcal{O}^n$.

It is easy to see that this system is a free basis of $\mathcal{O}^n$: this does not even depend on $\mathcal{O}$ being a local ring. Indeed, if $f_1, \dots, f_n$ is some free basis of $\mathcal{O}^n$ then $\overline{e}_i = \sum a_{ij} f_j$ and $f_i = \sum b_{ij} e_j$ with a_{ij} and $b_{ij} \in \mathcal{O}$. From the fact that $f_1, \dots, f_n$ is a free basis, it follows that the matrixes $A = (a_{ij})$ and $B = (b_{ij})$ satisfy $BA = 1$. But then also $AB = 1$, which means that $\overline{e}_1, \dots, \overline{e}_n$ is a free basis of $\mathcal{O}^n$. From the fact that $\overline{e}_1, \dots, \overline{e}_r$ are linearly independent over $\mathcal{O}$ if follows that $\mathcal{O}^n = \varphi(\mathcal{O}^r) \oplus N$, where N is the module generated by $\overline{e}_{r+1}, \dots, \overline{e}_n$. The proposition is proved. □

Now we can define the *quotient bundle* E/F of a vector bundle E by a subbundle $F \subset E$. As a set, of course,

$$E/F = \bigcup_{x \in X} E_x/F_x.$$

To give it a structure of variety, consider an open set U for which (6.7) holds, and identify $\bigcup_{x\in X} E_x/F_x$ with the algebraic variety G. It is easy to see that these structures are compatible on different open sets U and determine E/F as a vector bundle.

The proof of the proposition obviously remains valid for vector bundles over an arbitrary scheme X and leads to the definition of quotient bundle in this case.

The translation into the language of transition matrixes is obvious. If we choose a cover $X = \bigcup U_\alpha$ such that (6.7) holds for all U_α, the matrixes $C_{\alpha\beta}$ defining E can be expressed in the form

$$C_{\alpha\beta} = \begin{pmatrix} D_{\alpha\beta} & 0 \\ * & D'_{\alpha\beta} \end{pmatrix},$$

where $D_{\alpha\beta}$ defines the vector bundle F and $D'_{\alpha\beta}$ the vector bundle E/F. It follows at once from this that

$$\det E = \det F \otimes \det E/F. \tag{6.8}$$

Example 6.9 (The normal bundle $N_{X/Y}$) Let X be a nonsingular variety and $Y \subset X$ a nonsingular closed subvariety. We define the *normal bundle* $N_{X/Y}$ to Y in X. The definition used in differential geometry is not applicable in the algebraic situation, since it is based on the notion of the orthogonal complement $W^\perp$ of a vector subspace $W \subset V$. However, as a vector space, $W^\perp$ is determined by the fact that it is isomorphic to V/W. This is what we exploit.

Write Θ'_X for the restriction to Y of the tangent bundle Θ_X. It is defined as the pullback $j^*\Theta_X$, where $j\colon Y \hookrightarrow X$ is the closed embedding. The vector bundle Θ_Y is a subbundle of Θ'_X. Indeed, by definition $\Theta'_X = j^*\Theta_X = j^*((\Omega^1_X)^*) = (j^*\Omega^1_X)^*$. The restriction of differential forms from X to Y defines a surjective homomorphism $\varphi\colon j^*\Omega^1_X \to \Omega^1_Y$ and its dual $\varphi^*\colon \Theta_Y = (\Omega^1_Y)^* \to (j^*\Omega^1_X)^* = \Theta'_X$. By definition

$$N_{X/Y} = \Theta'_X/\Theta_Y.$$

We compute the transition matrix of the normal bundle. The homomorphism $\Theta'_X \to N_{X/Y}$ defines a dual homomorphism

$$\psi\colon N^*_{X/Y} \to j^*\Omega^1_X$$

of the dual vector bundles. It is easy to see that ψ defines a closed embedding, so that we can view $N^*_{X/Y}$ as a subbundle of $j^*\Omega^1_X$ and Ω^1_Y as the quotient bundle $(j^*\Omega^1_X)/N^*_{X/Y}$. It is enough to check these assertions on open sets on which our vector bundles are trivial, where they are obvious.

As we saw in Theorem 3.17, Corollary of Section 5.1, Chapter 3, forms $du_1, \dots, du_n$ are a basis of the $\mathcal{O}_X(U)$-module $\Omega^1_X[U]$ provided that the functions

$u_1, \dots, u_n$ define local parameters at each point $x \in U$. This basis defines a basis $\eta_1, \dots, \eta_n$ of the $\mathcal{O}_Y(U \cap Y)$-module of sections over $U \cap Y$ of the sheaf corresponding to the vector bundle $j^*\Omega^1_X$. Here $\varphi(\eta_i)$ is the restriction to Y of the form du_i.

Suppose that $n = \dim X$ and $m = \operatorname{codim}(Y \subset X)$. By Theorem 2.14 of Section 3.2, Chapter 2 we can choose the functions $u_1, \dots, u_n$ such that $u_1 = \dots = u_m = 0$ are the local equations of Y in U. By the same theorem, together with Theorem 3.17, Corollary of Section 5.1, Chapter 3, the restrictions of the forms $du_{m+1}, \dots, du_n$ define a basis of $\Omega^1_Y[U \cap Y]$, and hence $\eta_1, \dots, \eta_m$ is a basis of the $\mathcal{O}_Y(U \cap Y)$-module $N^*_{X/Y}(U \cap Y)$.

Suppose that U_α and U_β are two open sets in which $u_{\alpha,1}, \dots, u_{\alpha,n}$ and $u_{\beta,1}, \dots, u_{\beta,n}$ are systems of local parameters chosen as described. The transition matrix for the vector bundle Ω^1_X is determined by the expression

$$du_{\alpha,i} = \sum_{j=1}^{n} h_{ij} du_{\beta,j} \quad \text{for } i = 1, \dots, n, \tag{6.9}$$

where $h_{ij} \in \mathcal{O}_X(U)$ are the entries of the Jacobian matrix, that is, $h_{ij} = \partial u_{\alpha,i}/\partial u_{\beta,j}$, and the transition matrix of $j^*\Omega^1_X$ in the basis $\eta_1, \dots, \eta_n$ is obtained by restricting the entries of this matrix to $U \cap Y$.

Since $u_{\alpha,i} \in (u_{\beta,1}, \dots, u_{\beta,m})$ on $U_\alpha \cap U_\beta$ for $i = 1, \dots, m = \operatorname{codim} Y$, we have

$$u_{\alpha,i} = \sum_{j=1}^{m} f_{ij} u_{\beta,j} \quad \text{for } i = 1, \dots, m,$$

with $f_{ij} \in \mathcal{O}_X(U_\alpha \cap U_\beta)$. Hence for $i = 1, \dots, m$ we have

$$du_{\alpha,i} = \sum_{j=1}^{m} f_{ij} du_{\beta,j} + \sum_{j=1}^{m} u_{\beta,j} df_{ij}. \tag{6.10}$$

To reconcile this formula with (6.9) we would have to express the df_{ij} in terms of $du_1, \dots, du_n$. But we are only interested in formulas for the η_i, which are obtained by restricting to Y all the functions occurring in it. Since $u_{\beta,j} = 0$ on Y for $j = 1, \dots, m$, the second group of terms in (6.10) vanishes. Thus

$$\eta_{\alpha,i} = \sum_{j=1}^{m} \overline{f}_{ij} \eta_{\beta,j} \quad \text{for } i = 1, \dots, m,$$

where $\overline{f}_{ij}$ is the restriction of f_{ij} to $U_\alpha \cap U_\beta \cap Y$. As we have seen, these are the transition matrixes of the vector bundle $N^*_{X/Y}$. Those for $N_{X/Y}$ are obtained on transposing and taking the inverse; taking the inverse is equivalent to interchanging α and β. We finally arrive at the simple formulas

$$C_{\alpha\beta} = (h_{ij|Y}), \quad \text{for } i, j = 1, \dots, m \tag{6.11}$$

where $u_{\beta,j} = \sum h_{ij} u_{\alpha,i}$ in $U_\alpha \cap U_\beta$.

An important factor in practically all the constructions of this section is the possibility of specifying a vector bundle in an abstract way, without reference to an embedding into projective space. It can however be proved that a vector bundle over a quasiprojective variety is itself quasiprojective; we omit the proof.

1.4 Divisors and Line Bundles

In what follows, we do not assume that X is nonsingular, and consider locally principal divisors D (Section 1.2, Chapter 3). Corresponding to each divisor D on an irreducible variety X we have a vector space $\mathcal{L}(D)$ (Section 1.2, Chapter 3). This correspondence gives rise to a sheaf on X. To see this, note that the divisor D on X also defines a divisor on any open subset $U \subset X$, by restricting to U the local equations of D. We write D_U for the divisor thus obtained and set

$$\mathcal{L}_D(U) = \mathcal{L}(U, D_U),$$

where $\mathcal{L}(U, D_U)$ is the vector space corresponding to the divisor D_U on U. Obviously $\mathcal{L}_D(U) \subset k(X)$, and $\mathcal{L}_D(V) \subset \mathcal{L}_D(U)$ whenever $U \subset V$; write $\rho_U^V : \mathcal{L}_D(V) \hookrightarrow \mathcal{L}_D(U)$ for the inclusion map. The system $\{\mathcal{L}_D(U), \rho_U^V\}$ is a presheaf, and it is easy to see that it is a sheaf. We denote it by $\mathcal{L}_D$.

Multiplying elements $f \in \mathcal{L}_D(U)$ by $h \in \mathcal{O}_X(U)$ makes $\mathcal{L}_D$ into a sheaf of $\mathcal{O}_X$-modules. This sheaf is locally free. Indeed, if D is defined in an open set U_α by a local equation f_α then the elements $g \in \mathcal{L}_D(U_\alpha)$ are characterised by the condition $gf_\alpha \in \mathcal{O}_X(U_\alpha)$. This shows that the map $g \mapsto gf_\alpha$ defines an isomorphism

$$\varphi_\alpha : \mathcal{L}_{D|U_\alpha} \xrightarrow{\sim} \mathcal{O}_{X|U_\alpha}. \tag{6.12}$$

We saw in Section 1.3 that such a sheaf determines a vector bundle E_D; it follows from (6.12) that $\operatorname{rank} E_D = 1$. Vector bundles of rank 1 are called *line bundles* since their fibres are lines. We write out the transition functions for E_D. Since the isomorphism over U_α in (6.12) is given by multiplication by f_α, the automorphism $\varphi_\beta \circ \varphi_\alpha^{-1}$ over $U_\alpha \cap U_\beta$ is given by multiplication by f_β/f_α. Note that $f_\beta/f_\alpha \in \mathcal{O}_X(U_\alpha \cap U_\beta)$ by the compatibility of the f_α. Similarly $(f_\beta/f_\alpha)^{-1} = f_\alpha/f_\beta \in \mathcal{O}_X(U_\alpha \cap U_\beta)$. Thus in this case the transition matrix is the 1×1 matrix $\varphi_{\alpha\beta}$ given by

$$\varphi_{\alpha\beta} = f_\beta/f_\alpha. \tag{6.13}$$

If we replace the divisor D by a linearly equivalent divisor $D' = D + \operatorname{div} f$ with $f \in k(X)$ then multiplication by f defines an isomorphism of modules $\mathcal{L}(U, D_U) \to \mathcal{L}(U, D'_U)$. We verified this in Theorem 3.3 of Section 1.5, Chapter 3. In this way we obviously get an isomorphism of sheaves $\mathcal{L}_D \xrightarrow{\sim} \mathcal{L}_{D'}$. The two line bundles E_D and $E_{D'}$ actually have identical transition functions. Thus the sheaf $\mathcal{L}_D$ and the line bundle E_D both correspond to a whole divisor class.

Theorem 6.3 *The map $D \mapsto \mathcal{L}_D \to E_D$ defines a one-to-one correspondence between* (1) *linear equivalence classes of divisors,* (2) *isomorphism classes of sheaves of $\mathcal{O}_X$-modules locally isomorphic to $\mathcal{O}_X$, and* (3) *isomorphism classes of rank* 1 *vector bundles.*

Proof The correspondence between the sets (2) and (3) was established in Theorem 6.2. Thus we need only prove that $D \mapsto E_D$ defines a one-to-one correspondence between the sets (1) and (3). To do this we construct the inverse map.

Suppose that E is a line bundle defined in a cover $X = \bigcup U_\alpha$ by 1×1 transition matrixes $\varphi_{\alpha\beta}$, with $\varphi_{\alpha\beta}$ and $\varphi_{\alpha\beta}^{-1} \in \mathcal{O}_X(U_\alpha \cap U_\beta)$. It follows from the glueing conditions (6.2) that $\varphi_{\beta\alpha} = \varphi_{\alpha\beta}^{-1}$ and

$$\varphi_{\alpha\beta} = \varphi_{\gamma\alpha}^{-1}\varphi_{\gamma\beta} \quad \text{over } U_\alpha \cap U_\beta \cap U_\gamma. \tag{6.14}$$

The inclusion $\mathcal{O}_X(U_\alpha \cap U_\beta) \hookrightarrow k(X)$ allows us to consider the $\varphi_{\alpha\beta}$ as elements of $k(X)$, and (6.14) holds for these in the same way. Fix some subscript γ, say $\gamma = 0$. We substitute $\gamma = 0$ in (6.14) and set $f_\alpha = \varphi_{0\alpha}$. Then the system of elements f_α on U_α is compatible, since

$$f_\beta / f_\alpha = \varphi_{\alpha\beta}; \tag{6.15}$$

hence they define a certain divisor D. Comparing (6.13) and (6.15) shows that $E = E_D$.

Let us prove that the linear equivalence class of the divisor D depends only on the line bundle E and not on the choice of the cover or the transition matrixes. Two systems $\{U_\alpha, \varphi_{\alpha\beta}\}$ and $\{U'_\lambda, \varphi'_{\lambda\mu}\}$ can be compared on the cover $\{U_\alpha \cap U'_\lambda\}$ by setting

$$\widetilde{\varphi}_{\alpha\beta\lambda\mu} = \varphi_{\alpha\beta} \quad \text{and} \quad \widetilde{\varphi}'_{\alpha\beta\lambda\mu} = \varphi'_{\lambda\mu} \quad \text{on } U_\alpha \cap U_\beta \cap U'_\lambda \cap U'_\mu.$$

Therefore we can assume from the start that the two covers are the same, $X = \bigcup U_\alpha$. Then as shown in Section 1.2,

$$\varphi'_{\alpha\beta} = \psi_\alpha^{-1}\varphi_{\alpha\beta}\psi_\beta \quad \text{with } \psi_\alpha \text{ and } \psi_\alpha^{-1} \in \mathcal{O}_X(U_\alpha). \tag{6.16}$$

By definition of f_α and f'_α

$$f'_\alpha = \psi_0^{-1}\varphi_{0\alpha}\psi_\alpha = \psi_0^{-1} f_\alpha \psi_\alpha,$$

so that (6.16) gives $D' = D - \operatorname{div}(\psi_0)$.

Thus we have constructed a well-defined map from the set (3) to the set (1). An obvious check shows that it is the inverse of $D \mapsto E_D$. The theorem is proved. □

For any morphism $f \colon X \to Y$ we have the relation

$$f^*(E_D) = E_{f^*(D)}; \tag{6.17}$$

we leave the obvious verification to the reader.

The divisor class corresponding to a line bundle E under Theorem 6.3 is called the *characteristic class* of E and denoted by $c(E)$.

Example 6.10 If $\dim X = n$ and Ω^n is the line bundle introduced in Section 1.2 then $c(\Omega^n) = K$ is the canonical class of X.

Example 6.11 Let $X = \mathbb{P}^n$ and let D be a hyperplane of $\mathbb{P}^n$. The line bundle E_D corresponding to D under Theorem 6.3 is denoted by $\mathcal{O}(1)$. If D is given by $x_0 = 0$ then in the open set U_α where $x_\alpha \neq 0$ it has the local equation x_0/x_α. Hence the transition matrix for E_D is of the form $c_{\alpha\beta} = x_\alpha/x_\beta$. It follows that $\mathcal{O}(1)$ is the line bundle dual to the line bundle $\mathcal{L}$ of Example 6.2.

Let us find the sections of $\mathcal{O}(1)$. In U_α these are of the form $s_\alpha = P_\alpha/x_\alpha^k$, where P_α is a form of degree k; and they are related by $s_\beta = c_{\alpha\beta}s_\alpha$. It follows that $k = 1$ and that $P_\alpha = P_\beta$ is a form of degree 1 on $\mathbb{P}^n$. Similarly, the divisor mD corresponds to the line bundle denoted by $\mathcal{O}(m)$ with the transition matrix $c_{\alpha\beta} = (x_\alpha/x_\beta)^m$. The sections of $\mathcal{O}(m)$ are homogeneous polynomials of degree m. It is easy to see that $\mathcal{O}(m) = \mathcal{O}(1)^{\otimes m}$ is the mth tensor power of $\mathcal{O}(1)$. For a subvariety $X \subset \mathbb{P}^n$ we write $\mathcal{O}_X(m)$ for the restriction to X of the line bundle (or sheaf) $\mathcal{O}(m)$ on $\mathbb{P}^n$.

Example 6.12 Let X be a nonsingular variety and $Y \subset X$ a nonsingular hypersurface. In this case the normal bundle $N_{X/Y}$ is a line bundle. We compute its characteristic class.

Suppose that Y is given in an affine cover $X = \bigcup U_\alpha$ by local equations f_α. Then $f_\beta/f_\alpha = f_{\alpha\beta}$, where $f_{\alpha\beta}$ and $f_{\alpha\beta}^{-1} \in \mathcal{O}_X(U_\alpha \cap \mathcal{O}_\beta)$. By (6.11), the transition matrixes of $N_{X/Y}$ are of the form $f_{\alpha\beta|Y} = (f_\beta/f_\alpha)_{|Y}$. But we have just seen that f_β/f_α are the transition matrixes for the line bundle E_Y. Thus we have proved the formula

$$N_{X/Y} = E_{Y|Y}.$$

By (6.17) it follows from this that

$$c(N_{X/Y}) = \rho_Y(C_Y),$$

where C_Y is the divisor class on X containing Y and $\rho_Y \colon \operatorname{Cl} X \to \operatorname{Cl} Y$ the homomorphism of restriction to Y. Recall from Section 1.2, Chapter 3 the explicit description of ρ_Y: we must replace Y by a linearly equivalent divisor Y' not containing Y as a component, then restrict Y' to Y.

Since divisor classes form a group, the correspondence established in Theorem 6.3 defines a group law on the set of line bundles or sheaves locally isomorphic to $\mathcal{O}$. From (6.13) we see that addition of divisors corresponds to multiplication of the 1×1 transition matrixes. This operation is given more intrinsically by the tensor product of line bundles or sheaves (see Theorem 6.2). Here the sheaf $\mathcal{O}$ plays the role of the multiplicative identity, and the inverse of $\mathcal{L}_D$ is $\mathcal{L}_{-D}$. Because of this, locally free sheaves of $\mathcal{O}$-modules of rank 1 are also called *invertible sheaves*.

Although invertible sheaves and divisor classes are in one-to-one correspondence, it is often technically more convenient to use invertible sheaves. For example, the inverse image $f^*(\mathcal{F})$ can be defined in a natural way for any morphism f and any sheaf $\mathcal{F}$ (see for example Hartshorne [37, Section 5, Chapter II]). It is easy to check that if $\mathcal{F}$ is invertible then so is $f^*(\mathcal{F})$. The corresponding operation for divisor classes requires arguments concerned with moving the support of a divisor.

These technical advantages of invertible sheaves are related to matters of principle. In a closely related situation, in the theory of complex manifolds, the notions of invertible sheaf and divisor class are no longer equivalent, and there, invertible sheaves provide more information and lead to a more natural statement of the problems. For this, compare Exercises 6–8 of Section 2, Chapter 8.

For an arbitrary scheme X, a sheaf locally isomorphic to $\mathcal{O}_X$ is a natural analogue of a divisor class. Such sheaves form a group: multiplication is defined as tensor product, and the inverse of a sheaf $\mathcal{F}$ is its dual $\mathcal{H}om(\mathcal{F}, \mathcal{O}_X)$. This group is again denoted by $\operatorname{Pic} X$. In our case, the transition matrixes are invertible elements of the ring $\mathcal{O}_X(U_\alpha \cap U_\beta)$, the multiplication and inverse operations reduce to the same operations on transition matrixes (in our case, transition functions).

As an application of the ideas treated here we deduce the genus formula stated and used repeatedly in Section 2.3, Chapter 4.

Theorem 6.4 (Adjunction formula) *The genus g_Y of a nonsingular curve Y on a complete nonsingular surface X is given by the formula*

$$g_Y = \frac{1}{2}Y(Y+K)+1; \tag{6.18}$$

where K is the canonical class of X.

Proof Let X be a nonsingular variety and $Y \subset X$ an arbitrary nonsingular closed subvariety. By the definition of the normal bundle $N_{X/Y}$ and (6.8), we obtain

$$\rho_Y(\det \Theta_X) = \det \Theta'_X = \det \Theta_Y \otimes \det N_{X/Y}.$$

Since Θ_X is the dual of Ω^1_X and Θ_Y that of Ω^1_Y, we can apply the Corollary of Example 6.6, formula (6.5) to obtain

$$\rho_Y\bigl(c\bigl(\Omega^n_X\bigr)\bigr) = c\bigl(\Omega^m_Y\bigr)\cdot c(\det N_{X/Y})^{-1}.$$

It follows from (6.6) that $\det(E^*) = (\det E)^{-1}$ for any vector bundle. Since $\det \Omega^1_X = \bigwedge^n \Omega^1_X = \Omega^n_X$, we get

$$\rho_Y\bigl(c\bigl(\Omega^n_X\bigr)\bigr) = c\bigl(\Omega^m_Y\bigr)\cdot c(\det N_{X/Y})^{-1}, \tag{6.19}$$

with $\dim X = n$ and $\dim Y = m$. This formula holds for a nonsingular subvariety $Y \subset X$ of any dimension, and is usually called the *adjunction formula*.

Now suppose that $m = n-1$. We apply the results obtained in Examples 6.5–6.7. We arrive at the relation

$$\rho_Y(K_X) = K_Y - \rho_Y(C_Y). \tag{6.20}$$

Finally if $n = 2$ and $m = 1$, we deduce that the divisors on either side of (6.20) have equal degrees.

Note that in our case, the restriction of any divisor D on X is a divisor $\rho_Y(D)$ on Y, and it has a well-defined degree, equal to $\deg \rho_Y(D) = YD$. Now by Corollary 3.1, of Section 7, Chapter 3 we have $\deg K_Y = 2g_Y - 2$ and so

$$YK_X = 2g_Y - 2 - Y^2,$$

and the theorem follows from this based on simple properties of intersection numbers. □

1.5 Exercises to Section 1

1 Let k be an algebraically closed field. Define a *pseudovariety* over k to be a ringed space such that every point has a neighbourhood isomorphic to m-Spec A, where A is a finitely generated k-algebra with no nilpotents; the topology and structure sheaf on m-Spec A are defined exactly as in Chapter 5. Prove that taking a variety to its set of closed points defines an isomorphism of the category of varieties and pseudovarieties.

2 Define the product of pseudovarieties X and Y. Start by setting $X \times Y$ to be the set of pairs (x, y) with $x \in X$ and $y \in Y$, then construct an affine cover of this set based on affine covers of X and Y, using the definition of products of affine varieties given in Example 1.5 of Section 2.1, Chapter 1.

3 Prove that a variety is complete if and only if its irreducible components are complete.

4 We say that a fibration $X \to S$ is *locally trivial*, or is a *fibre bundle with fibre* F if every point $s \in S$ has a neighbourhood U such that the restriction of X over U is isomorphic to $F \times U$ as a scheme over U. Prove that if $X \to S$ is a locally trivial fibration with the base S and the fibre F both complete then X is also complete.

5 Determine the transition matrixes of the line bundle of Example 6.2, which corresponds to the cover of $\mathbb{P}^n$ by the sets $\mathbb{A}^n_i$ given by $x_i \neq 0$. Find the characteristic class of this line bundle.

6 Let D be an effective divisor on a variety X for which the vector space $\mathcal{L}(D)$ is finite dimensional, and $\mathcal{F} = \mathcal{F}_D$ the corresponding invertible sheaf. Let $f : X \to \mathbb{P}^n$ with $n = l(D) - 1$ be the rational map associated with $\mathcal{L}(D)$ as in Section 1.5, Chapter 3. Assume that the divisors $\operatorname{div} f$ of functions $f \in \mathcal{L}(D)$ have no common components. Prove that f is regular at a point $x \in X$ if and only if the stalk $\mathcal{F}_x$ of $\mathcal{F}$ is generated as an $\mathcal{O}_x$-module by the space $\rho_x \mathcal{L}(D)$.

7 Suppose that $X = \operatorname{Spec} A$ is a nonsingular affine variety. Prove that the A-module $\Theta_X(X)$ is isomorphic to the module of derivations of A (that is, k-linear maps $d\colon A \to A$ such that $d(xy) = d(x)y + xd(y)$ for $x, y \in A$).

8 Prove that the normal bundle to a line C in $\mathbb{P}^n$ is a direct sum of $n-1$ isomorphic line bundles E. Find $c(E)$.

9 Suppose that $n-1$ hypersurfaces $C_1, \dots, C_{n-1}$ in $\mathbb{P}^n$ of degrees $k_1, \dots, k_{n-1}$ intersect transversally in an irreducible curve X. Find the genus of X.

10 Let $f\colon E \to X$ be a vector bundle and $X = \bigcup U_\alpha$ a cover such that E is trivial over each U_α, that is, $E_{|U_\alpha} \cong U_\alpha \times k^n$. Embed k^n in $\mathbb{P}^n$ as the set of points with $x_0 \neq 0$, and glue the varieties $U_\alpha \times \mathbb{P}^n$ by means of the transition matrixes of E, now considered as matrixes of projective transformations of $\mathbb{P}^n$. Prove that in this way we obtain a variety $\overline{E}$ containing E as an open set, and $\overline{E}$ is nonsingular; moreover, $f\colon \overline{E} \to X$ is a regular map and its fibres are isomorphic to $\mathbb{P}^n$.

11 In the notation of Exercise 10, suppose that $X = \mathbb{P}^1$, and for $n \geq 0$ let E_n be the vector bundle of rank 1 corresponding to the divisor nx_∞ on $\mathbb{P}^1$. Prove that $\overline{E}_n \setminus E_n = C_\infty$ is a curve mapped isomorphically to $\mathbb{P}^1$ by f. Let C_0 be the zero section of E_n, which is obviously also contained in $\overline{E}_n$, and write F for the fibre of $\overline{E}_n \to \mathbb{P}^1$. Prove that $C_0 - C_\infty \sim nF$ on the surface $\overline{E}_n$, and determine C_0^2 and C_∞^2.

12 In the notation of Exercise 11, prove that the restriction of divisors $D \in \operatorname{Div} \overline{E}_n$ to a general fibre defines a homomorphism $\operatorname{Cl} \overline{E}_n \to \mathbb{Z}$ whose kernel is $\mathbb{Z} \cdot F$. Prove that $\operatorname{Cl} \overline{E}_n$ is a free Abelian group with the two generators C_0 and F.

13 In the notation of Exercises 11–12, find the canonical class of the surface $\overline{E}_n$.

14 Prove that the surfaces $\overline{E}_n$ corresponding to distinct $n \geq 0$ are not isomorphic. [Hint: Prove that $\overline{E}_n$ contains a unique irreducible curve with negative selfintersection, and this selfintersection is $-n$.]

15 Let X be a nonsingular affine variety and $A = k[X]$ its affine coordinate ring. Prove that the module $\Theta(X)$ is isomorphic to $\operatorname{Der}_k(A, A)$ (often written simply as $\operatorname{Der}_k(A)$. For the definition of $\operatorname{Der}_k(A, A)$, see Exercise 24, Section 1.6, Chapter 2; compare Exercise 12, Section 5.5, Chapter 3.

2 Abstract and Quasiprojective Varieties

2.1 Chow's Lemma

We prove a result that sheds some light on the relation between complete and projective varieties. Of course, every irreducible variety is birational to a projective variety, for example, the projective closure of any affine open subset. However, one can prove considerably more in this direction.

Theorem (Chow's lemma) *For any complete irreducible variety X, there exists a projective variety $\overline{X}$ and a surjective birational morphism $f\colon \overline{X} \to X$.*

Proof The idea of the proof is the same as that used to construct the projective embedding of the normalisation of a curve (Theorem 2.23 of Section 5.3, Chapter 2).

Let $X = \bigcup U_i$ be a finite affine cover. For each affine variety $U_i \subset \mathbb{A}^{n_i}$, write Y_i for the closure of U_i in the projective space $\mathbb{P}^{n_i} \supset \mathbb{A}^{n_i}$. The variety $Y = \prod Y_i$ is obviously projective.

Set $U = \bigcap U_i$. The inclusions $\psi\colon U \hookrightarrow X$ and $\psi_i\colon U \hookrightarrow U_i \hookrightarrow Y_i$ define a morphism

$$\varphi\colon U \to X \times Y, \quad \text{with } \varphi = \psi \times \prod \psi_i.$$

Write $\overline{X}$ for the closure of $\varphi(U)$ in $X \times Y$. The first projection $p_X\colon X \times Y \to X$ defines a morphism $f\colon \overline{X} \to X$. We prove that it is birational. For this it is enough to check that

$$f^{-1}(U) = \varphi(U). \tag{6.21}$$

Indeed, $p_X \circ \varphi = 1$ on U, and in view of (6.21), f coincides on $f^{-1}(U)$ with the isomorphism φ^{-1}. Now (6.21) is equivalent to

$$(U \times Y) \cap \overline{X} = \varphi(U), \tag{6.22}$$

that is, to $\varphi(U)$ closed in $U \times Y$. But this is obvious, since $\varphi(U)$ in $U \times Y$ is just the graph of the morphism $\prod \psi_i$. The morphism f is surjective, since $f(\overline{X}) \supset U$, and U is dense in X.

It remains to prove that $\overline{X}$ is projective. For this, we use the second projection $g\colon X \times Y \to Y$, and prove that its restriction $\overline{g}\colon \overline{X} \to Y$ is a closed embedding. Since to be a closed embedding is a local property, it is enough to find open sets $V_i \subset Y$ such that $\bigcup g^{-1}(V_i) \supset \overline{X}$ and $\overline{g}\colon \overline{X} \cap g^{-1}(V_i) \to V_i$ is a closed embedding. We set

$$V_i = p_i^{-1}(U_i),$$

where $p_i\colon Y \to Y_i$ is the projection. First of all, the $g^{-1}(V_i)$ cover $\overline{X}$. For this it is enough to prove that

$$g^{-1}(V_i) = f^{-1}(U_i), \tag{6.23}$$

since $\bigcup U_i = X$ and $\bigcup f^{-1}(U_i) = \overline{X}$. In turn, (6.23) will follow from

$$f = p_i \circ g \quad \text{on } f^{-1}(U). \tag{6.24}$$

But it is enough to prove (6.24) on some open subset $W \subset f^{-1}(U_i)$. We can in particular take $W = f^{-1}(U) = \varphi(U)$ (according to (6.21)), and then (6.24) is obvious.

Thus it remains to prove that

$$\overline{g}\colon \overline{X} \cap g^{-1}(V_i) \to V_i$$

defines a closed embedding. Now recall that

$$V_i = p_i^{-1}(U_i) = U_i \times \widehat{Y}_i, \quad \text{where } \widehat{Y}_i = \prod_{j \neq i} Y_j;$$

we get that

$$g^{-1}(V_i) = X \times U_i \times \widehat{Y}_i.$$

Write Z_i for the graph of the morphism $U_i \times \widehat{Y}_i \to X$, which is the composite of the projection to U_i and the embedding $U_i \hookrightarrow X$. The set Z_i is closed in $X \times U_i \times \widehat{Y}_i = g^{-1}(V_i)$, and its projection to $U_i \times \widehat{Y}_i = V_i$ is an isomorphism. On the other hand, $\varphi(U) \subset Z_i$, and since Z_i is closed, $\overline{X} \cap g^{-1}(V_i)$ is closed in Z_i. Hence the restriction of the projection to this set is a closed embedding. Chow's lemma is proved. □

Similar arguments prove the analogous statement for an arbitrary variety, when $\overline{X}$ is quasiprojective (see Exercise 7).

2.2 Blowup Along a Subvariety

Chow's lemma shows that arbitrary varieties are rather close to projective varieties. Nevertheless, the two notions do not always coincide. We construct simple examples of non-quasiprojective varieties in the following section. The construction uses a generalisation of the notion of blowup defined in Section 4.2, Chapter 2. The difference is that here we construct a morphism $\sigma\colon X' \to X$ such that the rational map σ^{-1} blows up a whole nonsingular subvariety $Y \subset X$ rather than just one point $x_0 \in X$. The construction follows closely that of Sections 4.1–4.3, Chapter 2.

(a) The Local Construction According to Theorem 2.14 of Volume 1, for any closed point of a nonsingular subvariety $Y \subset X$ of a nonsingular variety X, there exists a neighbourhood U and functions $u_1, \dots, u_m \in \mathcal{O}_X(U)$, where $m = \operatorname{codim}_X Y$, such that the ideal $\mathfrak{a}_Y \subset \mathcal{O}_X(U)$ is given by $\mathfrak{a}_Y = (u_1, \dots, u_m)$, and such that $d_x u_1, \dots, d_x u_m$ are linearly independent at every closed point $x \in U$. The final condition means that $u_1, \dots, u_m$ can be included in a system of local parameters at $x \in U$. If these conditions are satisfied, we say that $u_1, \dots, u_m$ are local parameters for Y in U.

Suppose that X is affine and $u_1, \dots, u_m$ are local parameters for Y everywhere in X. Consider the product $X \times \mathbb{P}^{m-1}$ and the closed subvariety $X' \subset X \times \mathbb{P}^{m-1}$ defined by the equations

$$t_i u_j(x) = t_j u_i(x) \quad \text{for } i, j = 1, \dots, m,$$

where $t_1, \dots, t_m$ are the homogeneous coordinates in $\mathbb{P}^{m-1}$. The projection $X \times \mathbb{P}^{m-1} \to X$ defines a morphism $\sigma \colon X' \to X$. Clearly, now $\sigma^{-1}(Y) = Y \times \mathbb{P}^{m-1}$, and σ defines an isomorphism

$$X' \setminus (Y \times \mathbb{P}^{m-1}) \xrightarrow{\sim} X \setminus Y.$$

Let $x' = (y, t)$ be a closed point of X', with $y \in X$ and $t \in \mathbb{P}^{m-1}$; suppose that $t = (t_1 : \cdots : t_m)$ with $t_i \neq 0$. Then in a neighbourhood of x', we have $u_j = u_i s_j$, where $s_j = t_j/t_i$. Let $v_1, \dots, v_{n-m}, u_1, \dots, u_m$ be a local parameter system at $y \in X$. Then the maximal ideal of $x' \in X'$ is of the form

$$\begin{aligned} \mathfrak{m}_{x'} &= \bigl(v_1, \dots, v_{n-m}, u_1, \dots, u_m, s_1 - s_1(x'), \dots, s_m - s_m(x')\bigr) \\ &= \bigl(v_1, \dots, v_{n-m}, s_1 - s_1(x'), \dots, \widehat{s_i - s_i(x')}, u_i, \dots, s_m - s_m(x')\bigr). \end{aligned}$$

It follows from this, as in Section 4.2, Chapter 2, that X' is nonsingular, n-dimensional and irreducible. As there, the following result holds.

Lemma *If $\tau \colon \overline{X} \to X$ is a blowup of the same subvariety $Y \subset X$ defined by a different local system of parameters $v_1, \dots, v_m$ of Y then there is an isomorphism $\varphi \colon X' \to \overline{X}$ for which the diagram*

$$\begin{array}{ccc} X' & \xrightarrow{\varphi} & \overline{X} \\ & {\scriptstyle\sigma}\searrow \quad \swarrow{\scriptstyle\tau} & \\ & X & \end{array}$$

commutes. The isomorphism φ is unique.

We have $\varphi = \tau^{-1} \circ \sigma$ on the open sets $X' \setminus \sigma^{-1}(Y)$ and $\overline{X} \setminus \tau^{-1}(Y)$, and the uniqueness of φ follows from this. By definition, in these sets

$$\begin{aligned} \varphi(x; t_1 : \cdots : t_m) &= \bigl(x; v_1(x) : \cdots : v_m(x)\bigr), \\ \psi\bigl(x; t'_1 : \cdots : t'_m\bigr) &= \bigl(x; u_1(x) : \cdots : u_m(x)\bigr), \end{aligned}$$

where $\psi = \varphi^{-1}$.

By assumption,

$$v_k = \sum_j h_{kj} u_j \quad \text{with } h_{kj} \in k[X]. \tag{6.25}$$

In the open set given by $t_i \neq 0$, we set $s_j = t_j/t_i$ and rewrite (6.25) in the form

$$v_k = u_i g_k \quad \text{with } g_k = \sum_j \sigma^*(h_{kj}) s_j. \tag{6.26}$$

Then define

$$\varphi(x; t_1 : \cdots : t_m) = (x; g_1 : \cdots : g_m). \tag{6.27}$$

The same simple verification as in the proof of the analogous lemma in Section 4.2, Chapter 2 shows that φ is a morphism, which is equal to that already constructed on $X' \setminus \sigma^{-1}(Y)$. The construction of ψ is similar.

(b) The Global Construction Let $X = \bigcup U_\alpha$ be an affine cover such that Y is defined in U_α by local parameters $u_{\alpha,1}, \dots, u_{\alpha,m}$. Over U_α we apply the construction of (a) to $Y \cap U_\alpha$; we get a system of varieties X'_α and morphisms $\sigma_\alpha : X'_\alpha \to U_\alpha$. Consider the subset $\sigma_\alpha^{-1}(U_\alpha \cap U_\beta) \subset X'_\alpha$ for all α and β; then by the lemma, there exist uniquely determined isomorphisms

$$\varphi_{\alpha\beta} : \sigma_\alpha^{-1}(U_\alpha \cap U_\beta) \to \sigma_\beta^{-1}(U_\alpha \cap U_\beta).$$

It is easy to check that these satisfy the glueing conditions and define a variety X' and a morphism $\sigma : X' \to X$. The morphism σ we have constructed is called the *blowup* of Y, or the blowup of X with centre in Y. It follows in an obvious way from the lemma that X' and σ are both independent of the cover $X = \bigcup U_\alpha$ and of the system of parameters $u_{\alpha,i}$.

(c) The Exceptional Locus The subvariety $\sigma^{-1}(Y)$ is known locally:

$$\sigma^{-1}(Y \cap U_\alpha) = (Y \cap U_\alpha) \times \mathbb{P}^{m-1}. \tag{6.28}$$

Globally, we are dealing with a fibre bundle of a new type: the fibre $\sigma^{-1}(y)$ over each $y \in Y$ is a projective space $\mathbb{P}^{m-1}$. Equation (6.28) shows the sense in which our fibre bundle is locally trivial.

With every vector bundle $p : E \to X$ we can associate a fibre bundle $\varphi : \mathbb{P}(E) \to X$ of this type. For this, we define $\mathbb{P}(E)$ as the set

$$\mathbb{P}(E) = \bigcup_{x \in X} \mathbb{P}(E_x),$$

where $\mathbb{P}(E_x)$ is the projective space of lines through 0 in the vector space E_x. To give $\mathbb{P}(E)$ the structure of an algebraic variety, consider a cover $X = \bigcup U_\alpha$ in which E is given by transition matrixes $C_{\alpha\beta}$. By fixing an isomorphism $p^{-1}(U_\alpha) \cong U_\alpha \times V$, where V is a vector space, we thus get a map

$$\bigcup_{x \in U_\alpha} \mathbb{P}(E_x) \to U_\alpha \times \mathbb{P}(V),$$

which allows us to give this set the structure of an algebraic variety. All the structures of this type are obviously compatible, and define a structure of algebraic variety on the whole of $\mathbb{P}(E)$. This variety is called the *projectivisation* of E.

More concretely, $\mathbb{P}(E)$ is obtained by glueing together open subsets

$$\varphi^{-1}(U_\alpha) \cong U_\alpha \times \mathbb{P}(V)$$

by means of the glueing law defined by automorphisms of $(U_\alpha \cap U_\beta) \times \mathbb{P}(V)$:

$$\varphi_{\alpha\beta}(u, \xi) = \big(u, \mathbb{P}(C_{\alpha\beta})\xi\big), \tag{6.29}$$

where $u \in U_\alpha \cap U_\beta$, $\xi \in \mathbb{P}(V)$ and $\mathbb{P}(C_{\alpha\beta})$ is the projective transformation with matrix $C_{\alpha\beta}$.

We return to the variety $\sigma^{-1}(Y)$ arising as the exceptional locus of a blowup $\sigma\colon X' \to X$. It is obtained by glueing together the open sets $(Y \cap U_\alpha) \times \mathbb{P}^{m-1}$, with the glueing law given by (6.25). This law is precisely of the type (6.29) if we take $C_{\alpha\beta}$ to be the matrix

$$C_{\alpha\beta} = \big((h_{ij})_{|Y}\big).$$

Here the functions h_{ij} are determined from (6.25), and a glance at the transition matrix of the normal bundle in (6.11) shows that $C_{\alpha\beta}$ corresponds to the vector bundles $N_{X/Y}$. Thus the result of our argument can be expressed by the simple formula

$$\sigma^{-1}(Y) \cong \mathbb{P}(N_{X/Y}).$$

(d) The Behaviour of Subvarieties Under a Blowup

Proposition *Let $Z \subset X$ be a closed irreducible nonsingular subvariety of X that is transversal to Y at every point of $Y \cap Z$, and let $\sigma\colon X' \to X$ be the blowup of Y. Then the subvariety $\sigma^{-1}(Z)$ consists of two irreducible components,*

$$\sigma^{-1}(Z) = \sigma^{-1}(Y \cap Z) \cup Z',$$

and $\sigma\colon Z' \to Z$ defines the blowup of Z with centre in $Y \cap Z$.

The subvariety $Z' \subset X'$ is called the *birational transform* of $Z \subset X$ under the blowup.

Proof The proof follows closely the arguments of Section 4.3, Chapter 2. The question is local, so that we can assume that $Y \subset X$ is defined by the local equations $u_1 = \cdots = u_a = u_{a+1} = \cdots = u_b = 0$, and $Z \subset X$ by the local equations $u_{a+1} = \cdots = u_b = u_{b+1} \cdots = u_c = 0$, so that the intersection $Y \cap Z$ is defined by $u_1 = \cdots = u_a = \cdots = u_b = \cdots = u_c = 0$; here $0 \le a < b < c \le d = \dim X$, and $u_1, \dots, u_d$ is a system of local parameters on X. Then X' is defined in $X \times \mathbb{P}^{b-1}$ by the equations

$$t_i u_j = t_j u_i \quad \text{for } i, j = 1, \dots, b. \tag{6.30}$$

Write $\overline{Z}$ for the closure of $\sigma^{-1}(Z \setminus (Y \cap Z))$. Then obviously, $\sigma^{-1}(Z) = \sigma^{-1}(Y \cap Z) \cup \overline{Z}$. Every point of $\sigma^{-1}(Z \setminus (Y \cap Z))$ has $u_{a+1} = \cdots = u_c = 0$ and at least one of $u_1, \dots, u_a \neq 0$; therefore

$$t_{a+1} = \cdots = t_c = 0 \quad \text{on } \overline{Z}.$$

Hence

$$\overline{Z} \subset Z \times \mathbb{P}^{a-1},$$

where $t_1, \dots, t_a$ are homogeneous coordinates of $\mathbb{P}^{a-1}$, and the relations

$$t_i u_j = t_j u_i \quad \text{for } i, j = 1, \dots, a.$$

hold on $\overline{Z}$. But these are just the equations defining the blowup $\sigma : Z' \to Z$ of Z with centre in $Y \cap Z$. We see that $\overline{Z} \subset Z'$, and therefore $\overline{Z} = Z'$, since both varieties have the same dimension and Z' is irreducible. The proposition is proved. □

We conclude this section with some remarks on the notion of blowup.

Remark 6.1 It can be shown that blowing up a quasiprojective variety does not take us outside the class of quasiprojective varieties; the proof is omitted.

Remark 6.2 The existence of blowups whose centres are not points creates a whole series of new difficulties in the theory of birational maps of varieties of dimension ≥ 3. In this connection, is not understood to what extent the results we obtained for birational maps of surfaces in Section 3.4, Chapter 4 can be carried over to higher dimensions. It is known that not every birational morphism $X \to Y$ can be expressed as a composite of blowups; the counterexample is due to Hironaka. It remains an open question whether every birational map is a composite of blowups and their inverses. On the other hand, the theorem on resolving the locus of indeterminacy of a rational map by blowups holds in any dimension, if the ground field k has characteristic 0; this is also a theorem of Hironaka.

2.3 *Example of Non-quasiprojective Variety*

The variety that we now construct to give an example of a non-quasiprojective variety will be complete. If a complete variety is isomorphic to a quasiprojective variety, then by the theorem on the closure of the image, it would be projective. Thus it is enough to construct an example of a complete nonprojective variety.

The proof of nonprojectivity will be based on the fact that intersection numbers on a projective variety has a specific property. We therefore start with some general remarks on intersection numbers.

We use notions which are a very special case of the cycle class ring mentioned in Section 6.2, Chapter 4. In our particular case, we can easily give the definitions from first principles. Let X be a complete nonsingular 3-fold, $C \subset X$ an irreducible curve and D a divisor on X. Suppose that $C \not\subset \operatorname{Supp} D$. Then the restriction $\rho_C(D)$ defines a locally principal divisor on C (we do not assume that C is nonsingular), for which the intersection number is defined (see the remark in connection with the definition of intersection number in Section 1.1, Chapter 4). In this case the intersection number is denoted by $\deg \rho_C(D)$ and is also called the *intersection number* of the curve C and the divisor D:

$$CD = \deg \rho_C(D).$$

The arguments of Sections 1.2–1.3, Chapter 4 show that this intersection number is additive as a function of D and invariant under linear equivalence. In particular, the intersection number $C\Delta$ is defined, where Δ is the divisor class containing D. In any case, we only require this for the case C a nonsingular curve, when both these properties are obvious.

Consider the free Abelian group A^1 generated by all curves $C \subset X$. The intersection number $a\Delta$ is defined for $a \in A^1$ and $\Delta \in \operatorname{Cl} X$ by additivity. We introduce on A^1 the equivalence relation

$$a \equiv b \iff a\Delta = b\Delta \quad \text{for all } \Delta \in \operatorname{Cl} X.$$

If this holds we say that a and b are *numerically equivalent*.

We consider an example which is basic for what follows. Suppose that $a = \sum n_i C_i$ and $a' = \sum n'_j C'_j$, where all the curves C_i and C'_j lie on a nonsingular surface $Y \subset X$, and $a \sim b$ are linearly equivalent as divisors on Y; then $a \approx b$. Indeed, for any divisor D on X the operation $\rho^X_{C_i}(D)$ of restriction to C_i can be carried out in two steps:

$$\rho^X_{C_i} = \rho^Y_{C_i} \circ \rho^X_Y,$$

and hence for $a \in \operatorname{Div} Y$

$$(aD)_X = \left(a\rho^X_Y(D)\right)_Y.$$

Therefore our assertion follows from the fact that intersection numbers of divisors on Y are invariant under linear equivalence of divisors.

The preceding considerations apply to any complete variety X. The assumption that X is projective has an important consequence for X: if $a = \sum n_i C_i$ with $n_i > 0$ then $a \not\approx 0$. Indeed, when we intersect an irreducible curve C with a hyperplane section H of X we obviously have the equality

$$CH = \deg C,$$

and in particular $CH > 0$. Hence also $aH = \sum n_i C_i H > 0$.

Before we start on the construction of the example, we consider an auxiliary construction. Suppose C_1 and C_2 are two nonsingular curves in a nonsingular 3-fold V intersecting transversally at a point x_0. We assume that C_1 and C_2 are rational; our results hold independently of this assumption, but it somewhat simplifies the deduction. Let $\sigma\colon W \to V$ be the blowup of C_1 and $S_1 = \sigma^{-1}(C_1)$ the exceptional surface. The restriction

$$\sigma_{|S_1}\colon S_1 \to C_1,$$

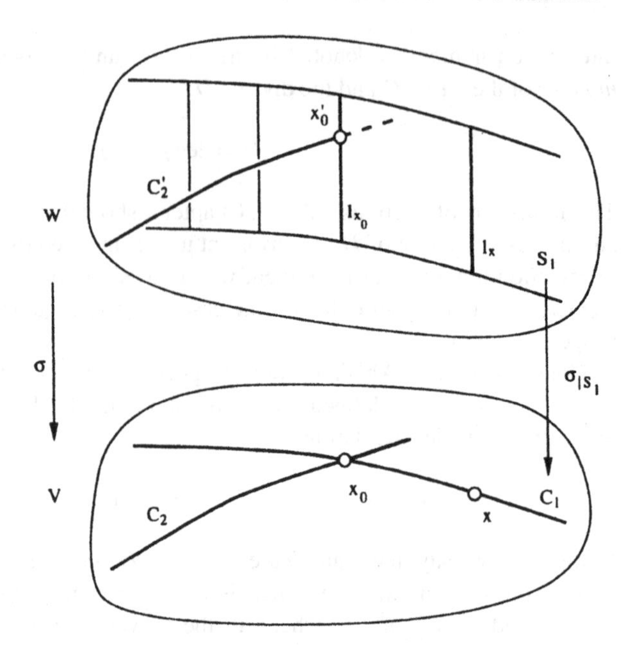

Figure 25 The first blowup

is a $\mathbb{P}^1$-bundle by Section 2.2, (c), and we write l_x for the fibre over $x \in C_1$. By Proposition 6.2, $\sigma^{-1}(C_2)$ consists of two components:

$$\sigma^{-1}(C_2) = l_{x_0} \cup C_2';$$

here $\sigma : C_2' \to C_2$ is the blowup of C_2 with centre in x_0, and in our case is therefore an isomorphism. As a very simple exercise in the formulas defining a blowup, we leave the reader to check that S_1 and C_2' intersect in a single point x_0' with $\sigma(x_0') = x_0$, and are transversal there. We arrive at the situation of Figure 25.

Since we have assumed that the curve C_1 is rational, all its points are linearly equivalent $x_1 \sim x_2$, and hence

$$l_{x_1} \sim l_{x_2} \quad \text{on } S_1 \quad \text{for all } x_1, x_2 \in C_1.$$

Now consider a second blowup, the blowup $\tau : X \to W$ of W with centre in C_2'. The inverse image $\tau^{-1}(S_1)$ of S_1 is irreducible: by Section 2.2, (d), Proposition, $\tau^{-1}(S_1) = \tau^{-1}(x_0') \cup S_1'$, where $\tau : S_1' \to S_1$ is the blowup of S_1 centred in x_0'. It follows that $\tau^{-1}(x_0') \subset S_1'$. On S_1' we have $\tau^{-1}(l_{x_0}) = L + L'$, where $L = \tau^{-1}(x_0')$ and $\tau : L' \to l_{x_0}$ is an isomorphism. For $x \neq x_0$, the fibre $\tau^{-1}(l_x)$ is irreducible; we denote it by L_x. By what we have said, we have

$$L_x \sim L + L' \quad \text{as divisors on } S_1'. \tag{6.31}$$

Now write S_2 for the surface $\tau^{-1}(C_2')$. In this same way as S_1, it is a $\mathbb{P}^1$-bundle $S_2 \to C_2'$ with fibre L_y over $y \in C_2'$, and on S_2

$$L_{y_1} \sim L_{y_2} \quad \text{for all } y_1, y_2 \in C_2, \text{ and } L_{x_0'} = L. \tag{6.32}$$

Figure 26 The second blowup

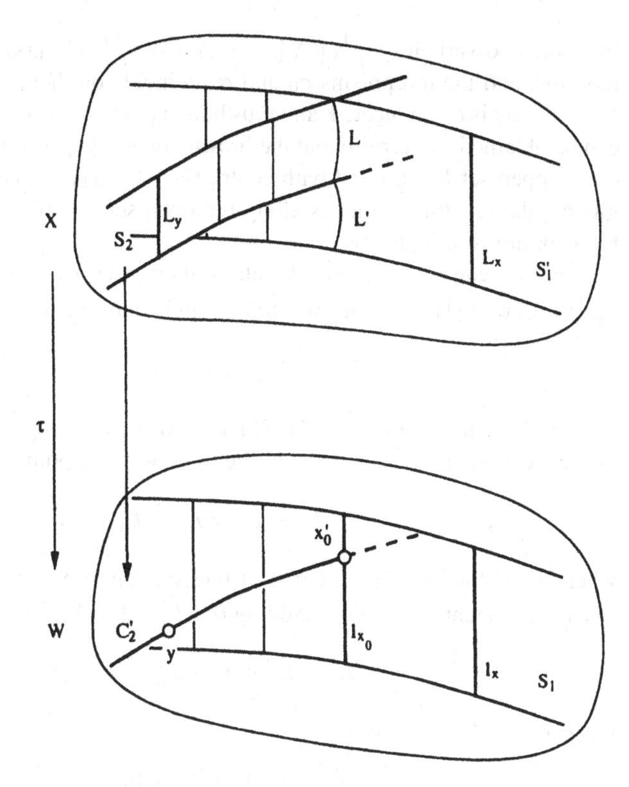

The two surfaces S_1' and S_2 intersect along the line L as shown in Figure 26.

We go over to numerical equivalence on X. Substituting (6.31) in (6.32) we get that

$$L_x \approx L + L' \approx L_y + L'. \tag{6.33}$$

The basic feature of this relation is its lack of symmetry with respect to the fibres L_x and L_y of the two ruled surfaces S_1' and S_2, arising from the order in which the blowups were performed. This is what we exploit in the example, the construction of which we now embark on.

Consider a nonsingular 3-fold V and two nonsingular rational curves $C_1, C_2 \subset V$ that intersect transversally in two points x_0 and x_1 (for example, V could contain a copy of $\mathbb{P}^2$, with C_1 a line and C_2 a conic). In the 3-fold $V_0 = V \setminus x_1$ we blow up as above first $C_1 \setminus x_1$, then the birational transform of $C_2 \setminus x_1$; we get a morphism

$$\sigma_0 \colon X_0 \to V \setminus x_1.$$

In $V_1 = V \setminus x_0$ we blow up the two curves in the opposite order, first $C_2 \setminus x_0$ then the birational transform of $C_1 \setminus x_0$; we get a morphism

$$\sigma_1 \colon X_1 \to V \setminus x_0.$$

Now the two varieties $\sigma_0^{-1}(V \setminus \{x_0, x_1\})$ and $\sigma_1^{-1}(V \setminus \{x_0, x_1\})$ are obviously isomorphic, and the morphisms σ_0 and σ_1 coincide on them. Indeed, the curve $C_1 \cup C_2 \setminus \{x_0, x_1\}$ is disconnected, and thus both $\sigma_0^{-1}(V \setminus \{x_0, x_1\})$ and $\sigma_1^{-1}(V \setminus \{x_0, x_1\})$ can be obtained by carrying out the blowup of $V \setminus \{x_0, x_1\}$ with centre $C_1 \setminus \{x_0, x_1\}$ on the open set $V \setminus C_2$ and with centre $C_2 \setminus \{x_0, x_1\}$ on the open set $V \setminus C_1$, then glueing the resulting varieties along the open set $V \setminus \{C_1 \cup C_2\}$, over which both blowups are isomorphisms.

Thus we can glue X_0 and X_1 along their open subsets $\sigma_0^{-1}(V \setminus \{x_0, x_1\})$ and $\sigma_1^{-1}(V \setminus \{x_0, x_1\})$, obtaining a 3-fold X and a morphism

$$\sigma\colon X \to V.$$

In X we have the relation (6.33), which we deduced using the existence of the point of intersection x_0 of C_1 and C_2. In the same way the point x_1 leads to the relation

$$L_y \approx L_1 + L_1' \approx L_x + L_1' \tag{6.34}$$

where L_1 is the irreducible curve of intersection of S_1 and S_2' over x_1 and L_1' the other component of $\sigma^{-1}(x_1)$. Adding (6.33) and (6.34) gives

$$L_x + L_y \approx L' + L_1' + L_x + L_y,$$

whence

$$L' + L_1' \approx 0. \tag{6.35}$$

To get a contradiction to X projective, it remains to prove that it is complete. For an arbitrary variety Z the projection $X \times Z \to Z$ factors as a composite of the map $\sigma \times 1\colon X \times Z \to V \times Z$ and the projection $V \times Z \to Z$. Since V is projective, the image of a closed set under the second projection is closed, and we need only prove that the same holds for $\sigma \times 1$. We know that V is a union of two open sets $V \setminus x_0$ and $V \setminus x_1$, and since closed is a local property, it is enough to check that both the restrictions

$$\sigma \times 1\colon (\sigma \times 1)^{-1}\big((V \setminus x_i) \times Z\big) \to (V \setminus x_i) \times Z \quad \text{for } i = 0, 1$$

take closed sets to closed sets. Now σ over $V \setminus x_i$ is just a composite of blowups, and it remains to prove that for any blowup $\sigma\colon U' \to U$ and any Z the morphism

$$\sigma \times 1\colon U' \times Z \to U \times Z$$

takes closed sets to closed sets. Once more since the question is local, we can assume that σ is given by the local construction Section 2.2, (a), that is, $U' \subset U \times \mathbb{P}^{m-1}$ and σ is induced by the projection $U \times \mathbb{P}^{m-1} \to U$. But then our assertion follows from the fact that projective space is complete, Theorem 1.11 of Section 5.2, Chapter 1.

Thus if X were quasiprojective it would be projective, and this is a contradiction, since (6.35) is impossible in a projective variety.

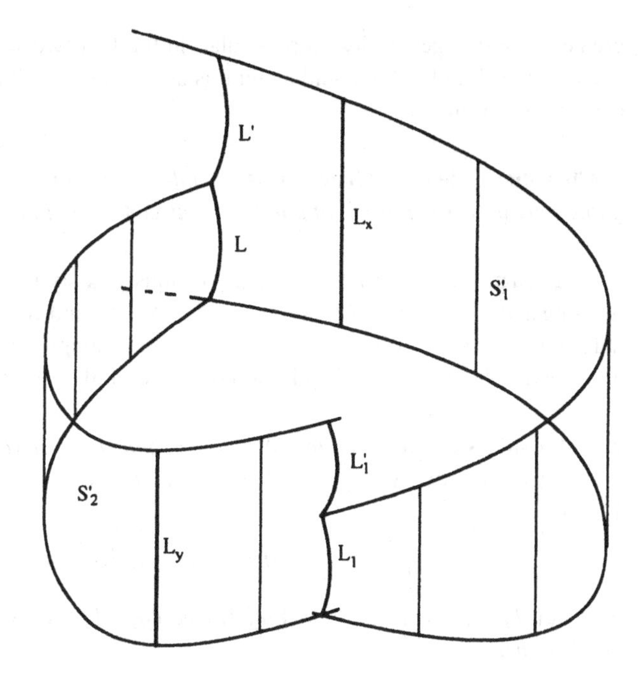

Figure 27 Hironaka's counterexample

The basic idea on which the example is built is of course the relations (6.33) and (6.34): they lead to (6.35), which cannot hold on a projective variety. These relations are perhaps clearer if we express them in a very primitive picture (Figure 27): the fibre L_{x_0} of the ruled surface S_1' is shown breaking up into two lines drawn as intervals L and L'.

Remark 6.3 It is no accident that this example has dimension 3. It can be proved that a 2-dimensional nonsingular complete variety is projective. On the other hand, there exist examples of complete nonprojective 2-dimensional varieties with singularities.

Remark 6.4 In the example we have constructed, consider an affine open set $U \subset X$. If both curves L' and L_1' in (6.35) had nonempty intersection with U, we would be able to find a divisor D such that $L'D > 0$, $L_1'D > 0$, contradicting (6.35); indeed, we could just take D to be the closure of a hyperplane section of the affine space containing U. Thus L' and L_1' are "very far apart" in X: if an affine open subset contains a point of L' then it must be disjoint from L_1'.

2.4 Criterions for Projectivity

To conclude this section, we discuss a number of criterions that characterise projective varieties among arbitrary complete varieties. We do not state them in the greatest possible generality. In particular, in the first two we assume that the varieties are nonsingular. We could avoid this assumption, but this would require some extra explanations.

1. Chevalley–Kleiman Criterion *A complete nonsingular variety X is projective if and only if every finite set of points of X is contained in an affine open subset.*

If X is a projective variety then there obviously exists a hyperplane section H not meeting any finite subset $S \subset X$, so that $S \subset X \setminus H$, and $X \setminus H$ is affine. Hence one half of the criterion is obvious. In the example of a nonprojective variety constructed at the end of Section 2.3, this criterion obviously fails (see Remark 6.4).

2. Nakai–Moishezon Criterion *A complete nonsingular variety X is projective if and only if there exists a divisor H on X such that for every irreducible subvariety $Y \subset X$,*

$$\left(\rho_Y(H)^m\right)_Y = H^m Y > 0, \quad \textit{where } m = \dim Y;$$

here $\rho_Y(H)$ is the restriction to Y of H and $(\rho_Y(H)^m)_Y$ its m-fold selfintersection number on Y.

If X is a projective variety then we can take H to be a hyperplane section. In this case

$$H^m Y = \deg Y.$$

Thus again the criterion obviously holds for projective varieties.

To state the final criterion, recall that projective space $\mathbb{P}^n$ has a line bundle $E \subset \mathbb{P}^n \times V$, where V is the vector space whose lines are represented by points of $\mathbb{P}^n$ (see Examples 6.10–6.11). Moreover, the projection $\mathbb{P}^n \times V \to V$ defines a morphism $E \to V$ that is the blowup of V centred in the origin. For this map, the unique exceptional subvariety is the zero section of E. Let $X \subset \mathbb{P}^n$ be a closed subvariety. The line bundle $E' = \rho_X(E)$, the restriction to X of E, is a closed subvariety of E, and the blowup $\sigma\colon E \to V$ defines a morphism $\sigma'\colon E' \to V$. The completeness of $\mathbb{P}^n$ implies that σ takes closed sets to closed sets. Hence $W = \sigma'(E')$ is an affine variety. In fact it is easy to see that W is the affine cone over $X \subset \mathbb{P}^n$ as in the proof of Theorem 6.7 (compare Exercise 8 of Section 4.5). Obviously the unique exceptional subvariety of σ' is the zero section of E'.

These arguments prove the "only if" part of the following criterion.

Grauert Criterion *A complete variety X is projective if and only if there exists a line bundle E over X, and a morphism $f\colon E \to V$ to an affine variety V such that f is birational, and the unique exceptional subvariety of f is the zero section of E. A shorter way of stating the condition is that the zero section of the line bundle E can be contracted to a point.*

2.5 Exercises to Section 2

1 Give an alternative proof of Theorem 6.1 using Chow's lemma and a reduction to Theorem 2.12 of Section 3.1, Chapter 2.

2 If X is a complete variety and $\sigma\colon X' \to X$ a blowup, prove that X' is also complete.

3 If E and E' are vector bundles such that $E' = E \otimes L$ for L a line bundle, and $\mathbb{P}(E)$, $\mathbb{P}(E')$ are as in Exercise 10 of Section 1.5, prove that $\mathbb{P}(E) \cong \mathbb{P}(E')$.

4 Suppose that X is a nonsingular complete variety with $\dim X = 3$, and $Y \subset X$ a nonsingular curve; let $\sigma\colon X' \to X$ be the blowup of X with centre Y, and $l = \sigma^{-1}(y_0)$ with $y_0 \in Y$. Prove that $\sigma^*(D)l = 0$, where D is any divisor on X and $\sigma^*(D)$ its pullback to X'.

5 Under the conditions of Exercise 4, set $S = \sigma^{-1}(Y)$. Prove that $Sl = -1$. [Hint: Consider a surface D on X containing Y and nonsingular at y_0, and apply the result of Exercise 4 to D. Compare the calculations of Section 3.2, Chapter 4.]

6 Prove that for any nonsingular projective 3-fold X there exists a complete non-projective variety birational to X.

7 Prove that for any irreducible variety X there exists a quasiprojective variety $\overline{X}$ and a surjective birational morphism $f\colon \overline{X} \to X$. There exists an embedding $\overline{X} \hookrightarrow \mathbb{P}^n \times X$ such that f is the restriction to $\overline{X}$ of the projection $\mathbb{P}^n \times X \to X$.

3 Coherent Sheaves

3.1 Sheaves of $\mathcal{O}_X$-Modules

Sheaves of modules over the sheaf of rings $\mathcal{O}_X$ have already appeared in Section 1.3 in connection with vector bundles. Sheaves of this type are an extraordinarily convenient tool in the study of algebraic varieties; we discuss one example of this in this section. But first we start with certain general properties of these sheaves.

Consider the most general situation: a ringed space, that is, a topological space X with a given sheaf of rings $\mathcal{O}$. In what follows we consider sheaves on X that are sheaves of modules over $\mathcal{O}$; we usually omit mention of this, speaking simply of sheaves of modules. Any sheaf of Abelian groups on a topological space X can obviously be viewed as a sheaf of modules over a sheaf of rings $\mathcal{O}$ by taking $\mathcal{O}$ to be the sheaf of locally constant $\mathbb{Z}$-valued functions.

The definition of a homomorphism $f\colon \mathcal{F} \to \mathcal{G}$ of sheaves of modules was given in Section 1.3. Recall that it is a system of $\mathcal{O}(U)$-module homomorphisms $f_U\colon \mathcal{F}(U) \to \mathcal{G}(U)$ satisfying certain compatibility requirements.

Example 6.13 Let X be a nonsingular algebraic variety over k, $\mathcal{O}_X$ the sheaf of regular functions, and Ω^1 the sheaf of regular differential 1-forms. Sending $f \in \mathcal{O}_X(U)$ to the differential $\mathrm{d}f \in \Omega^1(U)$ defines a homomorphism of sheaves

$$\mathrm{d}\colon \mathcal{O}_X \to \Omega^1.$$

This is a homomorphism of sheaves of modules over the sheaf of locally constant k-valued functions, but not over $\mathcal{O}_X$ (because, by Leibnitz' rule, d is not $\mathcal{O}_X$-linear).

Our immediate objective is to define the kernel and image of a homomorphism of sheaves of modules. The first definition is completely obvious. Let $f\colon \mathcal{F} \to \mathcal{G}$ be a homomorphism of sheaves of modules. Set $\mathcal{K}(U) = \ker f_U$. By definition of a homomorphism it follows that for $U \subset V$ we have $\rho_U^V(\mathcal{K}(V)) \subset \mathcal{K}(U)$. Hence the system $\{\mathcal{K}(U), \rho_U^V\}$ is a presheaf; an easy verification shows that it is a sheaf of modules. By definition this is the *kernel* of f.

The kernel of a homomorphism is an example of a *subsheaf* of a sheaf $\mathcal{F}$. This is a sheaf of modules $\mathcal{F}'$ such that $\mathcal{F}'(U) \subset \mathcal{F}(U)$ for every open set $U \subset X$, and such that $\rho_{U,\mathcal{F}'}^V$ is the restriction of $\rho_{U,\mathcal{F}}^V$ to the submodule $\mathcal{F}'(V)$.

The image of a homomorphism $f\colon \mathcal{F} \to \mathcal{G}$ is a somewhat more complicated notion. The point is that the $\mathcal{O}(U)$-modules $\mathcal{I}(U) = \operatorname{im} f_U$, together with the homomorphisms $\rho_{U,\mathcal{G}}^V$, define a presheaf that is in general not a sheaf.

Example 6.14 Let X be a nonsingular irreducible curve, and $\mathcal{K}^*$ the constant sheaf with $\mathcal{K}^*(U) = k(X)^*$ the group of nonzero elements of $k(X)$ under multiplication; let $\mathcal{D}$ be the sheaf of local divisors, defined by $\mathcal{D}(U) = \operatorname{Div} U$, with the obvious restriction homomorphisms. The homomorphism $f\colon \mathcal{K}^* \to \mathcal{D}$ takes a function $u \in \mathcal{K}^*(U)$ into its divisor $\operatorname{div} u$ on U. Since every divisor is locally principal, for every $D \in \operatorname{Div} U$ and every point $x \in U$ there exists a neighbourhood V_x of x and a function $u \in \mathcal{K}^*(V_x)$ such that $f_{V_x}(u) = D$; in other words, $(\operatorname{im} f)(V_x) \ni \rho_{V_x}^U(D)$. However, it is not always the case that $D \in (\operatorname{im} f)(U)$. For example, if X is projective then not every divisor is principal. Thus $\operatorname{im} f$ does not satisfy condition (2) in the definition of a sheaf in Section 2.3, Chapter 5.

Thus it seems natural to define the image of a homomorphism $f\colon \mathcal{F} \to \mathcal{G}$ of sheaves of modules as follows. First define the presheaf $\mathcal{I}'$ by setting

$$\mathcal{I}'(U) = f_U\big(\mathcal{F}(U)\big) \quad \text{for } U \subset X;$$

then take the sheafication $\mathcal{I}$ of the presheaf $\mathcal{I}'$ as in Section 2.4, Chapter 5; it is called the *image* of f and is denoted by $\operatorname{im} f$.

Recalling the definition of the sheafication of a presheaf, we see that $\operatorname{im} f$ is a subsheaf of $\mathcal{G}$, and $(\operatorname{im} f)(U)$ consists of elements $a \in \mathcal{G}(U)$ such that every point $x \in U$ has a neighbourhood U_x for which

$$\rho_{U_x}^U(a) \in f_{U_x}\big(\mathcal{F}(U_x)\big).$$

Obviously, f defines a homomorphism

$$\mathcal{F} \to \operatorname{im} f.$$

It follows at once from the definition that a homomorphism $f: \mathcal{F} \to \mathcal{G}$ for which $\ker \mathcal{F} = 0$ and $\operatorname{im} f = \mathcal{G}$ is an isomorphism.

A sequence $\mathcal{F}_1 \xrightarrow{f_1} \mathcal{F}_2 \to \cdots \xrightarrow{f_n} \mathcal{F}_{n+1}$ of homomorphisms is called an *exact sequence* if $\operatorname{im} f_i = \ker f_{i+1}$ for $i = 1, \dots, n$. If $0 \to \mathcal{F} \xrightarrow{f} \mathcal{G} \xrightarrow{g} \mathcal{H} \to 0$ is an exact sequence then $\mathcal{F}$ can be viewed as a subsheaf of $\mathcal{G}$. Because of this,

$$(\operatorname{im} f)(U) = f\bigl(\mathcal{F}(U)\bigr);$$

that is, in constructing the image sheaf of an injective homomorphism f, passing to the sheafication is unnecessary. Hence the sequence

$$0 \to \mathcal{F}(U) \xrightarrow{f_U} \mathcal{G}(U) \xrightarrow{g_U} \mathcal{H}(U) \tag{6.36}$$

is exact for any open set U.

Example 6.14 shows that the sequence

$$0 \to \mathcal{F}(U) \xrightarrow{f_U} \mathcal{G}(U) \xrightarrow{g_U} \mathcal{H}(U) \to 0$$

is in general not exact (for example, for $U = X$). This phenomenon is the reason for the existence of a nontrivial theory of sheaf cohomology.

For any subsheaf $\mathcal{F}$ of a sheaf $\mathcal{G}$ one can construct a homomorphism $f: \mathcal{G} \to \mathcal{H}$ such that $\ker f = \mathcal{F}$ and $\operatorname{im} f = \mathcal{H}$. To obtain this, set

$$\mathcal{H}'(U) = \mathcal{G}(U)/\mathcal{F}(U)$$

and define homomorphisms $\rho^V_{U,\mathcal{H}'}$ as the maps induced on these quotient groups by the homomorphisms $\rho^V_{U,\mathcal{G}}$. We define $\mathcal{H}$ to be the sheafication of $\mathcal{H}'$.

It is easy to check that the stalks of this sheaf satisfy

$$\mathcal{H}_x = \mathcal{G}_x/\mathcal{F}_x.$$

Hence an element $a \in \mathcal{G}(U)$ defines elements $a_x \in \mathcal{H}_x$ for all points $x \in U$. An obvious verification shows that the set of all the $\{a_x\}$ specify an element $a' \in \mathcal{H}(U)$, and $f: a \mapsto a'$ defines a homomorphism with the required properties. The sheaf $\mathcal{H}$ is the *quotient sheaf* of $\mathcal{G}$ by $\mathcal{F}$. Obviously the sequence $0 \to \mathcal{F} \to \mathcal{G} \to \mathcal{H} \to 0$ is exact.

Example 6.15 Let X be an irreducible algebraic variety over a field k and $\mathcal{K}^*$ the sheaf of locally constant functions with values in the multiplicative group of $k(X)$. The sheaf $\mathcal{O}^*$ is defined by setting $\mathcal{O}^*(U)$ to be the set of invertible elements of $\mathcal{O}(U)$; here $\mathcal{K}^*$ and $\mathcal{O}^*$ are viewed as sheaves of Abelian groups. It is each to check that the quotient sheaf $\mathcal{D} = \mathcal{K}^*/\mathcal{O}^*$ has $\mathcal{D}(U)$ isomorphic to the group of locally principal divisors of U.

Definition The *support* of a sheaf $\mathcal{F}$ is the set $X \setminus W$, where W is the union of all open sets $V \subset X$ with $\mathcal{F}(U) = 0$ for all nonempty open set $U \subset V$. This is a closed set, and is denoted by $\operatorname{Supp} \mathcal{F}$.

Proposition *If S is the support of a sheaf $\mathcal{F}$ and $U \subset V$ are two open sets such that $U \cap S = V \cap S$ then the restriction $\rho_U^V : \mathcal{F}(V) \to \mathcal{F}(U)$ is an isomorphism.*

Proof Let $a \in \mathcal{F}(V)$ be such that $\rho_U^V(a) = 0$. By definition of S every point $x \in V$ with $x \notin S$ has a neighbourhood V_x, which we can assume to be contained in V, for which

$$\rho_{V_x}^V(a) = 0.$$

By the assumption, U is a neighbourhood with this property for points $x \in S$. It follows from the definition of a sheaf that $a = 0$, and thus ρ_U^V is injective.

Let $a \in \mathcal{F}(U)$. Consider a cover $V = \bigcup U_\alpha$ with $U_0 = U$ and $U_\alpha \cap S = \emptyset$ for $\alpha \neq 0$ (for example, this holds if U_α for $\alpha \neq 0$ are sufficiently small neighbourhoods of points $x \in V$ with $x \notin S$). Set $a_0 = a$ and $a_\alpha = 0$ for $\alpha \neq 0$. From the assumption of the proposition it follows that

$$\rho_{U_\alpha \cap U_\beta}^{U_\alpha}(a_\alpha) = \rho_{U_\alpha \cap U_\beta}^{U_\beta}(a_\beta).$$

Hence by the definition of a sheaf there exists an element $a' \in \mathcal{F}(V)$ such that

$$\rho_{U_\alpha}^V(a') = a_\alpha,$$

and in particular, when $\alpha = 0$, we have $\rho_U^V(a') = a$. Thus ρ_U^V is surjective, and the proposition is proved. □

It follows from the proposition that if S is the support of a sheaf $\mathcal{F}$ then the modules $\mathcal{F}(U)$ are canonically isomorphic for all open sets U whose intersection with S is a given subset. Therefore we can define a sheaf $\overline{\mathcal{F}}$ on S by setting

$$\overline{\mathcal{F}}(\overline{U}) = \mathcal{F}(U), \quad \text{where } U \cap S = \overline{U}$$

for open sets $\overline{U} \subset S$.

Example 6.16 Let X be a scheme and $Y \subset X$ a closed subscheme. Define a subsheaf $\mathcal{J}_Y$ of the structure sheaf $\mathcal{O}_X$ by the condition $\mathcal{J}_Y(U) = \mathfrak{a}_Y$ if U is an affine open set with $U = \operatorname{Spec} A$ and $\mathfrak{a}_Y \subset A$ is the ideal of the subscheme $Y \cap U$. Obviously if U is disjoint from Y then $\mathcal{J}_{Y|U} = \mathcal{O}_{X|U}$. Hence the sheaf $\mathcal{F} = \mathcal{O}_X / \mathcal{J}_Y$ is equal to 0 on such open sets, that is, its support is contained in Y. The corresponding sheaf $\overline{\mathcal{F}}$ on Y coincides with the structure sheaf $\mathcal{O}_Y$ on the subscheme Y.

Remark The definition of support just given is not the usually accepted one, but it is slightly more convenient for our purposes. In any case, in what follows, the two definitions coincide in the cases where they are applied.

3.2 Coherent Sheaves

Locally free sheaves have already appeared in Section 1.3 in connection with vector bundles. We now consider a class of sheaves that are to arbitrary finite modules as locally free sheaves are to finite free modules.

We now apply the notions introduced in Section 3.1 to the case that $X, \mathcal{O}_X$ is an arbitrary scheme. We start with local considerations, and suppose that $X = \operatorname{Spec} A$, where A is an arbitrary ring.

For any module M over a ring A and any multiplicative system S of elements of A we define the localisation of M with respect to S, setting

$$M_S = M \otimes_A A_S.$$

M_S can be described in the same way as the localisation A_S in Section 1.1, Chapter 5. It consists of pairs (m, s) with $m \in M$ and $s \in S$ with the same rules of identification, addition and multiplication by elements of A_S as in the case of rings; we write m/s for the pair (m, s). In particular, taking S to be the system of powers of an element $f \in A$ gives the module M_f over A_f.

The homomorphisms $A_S \to A_{S'}$ defined for $S \subset S'$ generate homomorphisms $M_S \to M_{S'}$. This allows us to associate with an A-module M a sheaf $\widetilde{M}$ on $\operatorname{Spec} A$. The definition mirrors exactly that of the sheaf $\mathcal{O}$, to which it reduces in the case $M = A$. In view of this, we omit some of the verifications, when these do not differ in the general case from those carried out in Section 2.2, Chapter 5.

For an open set of the form $U = D(f)$ with $f \in A$ we set

$$\widetilde{M}(U) = M_f.$$

For an arbitrary open set U we consider all $f \in A$ for which $D(f) \subset U$. Whenever $D(g) \supset D(f)$, we have homomorphisms

$$M_g \to M_f.$$

Using these, we can define the projective limit of the groups M_f. Set

$$\widetilde{M}(U) = \varprojlim M_f$$

where the limit runs over $f \in A$ such that $D(f) \subset U$. Then $\widetilde{M}(U)$ is a module over the ring $\mathcal{O}(U) = \varprojlim A_f$; this is a general property of projective limits. An inclusion $U \subset V$ defines a homomorphism $\rho_U^V : \widetilde{M}(V) \to \widetilde{M}(U)$ as in the case $M = A$. The system $\{\widetilde{M}(U), \rho_U^V\}$ defines a sheaf $\widetilde{M}$ of modules over the sheaf of rings $\mathcal{O}_X$.

A homomorphism of A-modules $\varphi : M \to N$ defines homomorphisms $\varphi_f : M_f \to N_f$ for all $f \in A$, and on passing to the limit, a homomorphism of sheaves $\widetilde{\varphi} : \widetilde{M} \to \widetilde{N}$. If $\varphi : M \to N$ and $\psi : N \to L$ are two such homomorphisms then

$$\widetilde{\varphi \circ \psi} = \widetilde{\varphi} \circ \widetilde{\psi}.$$

M can be recovered from $\widetilde{M}$. Namely, we have a generalisation of the relation proved in Section 2.2, Chapter 5:

$$\widetilde{M}(\operatorname{Spec} A) = M;$$

the proof is word-for-word the same. It follows that $M \mapsto \widetilde{M}$ is a one-to-one correspondence between modules M and sheaves of the form $\widetilde{M}$. Moreover, a simple check allows us to deduce that $\varphi \mapsto \widetilde{\varphi}$ is an isomorphism of groups

$$\operatorname{Hom}_A(M, N) \cong \operatorname{Hom}_{\mathcal{O}}(\widetilde{M}, \widetilde{N}),$$

from the group of A-module homomorphisms to the group of homomorphisms of sheaves of $\mathcal{O}_{\operatorname{Spec} A}$-modules.

We can now proceed to globalise these notions. Let X be a Noetherian scheme.

Definition A sheaf $\mathcal{F}$ on X is *coherent* if every point $x \in X$ has an affine neighbourhood U of the form $U = \operatorname{Spec} A$ with A a Noetherian ring, such that $\mathcal{F}_{|U}$ is isomorphic to a sheaf of the form $\widetilde{M}$ for some finite A-module M.

Proposition *If $X = \operatorname{Spec} A$ is an affine and Noetherian scheme, then any coherent sheaf $\mathcal{F}$ on X is of the form $\widetilde{M}$, where M is a finite A-module.*

Proof We set $\mathcal{F}(X) = M$ and prove that $\mathcal{F} = \widetilde{M}$.

Since open sets of the form $D(f)$ are a basis of the Zariski topology, there exist elements $f_i \in A$ such that $\bigcup D(f_i) = X$ and $\mathcal{F}$ is isomorphic over $D(f_i)$ to a sheaf $\widetilde{M}_i$, where M_i is a finite A_{f_i}-module. Since Spec A is compact we can assume that the f_i are finite in number.

For any $g \in A$, since $\mathcal{F}(D(g))$ is an A_g-module, the restriction homomorphism $\rho^X_{D(g)}\colon M = \mathcal{F}(X) \to \mathcal{F}(D(g))$ extends in a unique way to a homomorphism of A_g-modules.

$$\varphi_g\colon \widetilde{M}\big(D(g)\big) \to \mathcal{F}\big(D(g)\big).$$

One checks easily that this system of homomorphisms defines a homomorphism $\widetilde{M} \to \mathcal{F}$ of sheaves of modules.

Everything thus reduces to proving that the homomorphism φ_g is an isomorphism. For this, consider the sequence of homomorphisms

$$0 \to M \xrightarrow{\lambda} \bigoplus_i M_i \xrightarrow{\mu} \bigoplus_{i,j} M_{ij}, \tag{6.37}$$

where

$$M_{ij} = (M_i)_{f_j} = (M_j)_{f_i} = \mathcal{F}\big(D(f_i f_j)\big), \qquad \lambda(m) = \big(\dots, \rho^X_{D(f_i)}(m), \dots\big),$$

$$\text{and} \quad \mu(\dots, m_i, \dots, m_j, \dots) = \big(\dots, \big(\rho^{D(f_i)}_{D(f_i f_j)}(m_i) - \rho^{D(f_j)}_{D(f_i f_j)}(m_j)\big), \dots\big).$$

We view M_i and M_j as A-modules in (6.37). It follows from the definition of sheaf that (6.37) is exact. We now use a property of the functor $M \mapsto M_g$ which is important, although trivial to verify: it takes exact sequences into exact sequences. In particular,

$$0 \to M_g \xrightarrow[g]{\lambda} \bigoplus (M_i)_g \xrightarrow[g]{\mu} \bigoplus (M_{ij})_g$$

is exact. On the other hand, consider the sheaf $\mathcal{F}_{|D(g)}$. For it we have a similar exact sequence

$$0 \to \mathcal{F}\big(D(g)\big) \xrightarrow{\lambda'_g} \bigoplus_i \mathcal{F}\big(D(gf_i)\big) \xrightarrow{\mu'_g} \bigoplus_{ij} \mathcal{F}\big(D(gf_i f_j)\big).$$

But $\mathcal{F}(D(gf_i)) = (M_i)_g$ and $\mathcal{F}(D(gf_i f_j)) = (M_{ij})_g$. These isomorphisms induce an isomorphism $\varphi'_g \colon M_g \to \mathcal{F}(D(g))$. It is easy to check that $\varphi'_g = \varphi_g$ on the images of elements of M, and therefore on the whole of M_g. This proves that φ is an isomorphism and $\mathcal{F} = \widetilde{M}$.

It remains to prove that M is Noetherian; we know that the modules $M_i = M_{f_i}$ are Noetherian. Let M_n be an ascending chain of submodules of M. Then $(M_n)_{f_i} = (M_{n+1})_{f_i}$ for all f_i and for n sufficiently large. It follows from this that $M_n = M_{n+1}$. The proposition is proved. □

Example 6.17 The simplest example of a coherent sheaf is the structure sheaf $\mathcal{O}_X$. In the case $X = \operatorname{Spec} A$, this is the ring A viewed as a module over itself. A more general example is the sheaf $\mathcal{L}_E$ corresponding to a vector bundle over a scheme X as in Theorem 6.2.

Example 6.18 For any sheaf $\mathcal{F}$ on a scheme X, the *dual sheaf* $\mathcal{G} = \mathcal{H}om(\mathcal{F}, \mathcal{O}_X)$ is the sheafication of the presheaf $\mathcal{G}(U) = \operatorname{Hom}(\mathcal{F}(U), \mathcal{O}_X(U))$. If $X = \operatorname{Spec} A$ and $\mathcal{F} = \widetilde{M}$ then $\mathcal{H}om(\mathcal{F}, \mathcal{O}_X) = \widetilde{N}$, where $N = \operatorname{Hom}_A(M, A)$. If A is Noetherian and $M = Am_1 + \cdots + Am_r$ is finite then a homomorphism $M \to A$ is determined by its values on the generators m_i, so that $\operatorname{Hom}_A(M, A) \subset A^r$, and is therefore again finite. It follows from this that if X is a Noetherian scheme and $\mathcal{F}$ is coherent then $\mathcal{H}om(\mathcal{F}, \mathcal{O}_X)$ is again coherent.

Example 6.19 Let X be a scheme of finite type over k. We define for X the analogue of the cotangent sheaf Ω^1_X (Example 6.7). If $X = \operatorname{Spec} A$ then we constructed in Section 5.2, Chapter 3 an A-module Ω_A that coincides with $\Omega^1_X[X]$ for a nonsingular variety X. By construction, Ω_A is a finite A-module. For any scheme X of finite type over k and any affine open $U = \operatorname{Spec} A$ we set $\Omega(U) = \Omega_A$. The sheafication Ω of this subsheaf is coherent and is called the *cotangent sheaf*. The sheaf $\Theta = \mathcal{H}om(\Omega, \mathcal{O}_X)$ is also coherent and is called the *tangent sheaf.* If $X = \operatorname{Spec} A$ then $\Theta(X) = \operatorname{Der}_k(A, A)$ is the module of derivations of A (compare Exercise 10 of Section 5.5, Chapter 3). If X is nonsingular then, as we know, both sheaves Ω and Θ are locally free, and correspond to the cotangent and tangent bundles.

Example 6.20 Let X be a Noetherian scheme, Y a closed subscheme and $\mathcal{I}_Y$ the sheaf of ideals corresponding to Y (Example 6.16). Since $\mathcal{O}_X(U)$ is Noetherian by assumption, $\mathcal{I}_Y$ is a coherent sheaf.

Example 6.21 Under the assumptions of Example 6.20, the sheaf of modules $\mathcal{I}_Y/\mathcal{I}_Y^2$ is coherent. We prove that if X and Y are nonsingular then it is locally free. This is a local assertion, and it is enough to check it in the case $X = \operatorname{Spec} A$, $Y = \operatorname{Spec} B$ and $B = A/I$, and we can even assume that A is the local ring of a point $x \in X$. Since X and Y are nonsingular we can assume that $I = (u_1, \dots, u_m)$, where $u_1, \dots, u_n$ (with $n > m$) is a system of local parameters of the maximal ideal of the ring A. Obviously I/I^2 is generated as B-module by $u_1, \dots, u_m$, and we need only check that they are free. This means that if $\sum u_i a_i \in I^2$ then $a_i \in I$. Suppose that $\sum u_i a_i = \sum u_i v_i$ with $v_i \in I$. Then $\sum u_i a_i' = 0$ where $a_i' = a_i - v_i$. Hence $u_i a_i' \in (u_1, \dots, \widehat{u_i}, \dots, u_m)$, and since $u_1, \dots, u_n$ is a regular sequence (see Section 1.2, Chapter 4), it follows that $a_i' \in (u_1, \dots, \widehat{u_i}, \dots, u_m) \subset I$, and hence $a_i \in I$.

Thus in this case, the sheaf $\mathcal{I}_Y/\mathcal{I}_Y^2$ corresponds to some vector bundle on Y. The transition matrixes $C_{\alpha\beta}$ of this vector bundle are of the form $C_{\alpha\beta} = (h_{ij})$, where the h_{ij} are given as follows: if $u_{\alpha,1}, \dots, u_{\alpha,m}$ are local equations of Y in U_α and $u_{\beta,1}, \dots, u_{\beta,m}$ local equations of Y in U_β and $u_{\alpha,i} = \sum f_{ij} u_{\beta,j}$ then h_{ij} is the restriction to Y of f_{ij}. As we saw in Section 1.3, this is the transition matrix of the vector bundle $N^*_{X/Y}$, which is in this case the vector bundle corresponding to the sheaf $\mathcal{I}_Y/\mathcal{I}_Y^2$. In the general case (when X and Y are not assumed to be nonsingular), $\mathcal{I}_Y/\mathcal{I}_Y^2$ is the *conormal sheaf* to Y in X. If X and Y are nonsingular then the vector bundle $N_{X/Y}$ corresponds to the sheaf $\mathcal{H}om(\mathcal{I}_Y/\mathcal{I}_Y^2, \mathcal{O}_Y)$. This sheaf is called the *normal sheaf* of the subscheme $Y \subset X$ and denoted by $\mathcal{N}_{X/Y}$.

We give an interpretation in these terms of the sequence

$$0 \to \Theta_Y \to j^*\Theta_X \to N_{X/Y} \to 0, \tag{6.38}$$

where j^* is the restriction to Y. For the corresponding sheaves and affine varieties it gives

$$0 \to \operatorname{Der}_k(B, B) \to \operatorname{Der}_k(A, B) \to \operatorname{Der}_k(I, B) \to 0, \tag{6.39}$$

where $B = A/I$, and $\operatorname{Der}_k(P, Q)$ is the module of derivations $D\colon P \to Q$. It is easy to see that $D(I^2) = 0$ for $D \in \operatorname{Der}_k(I, B)$, so that $\operatorname{Der}_k(I, B) = \operatorname{Hom}_B(I/I^2, B)$. Hence the sequences (6.38) and (6.39) coincide.

3.3 Dévissage of Coherent Sheaves

We now discuss a method that allows us to reduce arbitrary coherent sheaves to free sheaves (admittedly, only in some very coarse respects).

Proposition 6.1 *For any coherent sheaf $\mathcal{F}$ over a Noetherian reduced irreducible scheme X, there exists a dense open set W such that $\mathcal{F}_{|W}$ is free.*

Proof The assertion is local in nature, so that we can restrict to the case $X = \operatorname{Spec} A$, where A is a Noetherian ring without nilpotents and $\mathcal{F} = \widetilde{M}$ for a finite A-module M. In addition, we can obviously assume that X is irreducible. Then X reduced and irreducible implies that A has no zerodivisors.

Recall that the rank of an A-module is the maximal number of linearly independent elements of M over A. By assumption, M has finite rank. Write r for the rank, and let $x_1, \dots, x_r \in M$ be linearly independent over A; by definition, they generate a free submodule $M' \subset M$. Let $y_1, \dots, y_m$ be a system of generators of M. Then for each i there exist elements $d_i \in A$ with $d_i \neq 0$ such that

$$d_i y_i \in M'. \tag{6.40}$$

Consider the open set $W = D(d)$, where $d = d_1 \cdots d_m$. The sheaf $\mathcal{F}_{|W}$ is isomorphic to $\widetilde{M}_d$. But $M_d = M'_d$ by (6.40), and hence

$$\mathcal{F}_{|W} \cong \widetilde{M}'_d.$$

Now M'_d is a free module over the ring A_d, since M' is free. The proposition is proved. □

Proposition 6.2 *For any coherent sheaf $\mathcal{F}$ over a Noetherian reduced irreducible scheme X, there exists a coherent sheaf $\mathcal{G}$ containing a free subsheaf $\mathcal{O}^r$, and a homomorphism $\varphi\colon \mathcal{F} \to \mathcal{G}$ such that the two sheaves* $\ker\varphi$ *and $\mathcal{G}/\mathcal{O}_r$ both have support distinct from X.*

As we will see, in the proof we construct a homomorphism $\varphi\colon \mathcal{F} \to \mathcal{G}$ such that both $\ker\varphi$ and $\mathcal{G}/\operatorname{im}\varphi$ have support distinct from the whole of X. Since $\mathcal{G}/\mathcal{O}^r$ also has support distinct from the whole of X, Proposition 6.2 shows that any coherent sheaf is "free modulo sheaves with support distinct from the whole of X".

Proof Let W be the open set and $f\colon \mathcal{F}_{|W} \xrightarrow{\sim} \mathcal{O}^r{}_{|W}$ the isomorphism whose existence was established in Proposition 6.1. We can assume that W is a principal open set, and will do so in what follows. Define the sheaf $\mathcal{G}$ by the condition

$$\mathcal{G}(U) = f_{U\cap W}\bigl(\rho^U_{U\cap W}\mathcal{F}(U)\bigr) + \rho^U_{U\cap W}\bigl(\mathcal{O}^r(U)\bigr). \tag{6.41}$$

Since $\rho^U_{U\cap W}(\mathcal{O}^r(U)) \subset \mathcal{O}^r(U \cap W)$ and $f_{U\cap W}(\rho^U_{U\cap W}\mathcal{F}(U)) \subset \mathcal{O}^r(U \cap W)$, both terms of the right-hand side of (6.41) are contained in the same group. We consider the sum of these subgroups, which obviously becomes an $\mathcal{O}(U)$-submodule of $\mathcal{O}^r(U \cap W)$ when we set

$$ax = \rho^U_{U\cap W}(a)x \quad \text{for } a \in \mathcal{O}(U) \text{ and } x \in \mathcal{O}^r(U \cap W).$$

Since $\mathcal{F}(U)$ and $\mathcal{O}^r(U)$ are finite $\mathcal{O}(U)$-modules, the same holds for $\mathcal{G}(U)$.

The definition of the homomorphisms $\rho^U_{V,\mathcal{G}}$ is self-explanatory. It follows at once from what we said above that the sheaf $\mathcal{G}$ we have constructed is coherent.

For the sheaf $\mathcal{O}^r$, the restriction $\rho^U_{U\cap W}$ is an inclusion. It is enough to verify this for an affine open set $U = \operatorname{Spec} A$. Consider a principal open set $D(f) \subset U \cap W$. The kernel of $\rho^U_{D(f)}$ consists of elements $x \in A$ such that $f^n x = 0$ for some $n \geq 0$; since X is irreducible, the ring A has no zerodivisors, and hence $x = 0$. A fortiori, $\ker \rho^U_{U\cap W} = 0$. Thus $\rho^U_{U\cap W}$ allows us to identify the sheaf $\mathcal{O}^r$ with a subsheaf of $\mathcal{G}$.

We define the homomorphism $\varphi \colon \mathcal{F} \to \mathcal{G}$ by the condition

$$\varphi_U = f_{U\cap W} \circ \rho^U_{U\cap W}.$$

If $U \subset W$ then

$$\mathcal{G}(U) = f_U\big(\rho^U_{U\cap W}(\mathcal{F}(U))\big) = f_U\big(\mathcal{F}(U)\big) = \mathcal{O}^r(U) = \rho^U_{U\cap W}\big(\mathcal{O}^r(U)\big),$$

and f_U is an isomorphism. Hence φ_U is an isomorphism, and $\mathcal{G}(U) = \mathcal{O}^r(U)$. This proves that the sheaves $\ker \varphi$ and $\mathcal{G}/\mathcal{O}^r$ are both 0 on W, and hence they have supports contained in $X \setminus W$. The proposition is proved. □

Proposition 6.2 leads us to the question of the structure of coherent sheaves whose support is distinct from the whole scheme. If the support of $\mathcal{F}$ is a closed set $Y \subset X$ then by the discussion at the end of Section 3.1, there is a sheaf $\overline{\mathcal{F}}$ on Y defined by the condition

$$\overline{\mathcal{F}}(\overline{U}) = \mathcal{F}(U), \quad \text{where } U \cap Y = \overline{U}$$

for open sets $\overline{U} \subset Y$.

We consider Y as a reduced closed subscheme of $\overline{X}$. Is $\overline{\mathcal{F}}$ a coherent sheaf on Y, or even a sheaf of $\mathcal{O}$-modules? This is false in general, as shown by the following example. Suppose that $X = \operatorname{Spec} \mathbb{Z}$, and let $\mathcal{F}$ be the coherent sheaf corresponding to the module $\mathbb{Z}/p^2\mathbb{Z}$, where p is a prime. The support of $\mathcal{F}$ is the prime ideal (p), and the corresponding reduced subscheme is $\operatorname{Spec}(\mathbb{Z}/p\mathbb{Z})$. It is obviously impossible to put a $(\mathbb{Z}/p)$-module structure on $\mathbb{Z}/p^2\mathbb{Z}$.

Nevertheless, we prove that there is a weaker sense in which the sheaf $\mathcal{F}$ can be reduced to coherent sheaves on Y.

Proposition 6.3 *A coherent sheaf $\mathcal{F}$ on a Noetherian scheme X with support $Y \neq X$ has a chain of subsheaves*

$$\mathcal{F} = \mathcal{F}_0 \supset \mathcal{F}_1 \supset \cdots \supset \mathcal{F}_m = 0$$

such that each quotient sheaf $\overline{\mathcal{F}}_i/\overline{\mathcal{F}}_{i+1}$ is a coherent sheaf of $\mathcal{O}_Y$-modules.

Proof In Example 6.16, we gave the example of the sheaf $\mathcal{I}_Y$ of ideals of the reduced subscheme Y. Obviously $\overline{\mathcal{F}}$ is a coherent sheaf of $\mathcal{O}_Y$-modules if

$$\mathcal{I}_Y \cdot \mathcal{F} = 0. \tag{6.42}$$

Indeed, under this assumption, all the $\mathcal{O}_X(U)$-modules $\mathcal{F}(U)$ are modules over $\mathcal{O}_X(U)/\mathcal{I}_Y(U) = \mathcal{O}_Y(\overline{U})$. Thus if $\mathcal{F}$ is of the form $\widetilde{M}$ on an affine open set

$U = \operatorname{Spec} A$ then $\mathfrak{a}_Y \cdot M = 0$, and M is therefore an $(A/\mathfrak{a}_Y)$-module. Moreover, if we now view M as an $(A/\mathfrak{a}_Y)$-module then $\overline{\mathcal{F}} = \widetilde{M}$.

We show that a slightly weaker statement always holds: there exists an integer $k > 0$ such that

$$\mathcal{I}_Y^k \cdot \mathcal{F} = 0. \tag{6.43}$$

Consider an affine open set $U = \operatorname{Spec} A$ such that $\mathcal{F}_{|U}$ is of the form $\widetilde{M}$ with M a finite A-module. Let $\mathfrak{a}_Y$ be the ideal of the subset $Y \cap U$. If $f \in \mathfrak{a}_Y$ then $D(f) \subset U \setminus (U \cap Y)$, and by assumption the restriction of $\mathcal{F}$ to $D(f)$ is zero. This means that $M_f = 0$, and hence for every $m \in M$ there exists $k(m) > 0$ such that $f^{k(m)}m = 0$. Since M is a finite A-module, it follows that $f^k M = 0$ for some $k > 0$. Since this relation holds for any $f \in \mathfrak{a}_Y$ and $\mathfrak{a}_Y$ has a finite basis, it follows that

$$\mathfrak{a}_Y^l \cdot M = 0 \tag{6.44}$$

for some $l > 0$. In other words, (6.43) holds on the open set U. Choosing a finite cover of X by open sets U as above, and taking k to be the maximum of the l for which (6.44) holds on each of the U, we get (6.43) on the whole of X.

Set $\mathcal{F}_i = \mathcal{I}_Y^i \cdot \mathcal{F}$ for $i = 0, \dots, k$ and $\mathcal{F} = \mathcal{F}_0$. Obviously the support of each of the $\mathcal{F}_i$ is contained in Y. Write $\overline{\mathcal{F}}_i$ for the sheaf on Y determined by $\mathcal{F}_i$ on X. Since

$$\mathcal{I}_Y \cdot (\mathcal{F}_i/\mathcal{F}_{i+1}) = 0,$$

the sheaf $\overline{\mathcal{F}}_i/\overline{\mathcal{F}}_{i+1}$ satisfies (6.42), and so is a coherent sheaf of $\mathcal{O}_Y$-modules. This proves Proposition 6.3. □

To conclude, we show how the methods used throughout this section allow us to reduce the study of sheaves to the case of irreducible schemes.

Proposition 6.4 *Let X be a Noetherian reduced scheme with $X = \bigcup X_i$ its decomposition as a union of irreducible components, and suppose that $\mathcal{F}$ is a coherent sheaf on X. There exist coherent sheaves $\mathcal{F}_i$ on X and a homomorphism $\varphi\colon \mathcal{F} \to \bigoplus \mathcal{F}_i$ such that the support of $\mathcal{F}_i$ is contained in X_i, the sheaf $\overline{\mathcal{F}}_i$ defined on X_i by $\mathcal{F}$ is coherent, and the kernel of φ has support contained in $\bigcup_{i \neq j} X_i \cap X_j$.*

Proof Set $\mathcal{F}_i = \mathcal{F}/(\mathcal{I}_{X_i} \cdot \mathcal{F})$, and let $\varphi_i\colon \mathcal{F} \to \mathcal{F}_i$ be the natural projection and $\varphi = \bigoplus \varphi_i$. We saw in Section 3.2 that the support of $\mathcal{F}_i$ is contained in X_i, and $\overline{\mathcal{F}}_i$ is a coherent sheaf of $\mathcal{O}_{X_i}$-modules since $\mathcal{I}_{X_i} \cdot \mathcal{F}_i = 0$.

Consider the open set

$$U_i = X_i \setminus \bigcup_{i \neq j} X_i \cap X_j.$$

On U_i we have $\mathcal{I}_{X_j} = \mathcal{O}_X$ for $j \neq i$ and $\mathcal{I}_{X_i} = 0$, so that $\mathcal{F}_{j|U_i} = 0$ for $j \neq i$ and $\mathcal{F}_{i|U_i} = \mathcal{F}_{|U_i}$. Therefore on U_i we have $\varphi_j = 0$ for $j \neq i$, and $\varphi = \varphi_i$ is an isomorphism. Hence the kernel of φ equals 0 on $\bigcup U_i$, as required to prove. □

3.4 The Finiteness Theorem

Theorem *If X is a complete variety over a field k and $\mathcal{F}$ a coherent sheaf on X, the vector space $\mathcal{F}(X)$ is finite dimensional over k.*

Proof The essence of the proof is the following remark. Given a homomorphism $\varphi\colon \mathcal{F} \to \mathcal{G}$ of sheaves over X, set $\mathcal{H} = \ker \varphi$; then

$$\mathcal{H}(X) \text{ and } \mathcal{G}(X) \text{ finite dimensional} \quad \Longrightarrow \quad \mathcal{F}(X) \text{ is finite dimensional.} \tag{6.45}$$

This follows since $\mathcal{H}(X) = \ker\{\varphi_X\colon \mathcal{F}(X) \to \mathcal{G}(X)\}$, by definition of the kernel. From this, we deduce by induction that $\mathcal{F}(X)$ is finite dimensional if there exist subsheaves

$$\mathcal{F} = \mathcal{F}_0 \supset \mathcal{F}_1 \supset \cdots \supset \mathcal{F}_m = 0 \tag{6.46}$$

such that each vector space $\mathcal{F}_i/\mathcal{F}_{i+1}(X)$ is finite dimensional.

We prove the theorem by induction on the dimension of X. If $\dim X = 0$ then X consists of a finite number of points, and a coherent sheaf $\mathcal{F}$ on X is by definition a finite dimensional vector space over k, so that the theorem is obvious.

Suppose that the theorem holds for complete varieties of dimension less than $\dim X$. Let us prove that this implies the theorem for all sheaves $\mathcal{F}$ on X having support contained in a closed subvariety $Y \subset X$ with $\dim Y < \dim X$.

Indeed, by definition, the sheaf $\overline{\mathcal{F}}$ on Y has $\mathcal{F}(X) = \overline{\mathcal{F}}(Y)$, and we can apply the assertion of the theorem to coherent sheaves on Y. Here we run into the difficult that $\overline{\mathcal{F}}$ is not in general coherent on Y, but Proposition 6.3 saves the day. It provides a sequence

$$\overline{\mathcal{F}} = \overline{\mathcal{F}}_0 \supset \overline{\mathcal{F}}_1 \supset \cdots \supset \overline{\mathcal{F}}_m = 0$$

such that the quotient sheaves $\overline{\mathcal{F}}_i/\overline{\mathcal{F}}_{i+1}$ are coherent on Y and hence we can apply the inductive assumption to them. We get the existence of a sequence of sheaves (6.46), from which the finite dimensionality of $\overline{\mathcal{F}}(Y)$ follows, and hence also that of $\mathcal{F}(X)$.

The next step of the proof consists of reducing the assertion to the case of an irreducible variety. Suppose that $X = \bigcup X_i$ is a decomposition into irreducible components. Now we can apply Proposition 6.4. The homomorphism φ constructed there has kernel supported in the subvariety $\bigcup_{i\neq j} X_i \cap X_j$, which has dimension less than $\dim X$. Hence it is enough to prove that $(\bigoplus \mathcal{F}_i)(X)$ is finite dimensional. But

$$\Bigl(\bigoplus \mathcal{F}_i\Bigr)(X) = \bigoplus \overline{\mathcal{F}}_i(X_i),$$

and since $\overline{\mathcal{F}}_i$ is a coherent sheaf on X_i, this reduces the assertion to the case of the irreducible varieties X_i.

Finally we can proceed with the central step of the proof, assuming that X is irreducible. Here we build on the foundation of Proposition 6.2. Since X is complete,

$\mathcal{O}(X) = k$ by the discussion in Section 1.1, so that $\dim \mathcal{O}^r(X) = r$. Since the support of $\mathcal{G}/\mathcal{O}^r$ is distinct from X, the theorem holds for $\mathcal{G}/\mathcal{O}^r$, and hence for $\mathcal{G}$ we have a homomorphism $\psi: \mathcal{G} \to \mathcal{G}/\mathcal{O}^r$ satisfying conditions (6.45). Hence $\mathcal{G}(X)$ is finite dimensional. On the other hand, the homomorphism $\varphi: \mathcal{F} \to \mathcal{G}$ constructed in Proposition 6.2 again satisfies (6.45), so that $\mathcal{F}(X)$ is finite dimensional, which is what the theorem asserts. The theorem is proved. □

The theorem we have proved has many important applications. Some of these have been mentioned earlier. First of all, in Section 1.4 we associated with each divisor D on a variety X a sheaf $\mathcal{L}_D$ such that $\mathcal{L}_D(X)$ is isomorphic to the space $\mathcal{L}(D)$ introduced in Section 1.5, Chapter 3. We saw in Section 3.4 that $\mathcal{L}_D$ is locally free of rank 1, and therefore coherent. Thus our theorem is applicable to it, and we obtain the result that we have already used many times:

Corollary 6.1 *The dimension $l(D)$ of a locally principal divisor D on a complete variety is finite.*

Applying the theorem to the sheaf corresponding to the cotangent sheaf Ω^1 and its exterior powers Ω^p we get the following result.

Corollary 6.2 *On a complete nonsingular variety X, the dimension h^p of the space $\Omega^p[X]$ of regular differential p-forms is finite.*

This result was also stated in Section 6.1, Chapter 3, where we saw that it provides a series of birational invariants of varieties.

As a further example, consider the sheaf $\mathcal{T}$ corresponding to the tangent bundle. An element of $\mathcal{T}(X)$ is called a *regular vector field* on X. It can be viewed as a function taking each point $x \in X$ to a tangent vector $t_x \in \Theta_x$ at x. In this case our theorem gives the next result.

Corollary 6.3 *The space of regular vector fields on a complete nonsingular variety is finite dimensional.*

3.5 Exercises to Section 3

1 In this question, X is assumed to be irreducible. A coherent sheaf $\mathcal{F}$ is a *torsion sheaf* if $\mathcal{F}(U)$ is a torsion module over $\mathcal{O}_X(U)$ for any open set U. Prove that $\mathcal{F}$ is a torsion sheaf if and only if its support is distinct from X.

2 Find the general form of torsion sheaves on a nonsingular curve.

3 Let $E \to X$ be a vector bundle over an affine variety $X = \operatorname{Spec} A$. Prove that the set M_E of sections of E is a finite A-module.

4 Prove that the module M_E introduced in Exercise 3 is a projective A-module. (For the definition of a projective module, see for example Bourbaki [18, Section 3.2.2, Chapter II] or Matsumura [57, Appendix B].)

5 Under the assumptions of Exercise 3, prove that the modules M_E and $M_{E'}$ are isomorphic if and only if E and E' are isomorphic vector bundles.

6 Prove that every vector bundle over the affine line $\mathbb{A}^1$ is trivial.

7 Let $E \to X$ be a vector bundle over a complete variety X. Prove that the set of sections of E is a finite dimensional vector space.

8 Prove that the set of morphisms $f\colon E_1 \to E_2$ between vector bundles $E_i \to X$ (for $i = 1, 2$) over a complete variety X is a finite dimensional vector space.

9 Suppose that A is a 1-dimensional regular local ring with field of fractions K, and $X = \operatorname{Spec} A$; let $x \in X$ be the generic point and $U = \{x\}$. A sheaf $\mathcal{F}$ of $\mathcal{O}$-modules on X is given by an A-module M, a K-vector space L and a restriction map $\varphi\colon M \to L$ which is an A-module homomorphism. Express in terms of M, L and φ what it means for $\mathcal{F}$ to be a coherent sheaf. Construct an example of a subsheaf of a coherent sheaf which is not coherent.

10 Let X be an irreducible variety and $x_0 \in X$ a closed point. Define a sheaf $\mathcal{F}$ on X by setting $\mathcal{F}(U) = \mathcal{O}(U)$ if $U \not\ni x_0$ and $\mathcal{F}(U) = 0$ if $U \ni x_0$. Prove that $\mathcal{F}$ is a sheaf, that it is a subsheaf of $\mathcal{O}$, and that it is not coherent.

4 Classification of Geometric Objects and Universal Schemes

4.1 Schemes and Functors

A phenomenon that has already occurred several times is that a set of certain geometric objects depends on parameters, and more precisely, is parametrised by the points of some algebraic variety. For example, lines in the projective space $\mathbb{P}^3$ are parametrised by points of the 4-dimensional Plücker quadric (Section 4.1, Chapter 1). What is the precise meaning of this assertion? What meaning at all? We indicated the construction of the Plücker coordinates of a line, and showed that it defines a one-to-one correspondence between lines of $\mathbb{P}^3$ and points of the Plücker quadric. But there is no guarantee that this construction is unique; that is, that we might not be able to establish some other equally natural one-to-one correspondence between lines of $\mathbb{P}^3$ and points of some other variety, perhaps even of a different dimension. After all, as far as set theory goes, the set of lines has only one invariant, its cardinality. At the same time, it is obviously very important to be able to define some natural variety (or a more general notion) classifying geometric objects of a given type: its properties, such as dimension, rationality or unirationality and so on, give important characteristics of the whole set of these objects. We describe one approach that in many cases allows us to determine what precisely it means to say that a given set of objects is parametrised by the points of a given variety or scheme.

Since we are talking about geometric objects, the notion of an algebraic family of objects is usually well defined. For example, if we are talking about r-dimensional linear subspaces of a given vector space V, an algebraic family of these with base S is a vector bundle $E \to S$ of rank r which is a vector subbundle of the direct product $S \times V$. In exactly the same way, since we study objects modulo a well-defined equivalence relation, this equivalence relation carries over also to families over any base. For example, in studying the subspaces of a given space V, we naturally consider two vector subbundles $E \to S$ and $E' \to S$ in $S \times V$ as the same if they are equal as subschemes of $S \times V$. Or if we are interested, say, in the classification of nonsingular complete curves of genus g, then by a family of these curves we mean a scheme $C \to S$ all of whose (scheme-theoretic) fibres over closed points of S are nonsingular complete curves of genus g. An isomorphism between two families $C \to S$ and $C' \to S$ is an isomorphism of schemes $f\colon C \to C'$ commuting with the projection to S, that is, such that the diagram

$$\begin{array}{ccc} C & \xrightarrow{f} & C' \\ & \searrow \quad \swarrow & \\ & S & \end{array}$$

is commutative.

Suppose that for some type of geometric objects we have found a "natural" variety (or scheme) X classifying them. Let's try to clarify this idea of "naturality". Obviously, to each object there should correspond a definite closed point of X. Let $\varphi\colon Y \to S$ be an algebraic family of our objects over a base which is a variety. Then to each fibre $\varphi^{-1}(s)$ for $s \in S$ there corresponds some point of X, and this defines a map $f\colon S \to X$. In the notion of "naturality" it is first of all reasonable to include the requirement that this map of points be a morphism, and even to require that the same type of morphism exists for families whose base is a scheme (with certain conditions: over a field k, of finite type, and so on). Moreover, it is reasonable to suppose that two families $Y \to S$ and $Y' \to S$ determine the same morphism $f\colon S \to X$ if and only if they are equivalent in the sense of the equivalence defined for our objects. Finally, the "naturality" of X should include the requirement that every point of X corresponds to some object of our type. Then any map $f\colon S \to X$ of a variety S to X will determine over each point $s \in S$ the object which the point $f(s) \in X$ parametrises; in other words, set-theoretically, it will determine a "family" of objects parametrised by points $s \in S$. It is also reasonable to include in the notion of "naturality" the requirement that if f is a morphism then we obtain in this way an algebraic family of objects.

All of these conditions are summed up very simply in the single statement that there should exist a one-to-one correspondence between algebraic families $Y \to S$

of our objects (the base S may satisfy some restrictions such as being a Noetherian scheme), considered up to equivalence, and all morphisms $S \to X$.

We now formulate the definition we have arrived at. A scheme X is *universal* for some type of objects if for any scheme S (possibly with certain restrictions) there exists a one-to-one correspondence f_S between the set $\Phi(S)$ of all algebraic families $Y \to S$ of objects of the given type, considered up to equivalence, and the set $\mathcal{M}(S, X)$ of morphisms $S \to X$. The correspondence $f_S\colon \Phi(S) \to \mathcal{M}(S, X)$ should satisfy the following condition: for any morphism $\varphi\colon S \to S'$, the diagram

$$\begin{array}{ccc} \Phi(S') & \xrightarrow{f_{S'}} & \mathcal{M}(S', X) \\ {\scriptstyle g}\downarrow & & \downarrow{\scriptstyle h} \\ \Phi(S) & \xrightarrow[f_S]{} & \mathcal{M}(S, X) \end{array} \tag{6.47}$$

is commutative, where g is defined by taking the inverse image of families under φ (that is, their fibre product or pullback by φ), and h by composing a morphism $S' \to X$ with the morphism $\varphi\colon S \to S'$.

In the language of categories, an operation that sends a scheme S to a set $\Phi(S)$ and a morphism $\varphi\colon S \to S'$ to a map $\Phi(\varphi)\colon \Phi(S') \to \Phi(S)$ is called a *functor* if for two morphisms $\varphi\colon S \to S'$ and $\psi\colon S' \to S''$ we have $\Phi(\psi \circ \varphi) = \Phi(\varphi) \circ \Phi(\psi)$. In particular, if $\Phi(S)$ is the set of all algebraic families of objects of our type, and for a morphism $\varphi\colon S \to S'$ the map $\Phi(\varphi)\colon \Phi(S') \to \Phi(S)$ is defined by taking inverse image of families, then Φ is a functor. A trivial example of a functor $\Psi_X(S)$ is determined by an arbitrary scheme X: here $\Psi_X(S) = \mathcal{M}(S, X)$ is the set of all morphisms $S \to X$ to X and, if $\varphi\colon S \to S'$ is a morphism, the map $\Psi_X(\varphi)\colon \Psi_X(S') \to \Psi_X(S)$ sends $f\colon S' \to X$ into the composite $f \circ \varphi\colon S \to X$. Diagram (6.47) in the definition of universal scheme means that the functor Φ is isomorphic to the functor Ψ_X for some scheme X; in the theory of categories, Φ is then called a *representable functor*. Thus the question of the existence of a universal scheme is the question of the representability of the functor Φ of families of objects of the given type.

Note that our definition does not in any way guarantees the existence of a universal scheme: we will soon see that it does not always exist. For the moment, we assume that a universal scheme exists for objects of some type, and note some properties that support the naturality of the definition.

First of all, a universal scheme X is unique if it exists. Indeed, if Y is a second such scheme then by definition, we have isomorphisms $u\colon \mathcal{M}(X, X) \cong \Phi(X) \cong \mathcal{M}(X, Y)$ and $v\colon \mathcal{M}(Y, Y) \cong \Phi(Y) \cong \mathcal{M}(Y, X)$; and a morphism $\varphi\colon X \to Y$ gives rise to a commutative diagram

$$\begin{array}{ccc} \Phi(X) \cong \mathcal{M}(X, X) & \xrightarrow{u} & \mathcal{M}(X, Y) \\ {\scriptstyle g}\uparrow & & \uparrow{\scriptstyle h} \\ \Phi(Y) \cong \mathcal{M}(Y, X) & \xrightarrow[v]{} & \mathcal{M}(Y, Y) \end{array} \tag{6.48}$$

where $g(\xi) = \xi \circ \varphi$ and $h(\eta) = \eta \circ \varphi$. Let $1_X : X \to X$ and $1_Y : Y \to Y$ be the identity morphisms, and set $u(1_X) = \alpha$ and $v^{-1}(1_Y) = \beta$. Consider the diagram (6.48) for $\varphi = \alpha$, and apply it to $\beta \in \mathcal{M}(Y, X)$; then $u(\beta \circ \alpha) = \alpha = u(1_Y)$, and since u is a bijection, $\beta \circ \alpha = 1_Y$. Similarly one proves that $\alpha \circ \beta = 1_X$. Therefore α is an isomorphism.

But we can get even more. In view of the one-to-one correspondence $\Phi(X) \cong \mathcal{M}(X, X)$, the identity morphism $1_X \in \mathcal{M}(X, X)$ determines an element $\varepsilon_X \in \Phi(X)$ called the *universal family* over X. It follows from the definition that any family $\xi \in \Phi(S)$ not only determines a morphism $f : S \to X$, but is determined by it, as the inverse image of the universal family ε_X under f, that is, as the fibre product $\varepsilon_X \times_X S$.

Finally, suppose that all the objects and schemes are defined over an algebraically closed field k. Consider some individual object ξ, that is, a family $\xi \to \operatorname{Spec} k$. Then ξ is an element of the set $\Phi(\operatorname{Spec} k)$, which by definition is in one-to-one correspondence with the set $\mathcal{M}(\operatorname{Spec} k, X)$, that is, with the closed points of X. Therefore our object ξ determines a closed point of the scheme X, and all objects, up to equivalence, are in one-to-one correspondence with these points. Thus in this sense the objects under consideration are parametrised by points of X.

Example 6.22 Let us see that the Grassmannian $\operatorname{Grass}(r, V)$ really is a universal scheme for r-dimensional subspaces of a vector space V. We consider schemes over an algebraically closed field k. For a k-scheme S, we define $\Phi(S)$ as the set of vector bundles $E \to S$ that are vector subbundles of the direct product $S \times_k V$. For a morphism $\varphi : S' \to S$, we define $\Phi(\varphi) : \Phi(S) \to \Phi(S')$ as the inverse image map $E \mapsto E \times_S S'$. We need to determine a one-to-one correspondence $f_S : \Phi(S) \to \mathcal{M}(S, \operatorname{Grass}(r, V))$ which is functorial (that is, gives commutative diagrams (6.47)). These maps f_S are an exact analogue of writing down the Plücker coordinates (see Example 1.24 of Section 4.1, Chapter 1). Let $E \to S$ be a vector bundle of rank r, and $S = \bigcup U_\alpha$ a cover such that $E_{|U_\alpha} \cong U_\alpha \times \mathbb{A}^r$. We choose a basis $f_1, \dots, f_r$ in $\mathbb{A}^r$ and a basis $e_1, \dots, e_n$ in V. The embedding $E \hookrightarrow S \times_k V$ allows us to express the f_i as $f_i = \sum a_{ij} e_j$ with $a_{ij} \in \mathcal{O}(U_\alpha)$, and $f_1 \wedge \cdots \wedge f_r$ as $\sum p_{j_1 \dots j_r} e_{j_1} \wedge \cdots \wedge e_{j_r}$ with $p_{j_1 \dots j_r} \in \mathcal{O}(U_\alpha)$. This gives the morphism $U_\alpha \to \bigwedge^r V$ determined by the functions $p_{j_1 \dots j_r}$. From it we get a morphism $U_\alpha \to \mathbb{P}(\bigwedge^r V)$, which does not depend on the choice of the basis $f_1, \dots, f_r$. Obviously $p_{j_1 \dots j_r}$ satisfy the Plücker equations of the Grassmannian, so that we have a morphism $U_\alpha \to \operatorname{Grass}(r, V)$. Since these morphisms for different α are defined invariantly, they glue together to give a global morphism $S \to \operatorname{Grass}(r, V)$ that we take for $f_S(E)$. The inverse map $\mathcal{M}(S, \operatorname{Grass}(r, V)) \to \Phi(S)$ is obtained by taking the inverse image under any map $\varphi : S \to \operatorname{Grass}(r, V)$ of the universal bundle over $\operatorname{Grass}(r, V)$ (see Example 6.4). It is trivial to check that these two maps are inverse to one another.

Example 6.23 We now give an example of a situation where the universal scheme does not exist. This is an extremely important case, nonsingular curves of given genus g. The reason for nonexistence is already present most vividly in the most trivial case, curves of genus 0. We know that all such curves are isomorphic to $\mathbb{P}^1$. Therefore, if the universal scheme X exists, it must have a single closed point; that is, it would be an affine scheme Spec A, where A is a local ring. Now consider a concrete family of curves of genus 0. For this, consider the plane $\mathbb{P}^2$ with coordinates $(x_0 : x_1 : x_2)$ and the rational map $\mathbb{P}^2 \to \mathbb{P}^1$ given by $(x_0 : x_1 : x_2) \mapsto (x_1 : x_2)$. This has a single point of indeterminacy $(1 : 0 : 0)$. Blowing up this point we get a surface V and a morphism $\varphi \colon V \to \mathbb{P}^1$ (see the example at the end of Section 3.3 of Chapter 4). The fibres of φ are all isomorphic to the projective line, so φ is precisely a family of curves of genus 0 over $\mathbb{P}^1$, that is, an element of the set $\Phi(\mathbb{P}^1)$. If a universal scheme X existed then our family would be the inverse image of the universal family over $X = \operatorname{Spec} A$ under some morphism $f \colon \mathbb{P}^1 \to X$, and f must map $\mathbb{P}^1$ to the single closed point of X. However, f then corresponds to another element of $\Phi(\mathbb{P}^1)$, the direct product $\mathbb{P}^1 \times \mathbb{P}^1$. To nail down the contradiction, it remains to see that the family $V \to \mathbb{P}^1$ is not isomorphic to $\mathbb{P}^1 \times \mathbb{P}^1$. This follows for example from the fact that the selfintersection of any divisor on $\mathbb{P}^1 \times \mathbb{P}^1$ is even: if $C_1 = \mathbb{P}^1 \times x$ and $C_2 = y \times \mathbb{P}^1$ then any divisor D on $\mathbb{P}^1 \times \mathbb{P}^1$ is linearly equivalent to $n_1C_1 + n_2C_2$, so that $D^2 = 2n_1n_2$. On the other hand, V contains the curve L obtained by blowing up $(1 : 0 : 0) \in \mathbb{P}^2$, and $L^2 = -1$ (compare Exercise 14 of Section 1.5).

The family constructed above is locally trivial: it is easy to see that if $U_1 = \mathbb{P}^1 \setminus \infty$ and $U_2 = \mathbb{P}^1 \setminus 0$ then $\varphi^{-1}(U_1) \cong U_1 \times \mathbb{P}^1$ and $\varphi^{-1}(U_2) \cong U_2 \times \mathbb{P}^1$. But this is not necessarily the case: the family in $\mathbb{P}^2 \times \mathbb{A}^2$ given by $\xi_0^2 = u\xi_1^2 + v\xi_2^2$, where $\mathbb{A}^2$ has coordinates u, v and $\mathbb{P}^2$ has coordinates $(\xi_0 : \xi_1 : \xi_2)$ is not isomorphic to a trivial family over any open subset $U \subset \mathbb{A}^2$. This follows from the fact that it has no rational section: there do not exist polynomials $p_0, p_1, p_2 \in k[u, v]$ such that $p_0^2 = up_1^2 + vp_2^2$. Indeed, we can suppose that p_0, p_1, p_2 do not have any common factors. Setting $u = 0$ we get $p_0(0, v)^2 = vp_2(0, v)^2$, which is only possible if $p_0(0, v) = p_2(0, v) = 0$, that is, both p_0 and p_2 are divisible by u. Then p_1 would also be divisible by u.

Of course, similar examples can be constructed for curves of genus $g > 0$. Nevertheless, the notion of universal scheme can be modified in such a way that it does exist for curves of any genus. This can be done in two different ways. One can either drop the requirement in the definition of universal scheme X that the correspondence between families over S and morphisms $S \to X$ be one-to-one, and require only that every family defines a morphism: then the universal object will exist as a variety. Or one can insist on having a one-to-one correspondence, but allow the universal object to be something more general than a scheme, a so-called *topology* or *algebraic stack*. See Mumford and Fogarty [64] and Mumford [63].

The interpretation of a scheme as a functor has already appeared in a slightly different context. In Section 3.4, Chapter 5 we showed that if $x \in X$ is any closed point of a scheme X over a field k, we can describe the tangent space $\mathcal{O}_{X,x}$ as the set of morphisms $\mathcal{M}_x(\operatorname{Spec} D, X)$, where $D = k[\varepsilon]/(\varepsilon^2)$, and we allow in $\mathcal{M}_x$ only the morphisms that map the closed point of Spec D to the point $x \in X$. This interpretation of the tangent space gives a convenient method of describing it if the scheme

X itself is a universal scheme for some type of objects. Putting together Proposition of Section 3.4, Chapter 5 with the definition of a universal scheme shows that in this case $\Theta_{X,x}$ coincides with the set $\mathcal{M}_x(\operatorname{Spec} D, X) \cong \Phi_x(\operatorname{Spec} D)$ of families over $\operatorname{Spec} D$ with given fibre over the point 0. This provides grounds for the intuition that the tangent vector to a universal scheme is a first order infinitesimal deformation of a given object.

Example 6.24 (The tangent space to the Grassmannian $\operatorname{Grass}(r, V)$) Suppose that a point $x \in \operatorname{Grass}(r, V)$ corresponds to a vector subspace E with basis $e_1, \dots, e_r$. By what we said above, Θ_x is isomorphic to the set of vector bundles over $\operatorname{Spec} D$ that are vector subbundles of $\operatorname{Spec} D \times V$ with fibre over 0 equal to E. Passing to the corresponding sheaves, we see that a vector bundle over $\operatorname{Spec} D$ is a module over D that is locally free, hence free. Hence the vector bundle is trivial and has basis $e_1 + \varepsilon u_1, \dots, e_r + \varepsilon u_r$. It remains to determine when two bases of this form give the same vector subbundle of $\operatorname{Spec} D \times V$. If the second basis is $e_1 + \varepsilon v_1, \dots, e_r + \varepsilon v_r$ then this will happen if and only if

$$e_i + \varepsilon v_i = \sum_j (c_{ij} + \varepsilon d_{ij})(e_j + \varepsilon u_j)$$

for $i = 1, \dots, r$. This implies $e_i = \sum c_{ij} e_j$, so that (c_{ij}) is the identity matrix. Next, $v_i = \sum c_{ij} u_j + \sum d_{ij} e_j = u_i + w_i$, where $w_i = \sum d_{ij} e_j$ is an arbitrary vector in E. Thus the vector subbundle of $\operatorname{Spec} D \times V$ is uniquely determined by the vectors u_i in V/E. Setting $\varphi(e_i) = u_i \bmod E$, we see that the required vector subbundles are uniquely specified by homomorphisms $\varphi\colon E \to V/E$, so that $\Theta_x \cong \operatorname{Hom}(E, V/E)$.

Example 6.25 (The scheme of associative algebras) (See Example 2.5 of Section 4.1, Chapter 1 and Example 5.20.) A closed point of this scheme is a multiplication $E \times E \to E$; if E has basis $e_1, \dots, e_n$, the multiplication is given by $e_i e_j = \sum c_{ij}^m e_m$. Tautologically, the scheme is universal for multiplication laws in $S \times_k E$, where now S is an arbitrary scheme and $c_{ij}^m \in \mathcal{O}(S)$. Hence if x is the closed point of this scheme corresponding to the structure constants $\{c_{ij}^m\}$, the tangent space Θ_x is isomorphic to the set of multiplication laws on $D \times E$ of the form $e_i e_j = \sum (c_{ij}^m + \varepsilon d_{ij}^m) e_m$ where $d_{ij}^m \in k$ are any elements for which this multiplication is associative. The associativity condition can be written out at once by comparing the coefficient of ε in $(e_i e_j) e_k$ and $e_i (e_j e_k)$:

$$\sum_m c_{ij}^m d_{mk}^l + \sum_m d_{ij}^m c_{mk}^l = \sum_m c_{jk}^m d_{im}^l + \sum_m c_{jk}^m d_{im}^l$$

for all i, j, k, l. These are the same equations as we obtained in Example 2.5 of Section 1.3, Chapter 2 by differentiating the associativity relation; but now they have acquired a transparent meaning, as the first order infinitesimal deformations of the structure constants.

4.2 The Hilbert Polynomial

The remainder of this section will be taken up with the description of the universal scheme for an extremely important type of object: closed subvarieties, and even subschemes of projective space $\mathbb{P}^N$. For the case of linear subspaces we already know the universal scheme, the Grassmannian.

Already in the example of linear subspaces we see that, rather than considering all subvarieties at the same time, we get the natural answer by breaking up the subvarieties into classes, and then considering these separately. In the case of the Grassmannian, we fixed the dimension r of the subspace and its degree 1. We now describe similar discrete invariants of projective schemes that one has to fix in order to arrive at the natural universal schemes; these are the so-called Hilbert polynomials.

With each projective subscheme $X \subset \mathbb{P}^N$ we associate an infinite sequence $a_r(X)$ of integers: $a_r(X)$ is the number of forms of degree r in the homogeneous coordinates of $\mathbb{P}^N$ that are linearly independent on X. To give a more formal definition, consider the homogeneous ideal $\mathfrak{a}_X$ of a projective scheme $X \subset \mathbb{P}^N$ (Section 3.3, Chapter 5, and compare Section 4.1, Chapter 1), and write $\mathfrak{a}_X^{(r)}$ for its homogeneous piece of degree r, that is, the space of forms of degree r in $\mathfrak{a}_X$. Write $S^{(r)}$ for the space of forms of degree r in the homogeneous coordinates of $\mathbb{P}^N$. Now set $a_r(X) = \dim_k S^{(r)}/\mathfrak{a}_X^{(r)}$. These numbers depend, of course, on the embedding $X \hookrightarrow \mathbb{P}^N$, and in this respect they are analogous to the degree.

The infinite sequence of numbers just constructed can be described in finite terms.

Theorem 6.5 *There exists a polynomial $P_X(T) \in \mathbb{Q}[T]$ such that $a_r(X) = P_X(r)$ for all sufficiently large integers r.*

The polynomial $P_X(T)$ whose existence is established in the theorem is obviously uniquely determined. It is called the *Hilbert polynomial* of X.

Proof The theorem is proved by induction on the dimension N, and, as often happens, it is convenient to prove a more general assertion. Consider a finite *graded module* M over the polynomial ring $S = k[\xi_0, \dots, \xi_N]$. This means that M is a module over S, with a fixed decomposition $M = \bigoplus M^{(r)}$ as a direct sum of k-vector subspaces such that

$$x \in M^{(r)} \quad \text{and} \quad f \in S^{(l)} \quad \Longrightarrow \quad fx \in M^{(r+l)}.$$

The subspaces $M^{(r)}$ are called the *homogeneous pieces* of M of degree r. Each subspace $M^{(r)}$ is finite dimensional over k: indeed, as a k-vector space, $M^{(r)} \cong M'_r = (\bigoplus_{i \geq r} M^{(i)})/(\bigoplus_{i>r} M^{(i)})$, where $\xi_i M'_r = 0$ for each i, so that M'_r is a finite module over k. We set $a_r(M) = \dim_k M^{(r)}$ and prove that the statement of the theorem holds for $a_r(M)$. The theorem itself is obtained by setting $M = S/\mathfrak{a}_X$.

We can set $S = k$ for $N = -1$ and assume that in this case a graded module over S is of the form $M = M_0$, with M_0 a finite dimensional graded k-vector space. From this point on, the theorem is proved by induction on N. Consider the homomorphism $\xi_N \colon M \to M$ consisting of multiplication by the variable ξ_N. Then $\xi_N M^{(r)} \subset M^{(r+1)}$, which implies that the kernel K and cokernel $C = M/\xi_N M$ are both graded modules: $K = \bigoplus K^{(r)}$ and $C = \bigoplus C^{(r)}$, where $K^{(r)} = M^{(r)} \cap K$ and $C^{(r)} = M^{(r)}/\xi_N M^{(r-1)}$. We have an exact sequence

$$0 \to K^{(r)} \to M^{(r)} \xrightarrow{\xi_N} M^{(r+1)} \to C^{(r+1)} \to 0. \tag{6.49}$$

Now by construction K and C are graded S-modules on which ξ_N acts by 0, so that we can view them as modules over $k[\xi_0, \dots, \xi_{N-1}]$ and assume by induction that the assertion holds for them. Write P_K and P_C for the polynomials corresponding to them. Then the exact sequence (6.49) implies that

$$a_{r+1}(M) - a_r(M) = P_C(r+1) - P_K(r).$$

for all sufficiently large r. Now it follows from very simple properties of polynomials (see Section 2, Appendix) that a sequence of integers satisfying this condition is given for all sufficiently large r as the values of some polynomial $P_M(T) \in \mathbb{Q}[T]$, that is, $a_r(M) = P_M(T)$, as asserted. □

Example 6.26 Let $X \subset \mathbb{P}^N$ be a 0-dimensional subscheme. Suppose that the underlying set X_{red} does not intersect the hyperplane $\xi_0 = 0$. Taking a homogeneous polynomial $F \in S^{(r)}$ to the polynomial $f = F/\xi_0^r \in k[x_1, \dots, x_N] = k[\mathbb{A}^N]$ where $x_i = \xi_i/\xi_0$, we see that $S^{(r)}/\mathfrak{a}^{(r)} \cong V^{(r)}/(V^{(r)} \cap I)$, where $V^{(r)} \subset k[x_1, \dots, x_N]$ is the space of polynomials of degree $\le r$ and I the ideal defining the subscheme $X \subset \mathbb{A}^N$. Since $\dim V^{(r)}/(V^{(r)} \cap I) \le \dim V^{(r+1)}/(V^{(r+1)} \cap I)$, the sequence of numbers $a_r(X)$ stabilises from some r onwards. It follows that $P_X(T) = \text{const.} = \dim k[\mathbb{A}^N]/I$. In other words,

$$X = \operatorname{Spec} A, \quad A = k\big[\mathbb{A}^N\big]/I \quad \text{and} \quad P_X(T) = \text{const.} = \dim_k A.$$

Since $S^{(r)} \ne \mathfrak{a}_X^{(r)}$ for any r (assuming X nonempty), the Hilbert polynomial cannot be identically zero. We now determine how it reflects two of the simplest invariants of a scheme X, the dimension and the degree. We only carry through the proof in the case that X is a nonsingular variety (possibly irreducible). The same result holds for arbitrary closed subschemes $X \subset \mathbb{P}^N$, but to prove it requires a little more commutative algebra (see Hartshorne [37, Section 7, Chapter I] or Fulton [29, Example 2.5.2]).

Theorem 6.6 *The Hilbert polynomial P_X of a nonsingular variety X has degree equal to the dimension of X. If X has dimension n and degree d then the leading term of P_X is $(d/n!)T^n$.*

Proof The proof is based on the same arguments as that of Theorem 6.5. We use induction on $n = \dim X$. If $n = 0$, the result follows from Example 6.26: X consists of d distinct points, and $A = k[\mathbb{A}^N]/I$ is a direct sum of d copies of k, and obviously $a_r(X) = d$ for $r \gg 0$. This proves the theorem in this case.

In the general case, choose a coordinate system such that the hyperplane $\xi_N = 0$ is transversal to X at all points of intersection. The fact that this is possible follows easily from the usual dimension count that we have used many times. Write $\mathbb{P}^*$ for the projective space of hyperplanes of $\mathbb{P}^N$. In $X \times \mathbb{P}^*$ we need to consider the subvariety $Z = \{(x, \lambda) \mid x \in X, \lambda \in \mathbb{P}^* \text{ and } \lambda \supset \Theta_{X,x}\}$. Considering the projection $Z \to X$ shows that $\dim Z \le N - 1$, and hence the image of the projection of Z to $\mathbb{P}^*$ is not the whole of $\mathbb{P}^*$. Hence there exists a hyperplane transversal to X at every point of intersection, and we can take this to be $\xi_N = 0$.

We now apply the argument from the proof of Theorem 6.5 to the module $M = S/\mathfrak{a}_X$ and determine K and C in this case. We prove that $K = 0$. Suppose that $F \in K^{(r)}$, that is, $\xi_N F = 0$ on X. Then for any $i < N$ the function $f = F/\xi_i^r$ satisfies $(\xi_N/\xi_i)f = 0$ on X. But $u_N = \xi_N/\xi_i$ is part of a local parameter system at every point of X_N at which $u_N = 0$, and we saw in Section 1.2, Chapter 4 that none of the local parameters can be a zerodivisor. A fortiori u_N is not a zerodivisor in a neighbourhood of points where $u_N \ne 0$. Therefore $f = 0$ on every component of X, that is, $F \in \mathfrak{a}_X$.

Let us determine the module C. In what follows we use the notation introduced after the definition of projective scheme in Section 3.3, Chapter 5. By definition $C = S/(\xi_N, \mathfrak{a}_X)$. The ideal $(\xi_N, \mathfrak{a}_X)$ consists of forms that vanish on X', the section of X with the hyperplane $\xi_N = 0$. We prove that $(\xi_N, \mathfrak{a}_X) = \mathfrak{a}_{X'}$. For this it is enough to check on each affine open set U_i given by $\xi_i \ne 0$ that $(x_N, \mathfrak{a}_i) = \mathfrak{a}_i'$, where $x_N = \xi_N/\xi_i$ and $\mathfrak{a}_i'$ is the ideal of functions that vanish on the intersection of $X \cap U_i$ with $x_N = 0$. It is enough to prove that $(x_N, \mathfrak{a}_i)/\mathfrak{a}_i = \mathfrak{a}_i'/\mathfrak{a}_i$ in $k[X \cap U_i]$. For this it is enough to prove that if $\varphi \in k[X \cap U_i]$ and $\varphi^\rho \in (x_N)$ then $\varphi \in (x_N)$. This property holds locally in the neighbourhood of any point $\alpha \in X \cap U_i$. Indeed, as usual, it is enough to check this in the local ring $\mathcal{O}_\alpha$ of a point α. We need to prove that if $\varphi^\rho \in (x_N)$ then $\varphi \in (x_N)$ for $\varphi \in \mathcal{O}_\alpha$. But this follows at once because $\mathcal{O}_\alpha$ is a UFD (Theorem 2.10 of Section 3.1, Chapter 2), together with the fact that x_N is prime, as an element of a local system of parameters. Passing to the global situation, we can cover $X \cap U_i$ by open sets of the form $D(f_\lambda)$ and assume that $\varphi \in (x_N, \mathfrak{a}_i')k[D(f_\lambda)]$ for every λ. Now it is enough to find for any arbitrarily large m functions $g_\lambda \in k[X \cap U_i]$ such that $\sum f_\lambda^m g_\lambda = 1$. Then $\varphi = \varphi \sum f_\lambda^m g_\lambda$ and we can assume that $\varphi f_\lambda^m \in (x_N, \mathfrak{a}_i')$ by the choice of m.

Thus in the sequence (6.49) we now have $K = 0$ and $C = S'/\mathfrak{a}_{X'}$ where $S' = k[\xi_0, \dots, \xi_{N-1}]$ and X' is nonsingular, $(n-1)$-dimensional and of degree d. Using induction we can assume that the theorem holds for X'. We have an exact sequence

$$0 \to S^{(r)}/\mathfrak{a}_X^{(r)} \xrightarrow{\xi_N} S^{(r+1)}/\mathfrak{a}_X^{(r+1)} \to S'^{(r+1)}/\mathfrak{a}_{X'}^{(r+1)} \to 0,$$

and hence for sufficiently large r we have

$$P_X(r+1) - P_X(r) = P_{X'}(r+1),$$

that is,

$$P_X(T+1) - P_X(T) = P_{X'}(T+1). \tag{6.50}$$

By induction, we can assume that $P_{X'}(T)$ has leading term $(d/(n-1)!)T^{n-1}$. Writing the leading term of $P_X(T)$ as aT^m we see from (6.50) that $m = n$ and $a = d/n!$, as asserted in the theorem. □

The Hilbert polynomial provides the most natural answer to the question discussed at the beginning of this section of dividing up all projective subschemes $X \subset \mathbb{P}^N$ into natural classes, the classes of X with given Hilbert polynomial.

4.3 Flat Families

We proceed to consider families of closed subschemes $X \subset \mathbb{P}^N$ with a given Hilbert polynomial. First of all, we have to determine when all the schemes of a family with irreducible base have the same Hilbert polynomial. The fact that this does not always happen is shown by the following examples.

Example 6.27 Let $\sigma: X \to Y$ be a blowup of a point $y_0 \in Y$ with $\dim X = \dim Y > 1$, and let $Z = \sigma^{-1}(y_0)$. Then for $y \in Y$, we have $\dim \sigma^{-1}(y) = 0$ if $y \neq y_0$ and $\dim \sigma^{-1}(y_0) > 0$. By Theorem 6.6, even the degree of the Hilbert polynomial changes.

Example 6.28 Let X be a curve with an ordinary double point x_0 and let X^ν be the normalisation of X. We consider the family $\nu\colon X^\nu \to X$ as a family of 0-dimensional schemes over the base X. Then for $x \neq x_0$ the fibre $\nu^{-1}(x)$ is a single point, and $\nu^{-1}(x_0)$ is two points, that is, $\nu^{-1}(x) \cong \operatorname{Spec} k$ and $\nu^{-1}(x_0) \cong \operatorname{Spec}(k \oplus k)$. By Example 6.26, we have $P_{\nu^{-1}(x)}(r) = \text{const.} = 1$ for $x \neq x_0$ but $P_{\nu^{-1}(x_0)}(r) = \text{const.} = 2$.

Example 6.29 Suppose that $\operatorname{char} k \neq 2$; let g be the automorphism of $X = \mathbb{A}^2$ of order 2 given by $g(x, y) = (-x, -y)$ and $S = X/G$ the quotient of X by the group $G = \{1, g\}$ (see Example 1.21 of Section 2.3, Chapter 1 and Section 2.1, Chapter 2).

Then $S \subset \mathbb{A}^3$ is given by $uv = w^2$, and the morphism $X \to S$ by $u = x^2$, $v = y^2$ and $w = xy$. We view $X \to S$ as a family of 0-dimensional subschemes of $\mathbb{A}^2$ with base S. For $s = (a, b, c) \in S$, the fibre $f^{-1}(s) = \operatorname{Spec} k[x, y]/I$ where I is the ideal $I = (x^2 - a, y^2 - b, xy - c)$. Multiplying $xy - c$ by x and by y, we see that $I \ni ay - cx$ and $bx - cy$. Thus if, say, $a \neq 0$, we have $I = (x^2 - a, y - (c/a)x)$, and $k[x, y]/I \cong k[x]/(x^2 - a)$, so that $\dim k[x, y]/I = 2$. By Example 6.27, this means that $P_{f^{-1}(s)}(r) = \text{const.} = 2$. The same holds if $b \neq 0$. However, if $s = (0, 0, 0)$ then $I = (x^2, xy, y^2)$ and $\dim k[x, y]/I = 3$, that is, $P_{f^{-1}(s)}(r) = \text{const.} = 3$.

Thus we can only expect the Hilbert polynomial of fibres to remain constant in a family under some condition of "continuity" or "fluidity" of the fibres of the family. There does indeed exist such a condition, reflecting perfectly the idea of "no jumping" of the fibres; it is the condition that the family is flat. The definition of flat may seem somewhat strange at first sight, since it is purely algebraic in nature. It is hard to lead the reader to this notion by pure logic; it is easier to define it first, then to show just how useful it is.

Definition A module M over a ring A is *flat* if for any ideal, $\mathfrak{a} \subset A$ the surjective map $\mathfrak{a} \otimes M \to \mathfrak{a}M$ defined by $\alpha \otimes m \mapsto \alpha m$ is an isomorphism. A family $f: X \to S$, where X and S are schemes, is *flat* if $\mathcal{O}_x$ is flat as a module over $\mathcal{O}_{f(x)}$ for every $x \in X$. We then also say that f is a *flat morphism*, or that X is *flat over* S.

To check that M is a flat A-module, it is enough to check that the homomorphism $\mathfrak{a} \otimes M \to M$ defined by $\alpha \otimes m \mapsto \alpha m$ has no kernel. In particular, if $\mathfrak{a} = (a)$ is a principal ideal and a is a non-zerodivisor, the condition reduces to saying that the only element of M killed by a is 0. Thus a flat module over an integral principal ideal domain is just a torsion-free module.

We note that an individual scheme over a field k (that is, $S = \operatorname{Spec} k$) is automatically flat; thus flatness is a dynamic property, reflecting the change of the schemes in a family over a base S.

We now enumerate a number of properties of flat morphisms that we neither prove nor make use of, and which characterise flat families as "families with no jumping". They are all geometric restatements of the corresponding properties of rings, and are proved in this form in Bourbaki [17].[2]

Proposition A *If X and S are irreducible schemes of finite type over a field k and $f: X \to S$ is a flat morphism, then all the fibres of f have the same dimension.* (*Compare Example* 6.27.)

Proposition B *A finite morphism $f: X \to S$ of Noetherian schemes is flat if and only if $\ell_{k(s)}(f^{-1}(s))$ is a locally constant function of $s \in S$. Here $\ell_{k(s)}(f^{-1}(s)) =$* $\dim A_s$, *where the fibre is $f^{-1}(s) = \operatorname{Spec}(A_s)$.* (*Compare Examples* 6.28–6.29.)

Proposition C *If X and S are nonsingular varieties and $f: X \to S$ a morphism such that $\mathrm{d}f: \Theta_{X,x} \to \Theta_{S,f(x)}$ is surjective for every $x \in X$ then f is flat.*

Proposition D *If $f: X \to S$ is a flat morphism and $S' \to S$ an arbitrary morphism then $f': X \times_S S' \to S'$ is again flat.*

Proposition E *For rings A and B and a homomorphism $f: A \to B$, the morphism $\varphi: \operatorname{Spec} B \to \operatorname{Spec} A$ is flat if and only if the ring B is flat over A.*

---

[2] Compare also Hartshorne [37, especially A: Proposition 9.5, B: Theorem 9.9, C: Proposition 10.4, D–E: Proposition 9.2, Chapter III].

In what follows, we need one very particular case of this final property.

Lemma *Let A be a principal ideal domain, and B an A-algebra; if* $\operatorname{Spec} B$ *is flat over* $\operatorname{Spec} A$ *then B is a flat A-algebra.*

Proof We need to show that a nonzero element $a \in A$ is a non-zerodivisor in B. The given information is that the localisation B_P of B at any prime ideal $P \subset B$ is flat over $A_{\mathfrak{p}}$, where $\mathfrak{p} = P \cap A$. Hence if $ab = 0$ for some $b \in B$ then $\varphi_P(b) = 0$, where $\varphi_P \colon B \to B_P$ is the localisation map (Section 1.1, Chapter 2). We prove that this implies $b = 0$; moreover, the conditions $\varphi_{\mathfrak{m}}(b) = 0$ for all maximal ideal of B is already sufficient. Indeed, it follows from this that for any maximal ideal $\mathfrak{m}$ there exists an element $c_{\mathfrak{m}} \in B$ such that $c_{\mathfrak{m}} \notin \mathfrak{m}$ and $bc_{\mathfrak{m}} = 0$. Then $bI = 0$, where I is the ideal generated by all the $c_{\mathfrak{m}}$. But I is not contained in any maximal ideal $\mathfrak{m}$, since it contains $c_{\mathfrak{m}} \notin \mathfrak{m}$. Hence $I = B$ and $b = 0$. The lemma is proved. □

For the questions we are interested in, the flat condition on a family is also related to "uniformity": the Hilbert polynomial is constant in a flat family of closed subschemes of $\mathbb{P}^n$ with a connected base S. Straightforward arguments reduce this assertion to the case that S is Spec of a 1-dimensional regular local ring. Namely, it is enough to prove the theorem for a 1-dimensional base S, since in the general case we need only join any two points of S by a chain of curves. Moreover, we can assume that S is irreducible and normal, since otherwise we need to pass to the normalisation S^ν and pullback our family to S^ν, that is, replace $X \to S$ by $X \times_S S^\nu \to S^\nu$. Finally, to prove that the Hilbert polynomial of the fibres over all points $s \in S$ coincide, it is enough to prove this for any closed point $s \in S$ and the generic point $\eta \in S$. We set $A = \mathcal{O}_s$ and pass to the family $X \times_S \operatorname{Spec} A$, thus reducing the assertion to the following: to prove that the Hilbert polynomial of the fibres over the closed and generic points of $\operatorname{Spec} A$ are equal. We now consider this case.

We will understand a family of closed subschemes of $\mathbb{P}^N$ over the base $S = \operatorname{Spec} A$ to mean a closed subscheme of $\mathbb{P}^N_A$. Since there is a canonical morphism $\mathbb{P}^N_A \to \operatorname{Spec} A$, a morphism $X \to \operatorname{Spec} A$ is defined for any closed subscheme $X \subset \mathbb{P}^N_A$, which allows us to view X as a family over the base $\operatorname{Spec} A$.

Theorem 6.7 *Let A be the local ring of a nonsingular point of a curve over an algebraically closed field, and $X \subset \mathbb{P}^N_A$ a closed subscheme such that the morphism $X \to$* $\operatorname{Spec} A$ *is flat. Then the fibres of X over the closed and generic points of* $\operatorname{Spec} A$ *have the same Hilbert polynomial.*

Proof Let

$$\mathfrak{a}_X = \bigoplus_{r \geq 0} \mathfrak{a}_X^{(r)} \subset \Gamma = A[T_0, \dots, T_N]$$

be the homogeneous ideal corresponding to the closed subscheme X. Set $B = \Gamma/\mathfrak{a}_X = \bigoplus_{r \geq 0} B^{(r)}$. Then each $B^{(r)}$ is a finite A-module. Let K be the field of fractions of A and $(\tau) \subset A$ the maximal ideal. The fibre $X \otimes_A K$ of X over the generic point of $\operatorname{Spec} A$ is defined by the ideal $\mathfrak{a}_X \otimes K \subset K[T_0, \dots, T_N]$; and the fibre $X \otimes_A k$ over the closed point is defined by the ideal $\mathfrak{a}_X/\tau\mathfrak{a}_X \subset k[T_0, \dots, T_N]$. Hence the Hilbert polynomial of the fibre over the generic point is defined by the dimensions of the K-vector spaces $B^{(r)} \otimes_A K$ and that of the fibre over the closed point by the dimensions of the k-vector spaces $B^{(r)}/\tau B^{(r)}$. Since $B^{(r)}$ is a finite A-module, the equality

$$\dim_K\left(B^{(r)} \otimes_A K\right) = \dim_k\left(B^{(r)}/\tau B^{(r)}\right)$$

just means that $B^{(r)}$ is torsion-free for all sufficiently large r, and, for this, it is enough to check that $\tau b = 0$ with $b \in B^{(r)}$ is only possible for $b = 0$.

The ring B defines an affine scheme $Z = \operatorname{Spec} B$. This is called the *affine cone* over X, and X is the *base* of the cone Z; compare Exercise 8 of Section 4.5. The intersection $\mathfrak{a}_X \cap A$ is an ideal of A. If this ideal were nonzero it would be of the form (τ^k) for some $k \geq 0$, and thus $\tau^k B = 0$; one sees easily that this would imply $\tau^k \mathcal{O}_X = 0$, so that $\mathcal{O}_X$ would not be flat over A. Hence $\mathfrak{a}_X \cap A = 0$, that is, $B^{(0)} = A$.

Write η_i for the images of the T_i in B, and I for the ideal $(\eta_0, \dots, \eta_N)$. By what we just said $B/I \cong A$, so that I is a prime ideal. Write ζ for the point of Z corresponding to this prime ideal. We call it the *vertex* of Z. Obviously the subscheme defined by $\eta_0, \dots, \eta_N$ is the closure of ζ, that is, $\bigcap_0^N V(\eta_i) = \overline{\zeta}$, and hence $Z \setminus \overline{\zeta} = \bigcup_0^N D(\eta_i)$.

Consider a set $D(\eta_i)$. By definition $D(\eta_i) = \operatorname{Spec}(B_{\eta_i})$, where B_{η_i} is the ring of fractions u/η_i^ν with $u \in B$ and $\nu \geq 0$. If $u = \sum u^{(r)}$ then u/η_i^ν can be written uniquely in the form $\sum \eta_i^{\nu_r}(u^{(r)}/\eta_i^r)$. Here $\nu_r = r - \nu$ is an integer, possibly negative, so that u/η_i^ν can be written as a polynomial in η_i and η_i^{-1} with coefficients of the form $u^{(r)}/\eta_i^r$. We have seen (Section 3.3, Chapter 5) that elements $u^{(r)}/\eta_i^r$ form a ring $C_i = A_i/\mathfrak{a}_i$ with $\operatorname{Spec} C_i = V_i \subset X$. Hence $B_{\eta_i} = C_i[\eta_i, \eta_i^{-1}]$. Since $\operatorname{Spec}(\mathbb{Z}[T, T^{-1}]) \cong \mathbb{A}^1 \setminus 0$, $D(\eta_i) \cong V_i \times (\mathbb{A}^1 \setminus 0)$. It is easy to see (we do not require this) that the projections $D(\eta_i) \to V_i$ glue together to a global morphism $Z \setminus \overline{\zeta} \to X$. That is, removing the origin, the cone has a projection to its base with fibre $\mathbb{A}^1 \setminus 0$ (because we removed the origin). Note that we have proved more: this is a locally trivial fibration—over each V_i it turns into a direct product. (See Figure 28.)

Thus $Z \setminus \overline{\zeta}$ is covered by $N + 1$ open sets each of which is isomorphic to $V_i \times (\mathbb{A}^1 \setminus 0)$ where $V_i \subset X$ are open sets. Since X is flat over A so are the schemes V_i. It follows that the $V_i \times (\mathbb{A}^1 \setminus 0)$ are also flat over A; since in our case flat is equivalent to torsion-free, this follows from the fact that $V_i = \operatorname{Spec} C_i$ and $V_i \times (\mathbb{A}^1 \setminus 0) = \operatorname{Spec} C_i[T, T^{-1}]$. Finally, since flat is a local condition, we conclude that $Z \setminus \overline{\zeta}$ is flat over A.

What does this mean from the point of view of B? If we recall the definition of the ring $\mathcal{O}(U)$ for an open set $U \subset \operatorname{Spec} B$ (Section 2.2, Chapter 5), the answer is as follows: suppose that $b \in B$ and $\tau b = 0$; then for any $f \in (\eta_0, \dots, \eta_N)$ the element b is zero on the open set $D(f)$, that is, $f^s b = 0$ for some $s > 0$. In particular $\eta_i^{s_i} b = 0$ for some s_i, and hence $I^t b = 0$ for $t \geq s_0 + \dots + s_N$.

Figure 28 The affine cone

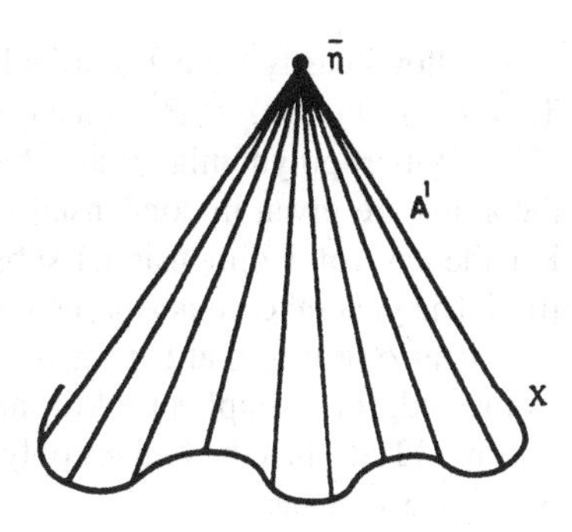

All elements $b \in B$ with $\tau b = 0$ form an ideal J, which, since B is Noetherian, has a finite basis, say, $J = a_1 B + \cdots + a_m B$. From the fact that $I^{t_i} a_i = 0$ for some $t_i > 0$, it follows that all components of $a_i B$ of sufficiently high degree are zero, and therefore the same holds for J. This means that $J \cap B^{(r)} = 0$ for all sufficiently large r; that is, $B^{(r)}$ is a torsion-free module, and this is what we had to prove. □

4.4 The Hilbert Scheme

We can now state the fundamental existence theorem. Let S be a scheme over a field k. A *family of closed subschemes* of $\mathbb{P}^N$ with *base* S is a closed subscheme $X \subset \mathbb{P}^N \times_k S$ with the natural projection morphism $X \to S$. Let $P \in \mathbb{Q}[T]$ be a polynomial. Consider the functor Ψ^P that sends a scheme S to the set $\Psi^P(S)$ of all flat families of closed subschemes of $\mathbb{P}^N$ with base S and Hilbert polynomial P. For a morphism $f\colon S' \to S$, we define $\Psi^P(f)$ to be the map $\Psi^P(S) \to \Psi^P(S')$ which sends a family $X \to S$ into the pullback family $X' = X \times_S S' \to S'$.

Theorem F *There exists a universal scheme* $\operatorname{Hilb}^P_{\mathbb{P}^N}$ *for the functor* Ψ^P*; it is a projective scheme over k, called the* Hilbert scheme *of* $\mathbb{P}^N$.

The proof of this theorem is not difficult, but we cannot give it here because it uses cohomological methods. Roughly speaking, one proves that for sufficiently large r, the homogeneous ideal $\mathfrak{a}_X$ of any flat family $X \to S$ with Hilbert polynomial P has every homogeneous component $\mathfrak{a}_X^{(t)}$ with $t \geq r$ generated by forms of degree r, that is, $\mathfrak{a}_X^{(t)} = \Gamma^{(t-r)} \cdot \mathfrak{a}_X^{(r)}$. For r sufficiently large, the codimension of $\mathfrak{a}_X^{(r)} \subset \Gamma^{(r)}$ equals $P(r)$, and it determines a point of the Grassmannian $\operatorname{Grass}(\binom{N+r}{r}, P(r))$. Conversely, this point determines $\mathfrak{a}_X^{(r)}$. One checks furthermore that, for sufficiently large r, the points of $\operatorname{Grass}(\binom{N+r}{r}, P(r))$ for which the corresponding space of forms $\mathfrak{a}^{(r)}$ generates a homogeneous ideal $\mathfrak{a}$ defining a closed subscheme with Hilbert polynomial P is itself a closed subscheme of $\operatorname{Grass}(\binom{N+r}{r}, P(r))$. This is the universal scheme $\operatorname{Hilb}^P_{\mathbb{P}^N}$.

It follows easily from Theorem F (or it can be proved directly in the same way as Theorem F) that if $Y \subset \mathbb{P}^N$ is a closed subscheme then closed subschemes of Y with given Hilbert polynomial P also have a universal family Hilb_Y^P. The proof of these theorems are given in condensed form in Grothendieck's Bourbaki seminars [35]. For the case of 1-dimensional subschemes of a surface Y it is given in Mumford [62]. The general case is worked out in Altman and Kleiman [5, 6].

It is also proved that for a given polynomial $P(T)$ the Hilbert scheme $\mathrm{Hilb}_{\mathbb{P}^N}^P$ is connected; for a simple proof of this theorem of Hartshorne, see Cartier's Bourbaki seminar [21]. Thus the Hilbert polynomial is a complete set of discrete invariants of projective schemes.

Applying Theorem F, we now show how one can find the tangent space to a point of $\mathrm{Hilb}_{\mathbb{P}^N}^P$.

Theorem 6.8 *Let $X \subset \mathbb{P}^N$ be a closed subscheme. The tangent space to the Hilbert scheme* $\mathrm{Hilb}_{\mathbb{P}^N}^{P_X}$ *at the point corresponding to X is isomorphic to the space $\mathcal{N}_{\mathbb{P}^N/X}(X)$ of sections of the normal sheaf $\mathcal{N}_{\mathbb{P}^N/X}$ (Example* 6.21).

Proof Write $x \in \mathrm{Hilb}_{\mathbb{P}^N}^{P_X}$ for the point corresponding under the universal property of the Hilbert scheme Hilb to the scheme X. The tangent space to Hilb, as for any scheme, equals $\mathcal{M}_x(\mathrm{Spec}\, D, \mathrm{Hilb}_{\mathbb{P}^N}^{P_X})$, where $D = k[\varepsilon]/(\varepsilon^2)$ (by Proposition of Section 3.4, Chapter 5). If we now use the universal property of the Hilbert scheme, this set can be given another interpretation: it equals the set of flat families of closed subschemes $\widetilde{X} \subset \mathbb{P}_D^N$ with base $\mathrm{Spec}\, D$ whose fibre over the closed point of $\mathrm{Spec}\, D$ coincides with X. We now describe this set.

We start with the analogous problem for affine schemes. Let A and B be algebras over k with $B = A/I$, so that $\mathrm{Spec}\, B \subset \mathrm{Spec}\, A$ is a closed subscheme. Write $\widetilde{A} = A \otimes_k D = A \oplus \varepsilon A$. A closed subscheme of $\mathrm{Spec}\, \widetilde{A}$ is of the form $\mathrm{Spec}\, \widetilde{B}$, where $\widetilde{B} = \widetilde{A}/\widetilde{I}$, and $\widetilde{I} \subset \widetilde{A}$ is an ideal such that $(\widetilde{I} + \varepsilon A)/\varepsilon A = I$. Since D has a unique nonzero ideal (ε), flatness over D means the isomorphism $\varepsilon \otimes \widetilde{B} \cong \varepsilon \widetilde{B}$. In other words, this means that for $\widetilde{b} \in \widetilde{B}$, we have $\varepsilon \widetilde{b} = 0 \iff \widetilde{b} = \varepsilon \widetilde{c}$. Or in terms of the ideal $\widetilde{I}$, if $\varepsilon \widetilde{a} \in \widetilde{I}$ for $\widetilde{a} \in \widetilde{A}$ then $\widetilde{a} \equiv \varepsilon \widetilde{x} \bmod I$; then $\widetilde{a} = \varepsilon \widetilde{x} + i$ for some $i \in I$ and $\varepsilon \widetilde{a} = \varepsilon i$. That is, $\widetilde{B}$ flat over D is the condition that

$$\varepsilon A \cap \widetilde{I} = \varepsilon I. \tag{6.51}$$

By assumption, $(\widetilde{I} + \varepsilon A)/\varepsilon A = I$, that is, any element $j \in \widetilde{I}$ is of the form $j = i + \varepsilon a$ with $a \in A$, and conversely, for any $i \in I$ one can find $a \in A$ such that $i + \varepsilon a \in \widetilde{I}$. By (6.51), $\varepsilon I \subset \widetilde{I}$, and hence a is only defined modulo I. But for given $i \in I$, it follows from (6.51) that the residue class modulo I consisting of elements a such that $i + \varepsilon a \in \widetilde{I}$ is uniquely determined. Thus by condition (6.51), that is, by the flatness of $\widetilde{B}$ over D, the ideal $\widetilde{I}$ is determined by a homomorphism $\varphi \colon I \to A/I = B$, and consists of elements $i + \varepsilon a$ such that $a \in \varphi(i)$. We see that the set of closed subschemes of $\mathrm{Spec}\, \widetilde{A}$ flat over $\mathrm{Spec}\, D$ which intersect the closed fibre in the given scheme $\mathrm{Spec}\, B$ is the set $\mathrm{Hom}_A(I, B)$. Since $IB = 0$, any $\varphi \in \mathrm{Hom}_A(I, B)$ has $\varphi(I^2) = 0$, so that $\mathrm{Hom}_A(I, B) = \mathrm{Hom}_A(I/I^2, B)$.

In the case of any scheme $\mathcal{P}$ (for example $\mathbb{P}^N$), closed subschemes of $\mathcal{P} \times \operatorname{Spec} D$ that are flat over D are described in an entirely analogous way. We have to cover $\mathcal{P}$ by affine pieces $U_\alpha = \operatorname{Spec} A_\alpha$. The closed subscheme $X \subset \mathcal{P}$ defines in U_α a subscheme $U_\alpha \cap X = U_\alpha \times_{\mathcal{P}} X = \operatorname{Spec}(A_\alpha/I_\alpha)$. A family $\widetilde{X} \subset \mathcal{P} \times \operatorname{Spec} D$ with closed fibre equal to X determines by what we said above homomorphisms $\varphi_\alpha \in \operatorname{Hom}_{A_\alpha}(I_\alpha/I_\alpha^2, A_\alpha/I_\alpha)$. These homomorphisms must be compatible on $U_\alpha \cap U_\beta$, and hence they define a global homomorphism $\varphi\colon \mathcal{I}_X/\mathcal{I}_X^2 \to \mathcal{O}_X$ of coherent sheaves on X, where $\mathcal{I}_X$ is the sheaf of ideals defining the subscheme $X \subset \mathcal{P}$. Conversely, any homomorphism of coherent sheaves $\varphi\colon \mathcal{I}_X/\mathcal{I}_X^2 \to \mathcal{O}_X$ defines flat subschemes $\widetilde{X}_\alpha \subset U_\alpha \times \operatorname{Spec} D$ that are compatible, that is, a subscheme $\widetilde{X} \subset \mathcal{P} \times \operatorname{Spec} D$.

We see that all families of the type we are interested in are described by homomorphisms $\varphi\colon \mathcal{I}_X/\mathcal{I}_X^2 \to \mathcal{O}_X$ of sheaves of $\mathcal{O}_X$-modules. The homomorphism φ is a section over X of the sheaf $\mathcal{H}om(\mathcal{I}_X/\mathcal{I}_X^2, \mathcal{O}_X)$. Since $\mathcal{H}om(\mathcal{I}_X/\mathcal{I}_X^2, \mathcal{O}_X) = \mathcal{N}_{\mathcal{P}/X}$ (see Example 6.21), the families under consideration are in one-to-one correspondence with elements of the set $\mathcal{N}_{\mathcal{P}/X}(X)$. By what we said at the start of the proof we thus establish a one-to-one correspondence between the set $\mathcal{N}_{\mathcal{P}/X}(X)$ and the tangent space to the Hilbert scheme $\operatorname{Hilb}_{\mathbb{P}^N}^{P_X}$. A routine verification shows that this correspondence is an isomorphism of vector spaces; we need to use the interpretation of the algebraic operations in the tangent space indicated after Proposition of Section 3.4, Chapter 5. The theorem is proved. □

Mumford [62, Lecture 22] gives an example (already known in different terminology to the ancient Italian geometers) of a nonsingular projective surface Y containing a curve C which does not move on Y, but for which the tangent space to the scheme $\operatorname{Hilb}_Y^{P_C}$ at the point ξ corresponding to C is 1-dimensional. That is, the reduced subscheme of $\operatorname{Hilb}_Y^{P_C}$ in a neighbourhood of ξ consists of the single point ξ, but the local ring of this point on $\operatorname{Hilb}_Y^{P_C}$ has nonzero nilpotent elements; in other words, this component of $\operatorname{Hilb}_Y^{P_C}$ is of the form $\operatorname{Spec} A$ where A is a finite dimensional k-algebra with radical $\mathfrak{m}$ and $A/\mathfrak{m} = k$. This result shows that the curve C on Y can be moved infinitesimally to first order, but not moved globally. It again demonstrates vividly that schemes with nilpotent elements appear naturally in entirely classical questions of algebraic geometry.

The Hilbert scheme plays a basic role not only in studying subschemes of $\mathbb{P}^N$, but also in the study of algebraic varieties in the "abstract" setting, that is, up to isomorphism. The reason, of course, is that one problem can be reduced to the other. Thus we saw in Section 7.1, Chapter 3 that for a nonsingular projective curve X of genus $g > 1$ the map φ_{3K} corresponding to the divisor class $3K$ is an isomorphic embedding $X \hookrightarrow \mathbb{P}^{5g-6}$. The images of curves of genus g under this embedding are curves of degree $6g - 6$, and their Hilbert polynomial is easily seen to be $P(T) = (6g-6)T - g + 1$. They are thus parametrised by points of the scheme $\operatorname{Hilb}_{\mathbb{P}^{5g-6}}^{P}$: more precisely, by points of the locally closed subset H_g corresponding to nonsingular curves for which the hyperplane section is in the class $3K$. Points $x, y \in H_g$ correspond to isomorphic curves if and only if the curves in $\mathbb{P}^{5g-6}$

parametrised by x and y are taken into one another by a projective transformation. Thus H_g has an action of the group G of projective transformations of $\mathbb{P}^{5g-6}$, and all nonsingular projective curves of genus g (up to isomorphism) are parametrised by the points of the quotient space H_g/G. A treatment of this theory is contained in Mumford and Fogarty [64].

4.5 Exercises to Section 4

1 Prove that for a closed subscheme $X \subset \mathbb{P}^N_k$ the power series $\sum_{r\geq 0} a_r(X)T^r$ represents a rational function.

2 Find the numbers $a_r(X)$ and the Hilbert polynomial $P_X(T)$ for a projective curve $X \subset \mathbb{P}^2$ of degree d. From what value of r is it true that $a_r(X) = P_X(r)$?

3 Find the Hilbert polynomial of a hypersurface of degree d in $\mathbb{P}^N$.

4 Find and prove a relation analogous to (6.50) in the case that X' is the intersection of X with a hypersurface of degree d transversal to X.

5 Find the Hilbert polynomial for the variety that is the intersection of two nonsingular transversal hypersurfaces of degree d_1 and d_2 in $\mathbb{P}^N$.

6 Is the ring $B = k[T]$ flat over its subring consisting of polynomials $F(T)$ such that $F'(T) = 0$?

7 Prove that a localisation A_S of any ring A is flat over A.

8 Prove that if $X \subset \mathbb{P}^N$ is a closed variety then the cone Z over it (introduced in proof of Theorem 6.7) is contained in $\mathbb{A}^{N+1}$.

9 Prove that if $a, b \in k$ with $4a^3 + 27b^2 \neq 0$ and $c(t) \in k[t]$ then the family of elliptic curves $y^2 = x^3 + ac(t)^2x + bc(t)^3$ has all the fibres over t with $c(t) \neq 0$ isomorphic. Prove that if $c(t)$ is not a perfect square in $k[t]$ then the family is not isomorphic to a direct product over any open set $U \subset \mathbb{A}^1$. Deduce from this that for elliptic curves there does not exist a universal family.

10 Find the Hilbert polynomial for the two curves of degree 2 in $\mathbb{P}^3$: a plane irreducible conic and a pair of skew lines.

11 Let $\varphi\colon X \to \mathbb{A}^1 = \operatorname{Spec} k[t]$ be a family of curves of degree 2 in $\mathbb{P}^3$ whose fibres over $t \neq 0$ are pairs of skew lines, and over $t = 0$ a pair of intersecting lines. Describe the scheme $\varphi^{-1}(0)$.

12 Prove the converse of Theorem 6.8: if $X \to \operatorname{Spec} A$ is a projective scheme over a 1-dimensional regular local ring A, and the fibres of X over the closed and generic points of $\operatorname{Spec} A$ have the same Hilbert polynomial, then X is flat over $\operatorname{Spec} A$.

Book 3: Complex Algebraic Varieties and Complex Manifolds

Chapter 7
The Topology of Algebraic Varieties

1 The Complex Topology

1.1 Definitions

We saw in Section 2.3, Chapter 2, that the set of complex points of an algebraic variety X defined over the field $\mathbb{C}$ of complex numbers is a topological space. In Section 2.3, Chapter 2, this was proved for quasiprojective varieties, at the time the only varieties known to us. But the same arguments are valid also for arbitrary varieties. We give here a general definition; the topology of X that comes from its structure of a scheme is called its *Zariski topology*.

We first introduce some notation. For a variety X defined over $\mathbb{C}$ we write $X(\mathbb{C})$ for its set of closed points. Consider an open set $U \subset X$ in the Zariski topology, a finite number of regular functions $f_1, \dots, f_m$ on U and a number $\varepsilon > 0$. Write $V(U; f_1, \dots, f_m; \varepsilon)$ for the set of points

$$V(U; f_1, \dots, f_m; \varepsilon) = \{x \in U(\mathbb{C}) \mid |f_i(x)| < \varepsilon \text{ for } i = 1, \dots, m\}.$$

We make $X(\mathbb{C})$ into a topological space by taking the $V(U; f_1, \dots, f_m; \varepsilon)$ as a basis for the open sets. The topology defined in this way is called the *complex topology*. We compare it with the Zariski topology considered earlier. If $Y \subset X$ is closed in the Zariski topology then $Y(\mathbb{C}) \subset X(\mathbb{C})$. It follows from the definition that $Y(\mathbb{C})$ is also closed in the complex topology, and that the complex topology of $Y(\mathbb{C})$ is the same as its topology as a subset of $X(\mathbb{C})$. However, not every closed set in the complex topology of $X(\mathbb{C})$ is of the form $Y(\mathbb{C})$. An example is the set of points $x \in \mathbb{A}^1(\mathbb{C})$ with $|t(x)| \le 1$, where t is a coordinate on $\mathbb{A}^1$. A morphism $f: X \to Y$ obviously defines a continuous map $f: X(\mathbb{C}) \to Y(\mathbb{C})$.

In certain respects the complex topology is simpler than the Zariski topology. As a very simple example, we show that $(X_1 \times X_2)(\mathbb{C})$ in its complex topology is the product of $X_1(\mathbb{C})$ and $X_2(\mathbb{C})$. Indeed, it is clear that

$$V(U_1; f_1, \dots, f_m; \varepsilon) \times V(U_2; g_1, \dots, g_m; \varepsilon) = V\big(U_1 \times U_2; p_1^*(f_i), p_2^*(g_i); \varepsilon\big),$$

I.R. Shafarevich, *Basic Algebraic Geometry 2*, DOI 10.1007/978-3-642-38010-5_3,

where $U_1 \subset X_1$ and $U_2 \subset X_2$ are Zariski open sets and p_1, p_2 the projections of $X_1 \times X_2$ to X_1 and X_2. Therefore a product of open sets of $X_1(\mathbb{C})$ and $X_2(\mathbb{C})$ is open in $(X_1 \times X_2)(\mathbb{C})$. To prove that they form a basis for the open sets, it is enough to do this for affine open sets of X_1 and X_2. Embedding these in affine spaces, we reduce the verification to $X_1 = \mathbb{A}^{n_1}_{\mathbb{C}}$ and $X_2 = \mathbb{A}^{n_2}_{\mathbb{C}}$, where it is obvious.

The space $X(\mathbb{C})$ in the complex topology is Hausdorff. Indeed, by definition of a variety, the diagonal Δ is closed in $X \times X$ in the Zariski topology, and hence $\Delta(\mathbb{C})$ is closed in $X(\mathbb{C}) \times X(\mathbb{C})$ in the complex topology. As we have just seen, $(X \times X)(\mathbb{C}) = X(\mathbb{C}) \times X(\mathbb{C})$, and $\Delta(\mathbb{C})$ is the diagonal of this space, the set of points of the form (x, x) with $x \in X(\mathbb{C})$. Now a topological space is Hausdorff if and only if its diagonal is closed (see any book on point set topology, for example, Bourbaki [16, I.8.1]).

The topological space $\mathbb{P}^n(\mathbb{C})$ is compact, and hence so are all its closed subspaces. This applies in particular to the space $X(\mathbb{C})$ where X is a projective variety. If X is a complete variety then using Chow's lemma from Section 2.1, Chapter 6, we construct a morphism $f\colon X' \to X$, where X' is a projective variety. This morphism is birational so that $f(X')$ is dense in X, and since X' is projective, $f(X') = X$. In particular $f(X'(\mathbb{C})) = X(\mathbb{C})$. Since f is a continuous map and $X'(\mathbb{C})$ is compact, it follows that $X(\mathbb{C})$ is compact. It can be shown that this property characterises complete varieties over $\mathbb{C}$. That is, if $X(\mathbb{C})$ is compact then X is complete (see Exercises 1–2 of Section 2.6). Obviously, for an arbitrary variety X, the space $X(\mathbb{C})$ is locally compact.

The arguments of Section 2.3, Chapter 2, can now be applied to study the complex topology of any nonsingular variety (not necessarily quasiprojective) defined over the complex field. They show that in this case $X(\mathbb{C})$ in its complex topology is a topological manifold of dimension $2 \dim X$.

The above definitions can be generalised as follows. Consider an arbitrary field k and write $\overline{k}$ for its algebraic closure. Let X be a scheme over k such that $X \times_k \operatorname{Spec} \overline{k}$ is an algebraic variety over $\overline{k}$. A scheme of this type is called an *algebraic variety defined over k*. An example is an affine or projective variety over $\overline{k}$ whose ideal has a basis consisting of polynomials with coefficients in k. If X is an algebraic variety over k then $X(k)$ denotes the set of closed points $x \in X$ such that $k(x) = k$.

If k is the real number field $\mathbb{R}$, or the p-adic number field $\mathbb{Q}_p$, then $X(k)$ can be given a topology in exactly the same way as for $k = \mathbb{C}$. If $k = \mathbb{R}$ and X is a nonsingular variety then $X(\mathbb{R})$ is a topological manifold of dimension $\dim X$. In the rest of the book, we consider only the topological space $X(\mathbb{C})$, except for Section 4 in which we study the space $X(\mathbb{R})$ in the case when X is a curve.

We always consider the space $X(\mathbb{C})$ with its complex topology. In the remainder of this section we consider $X(\mathbb{C})$ when X is nonsingular. We use a little more topological apparatus here than elsewhere in the book: we assume known the basic theory of homology and cohomology, Poincaré duality for manifolds, the theory of differential forms and their relation with cohomology (the Stokes–Poincaré theorem and de Rham's theorem). The reader can find a concise summary of the results we need in Mumford [60, Section 5C], and a more detailed exposition in de Rham [25].

1.2 Algebraic Varieties as Differentiable Manifolds; Orientation

Let X be a nonsingular n-dimensional variety over the field of complex numbers $\mathbb{C}$ and $x \in X$ a point (we consider only closed points from now on), and let $t_1, \dots, t_n$ be a system of local parameters at x. As we proved in Section 2.3, Chapter 2, there exists a neighbourhood U of x in $X(\mathbb{C})$ that is mapped homeomorphically to a domain of $\mathbb{C}^n$ by $t_1, \dots, t_n$. Because of this, any function on U can be considered as a function of the complex variables $t_1, \dots, t_n$ or of the real variables $u_1, \dots, u_n, v_1, \dots, v_n$, where $t_j = u_j + iv_j$.

Definition A real valued function on U is *smooth* or of *class* C^∞ if it is differentiable infinitely often as a function of $u_1, \dots, u_n, v_1, \dots, v_n$.

As we proved in Section 2.3, Chapter 2, if $t'_1, \dots, t'_n$ is another system of local parameters then $t'_1, \dots, t'_n$ are analytic functions of $t_1, \dots, t_n$. Hence the notion of smooth function is well defined, that is, does not depend on the choice of the local parameters. It is easy to check that our definition gives $X(\mathbb{C})$ the structure of a differentiable manifold (see de Rham [25, Chapter I]).

There is a natural relation between the properties of X as an algebraic variety and those of $X(\mathbb{C})$ as a differentiable manifold. Regular differential forms $\omega \in \Omega^p[X]$ are complex valued differential forms on $X(\mathbb{C})$. If $E \to X$ is a vector bundle then $E(\mathbb{C}) \to X(\mathbb{C})$ is a topological vector bundle. For this, we should forget that the fibres E_x are complex vector spaces and view them as real vector spaces of twice the dimension. Under this correspondence the tangent bundle $\Theta \to X$ corresponds to the tangent bundle of the differentiable manifold $X(\mathbb{C})$.

We now consider the question of the orientability of $X(\mathbb{C})$. We first recall the definitions. An orientation of a 1-dimensional vector space $\mathbb{R}$ is a choice of one of the two connected components of $\mathbb{R} \setminus 0$; an *orientation* of an n-dimensional vector space F is an orientation of the 1-dimensional space $\bigwedge^n F$. An orientation of a (locally trivial) vector bundle $f\colon E \to X$ is a collection of orientations of the fibres E_x such that each point has a neighbourhood U and an isomorphism $f^{-1}(U) \xrightarrow{\sim} U \times F$ taking the orientations ω_x of all the fibres E_x into the same orientation of F. An orientation of a differentiable manifold is an orientation of its tangent bundle.

Proposition *If X is a nonsingular variety over $\mathbb{C}$ then the differentiable manifold $X(\mathbb{C})$ has a canonical orientation.*

Proof The reason is very simple. It comes from the fact that if we consider an n-dimensional vector space F over $\mathbb{C}$ as a $2n$-dimensional vector space over $\mathbb{R}$ then it has a certain canonical orientation. To define it, choose a basis $e_1, \dots, e_n$ of F over $\mathbb{C}$. Then the vectors $\{u_1, \dots, u_{2n}\} = \{e_1, ie_1, \dots, e_n, ie_n\}$ form a basis of F over $\mathbb{R}$ and determine an orientation $u_1 \wedge \dots \wedge u_{2n}$ of this space. We check that this orientation is independent of the choice of the basis $e_1, \dots, e_n$. Let $f_1, \dots, f_n$ be another basis of F over $\mathbb{C}$. Write φ for the $\mathbb{C}$-linear map taking $e_1, \dots, e_n$ into

$f_1, \dots, f_n$, and $\widetilde{\varphi}$ for the same map viewed as an $\mathbb{R}$-linear map $F \to F$. We need to prove that $\det \widetilde{\varphi} > 0$; this will follow from the identity

$$\det \widetilde{\varphi} = |\det \varphi|^2. \tag{7.1}$$

For the proof, consider the space $\widetilde{F} = F \otimes_{\mathbb{R}} \mathbb{C}$ and the transformation I of $\widetilde{F}$ defined by

$$I(f \otimes \alpha) = if \otimes \alpha \quad \text{for } f \in F \text{ and } \alpha \in \mathbb{C}. \tag{7.2}$$

Then $\widetilde{F} = F_1 \oplus F_2$, where F_1 and F_2 are the eigenspaces of I corresponding to the eigenvalues i and $-i$. By analogy with (7.2), we extend $\widetilde{\varphi}$ to $\widetilde{F}$. Then of course the determinant does not change. It is easy to see that F_1 and F_2 are invariant under $\widetilde{\varphi}$, and that the transformation $\widetilde{\varphi}$ on F_1 has the same matrix as that of φ on F, and F_2 its complex conjugate. Equation (7.1) follows from this.

Now suppose that $f : \Theta \to X(\mathbb{C})$ is the tangent bundle and ω_x the canonical orientation on Θ_x for $x \in X$. We prove that this defines an orientation on X. If $U \subset X$ is such that

$$\psi : f^{-1}(U) \cong U \times F \tag{7.3}$$

is an isomorphism of algebraic vector bundles, then it is true a fortiori for the corresponding differentiable manifolds. But in (7.3)

$$\psi_x : \Theta_x \to F$$

is an isomorphism of complex vector spaces. Hence it takes the canonical orientation ω_x on Θ into the canonical orientation ω of F. The proposition is proved. □

The orientation just constructed is called the *canonical orientation* of $X(\mathbb{C})$.

We thus obtain a first restriction, showing that not every even dimensional manifold can be of the form $X(\mathbb{C})$ for X a nonsingular algebraic variety. For example, the real projective plane can not be represented in this way.

1.3 Homology of Nonsingular Projective Varieties

The orientability of a differentiable manifold can be expressed in terms of its homology. We recall this relation (see for example Husemoller [41, Section 4, Chapter 17]). An orientation ω of an n-dimensional vector space E over $\mathbb{R}$ determines an element of the relative homology group $\omega \in H_n(E, E \setminus 0, \mathbb{Z})$. If U is a coordinate chart around a point x of M and $\varphi : U \to E$ a diffeomorphism of U with a neighbourhood of 0 in E, then we have excision isomorphisms

$$H_n(U, U \setminus x, \mathbb{Z}) \to H_n(M, M \setminus x, \mathbb{Z}),$$

$$H_n\bigl(\varphi(U), \varphi(U) \setminus 0, \mathbb{Z}\bigr) \to H_n(E, E \setminus 0, \mathbb{Z}),$$

and an isomorphism

$$\varphi_*\colon H_n(U, U \setminus x, \mathbb{Z}) \to H_n\bigl(\varphi(U), \varphi(U) \setminus 0, \mathbb{Z}\bigr).$$

Finally, $\mathrm{d}_x\varphi$ is an isomorphism of the tangent spaces

$$\mathrm{d}_x\varphi\colon \Theta_x \xrightarrow{\sim} E.$$

Using this system of isomorphisms we can associate with the orientation ω_x of the tangent space Θ_x a homology class, for which we preserve the same notation

$$\omega_x \in H_n(M, M \setminus x, \mathbb{Z}).$$

The orientation of the compact manifold M then defines a class $\omega_M \in H_n(M, \mathbb{Z})$, which is uniquely characterised by the fact that it maps to the class ω_x under the homomorphism

$$H_n(M, \mathbb{Z}) \to H_n(M, M \setminus x, \mathbb{Z})$$

corresponding to any point $x \in M$. This class is a generator of $H_n(M, \mathbb{Z})$. We call ω_M the *orientation class* of M. In what follows, ω_M is denoted by $[M]$; we sometimes speak of the manifold M itself as an n-dimensional homology class, and then the class $[M] = \omega_M$ is intended.

Proposition of Section 1.2 shows that if X is a nonsingular complete algebraic variety then $X(\mathbb{C})$ has a canonically defined orientation class $[X] = \omega_{X(\mathbb{C})} \in H_{2n}(X(\mathbb{C}), \mathbb{Z})$, where $n = \dim X$. In what follows, we sometimes speak of the manifold X itself as a $2n$-dimensional homology class, and then the class $[X] = \omega_{X(\mathbb{C})}$ is intended.

The preceding arguments construct a class $\omega_{X(\mathbb{C})} \in H_{2n}(X(\mathbb{C}), \mathbb{Z})$, which is obviously nonzero, because for any point $x \in X$, it defines a nonzero class in $H_{2n}(X(\mathbb{C}), X(\mathbb{C}) \setminus x, \mathbb{Z})$. For the same reason this class is of infinite order. Thus for a nonsingular complete variety X,

$$H_{2n}\bigl(X(\mathbb{C}), \mathbb{C}\bigr) \neq 0.$$

This is a particular case of the following more general result.

Proposition *For an n-dimensional nonsingular projective variety X, we have*

$$H_{2k}\bigl(X(\mathbb{C}), \mathbb{C}\bigr) \neq 0 \quad \textit{for } k \leq n.$$

Proof We exhibit a $2k$-dimensional cycle on $X(\mathbb{C})$ not homologous to 0. For this, consider a nonsingular k-dimensional subvariety $Y \subset X$, for example, the intersection of $X \subset \mathbb{P}^N$ with a linear subspace of $\mathbb{P}^N$. Write j for the inclusion map $Y \hookrightarrow X$, and also for the inclusion $Y(\mathbb{C}) \hookrightarrow X(\mathbb{C})$. The homology class we consider is $j_*\omega_Y$. When there is no fear of confusion, we write simply $[Y]$ for this. We prove that it is not homologous to 0. More intuitively, if somewhat less precisely, we can express this by saying that nonsingular subvarieties are not homologous to 0 in their ambient space.

Recall that if M is a compact n-dimensional oriented manifold then the multiplication $H^p(M,\mathbb{C}) \otimes H^{n-p}(M,\mathbb{C}) \to H^n(M,\mathbb{C})$ defines a duality between $H^p(M,\mathbb{C})$ and $H^{n-p}(M,\mathbb{C})$. Since $H^p(M,\mathbb{C})$ is dual to $H_p(M,\mathbb{C})$, the spaces $H_p(M,\mathbb{C})$ and $H_{n-p}(M,\mathbb{C})$ are also dual to one another. The corresponding scalar product $H_p(M,\mathbb{C}) \otimes H_{n-p}(M,\mathbb{C}) \to \mathbb{C}$ is called the *intersection number* or *Kronecker pairing* of two cycles. Suppose that V and W are two nonsingular oriented submanifolds of dimension p and $n-p$ in M intersecting transversally. This means that their intersection consists of a finite number of points, and that at each point $x \in V \cap W$ we have $\Theta_{M,x} = \Theta_{V,x} \oplus \Theta_{W,x}$. In this case we can consider the embeddings $j_V \colon V \to M$ and $j_W \colon W \to M$ and give the simple formula

$$[V]\cdot[W] = j_{V*}(\omega_V)\cdot j_{W*}(\omega_W) = \sum_{x\in V\cap W} c(V,W,x), \tag{7.4}$$

for the intersection number $j_{V*}(\omega_V)\cdot j_{W*}(\omega_W)$, where the $c(V,W,x)$ are equal to $+1$ or -1 according as to whether the natural orientation of $\Theta_{V,x} \oplus \Theta_{W,x}$ is equal or opposite to that of $\Theta_{M,x}$. (For these results, see Pham [65, II.7].)

Now let $M = X(\mathbb{C})$, $V = Y(\mathbb{C})$ and $W = Z(\mathbb{C})$, where X is a nonsingular complete algebraic variety and $Y, Z \subset X$ are nonsingular complete subvarieties that intersect transversally. Then all the summands c in (7.4) are $+1$. Indeed, in this case $\Theta_{X,x}$, $\Theta_{Y,x}$ and $\Theta_{Z,x}$ are complex vector spaces. Passing from a complex basis of $\Theta_{Y,x} \oplus \Theta_{Z,x}$ to a complex basis of $\Theta_{X,x}$ is realised by a complex linear transformation. The corresponding real transformation taking a real basis of $\Theta_{Y,x} \oplus \Theta_{Z,x}$ to a real basis of $\Theta_{X,x}$ has positive determinant, as we saw in Section 1.2. Hence

$$c\bigl(Y(\mathbb{C}), Z(\mathbb{C}), x\bigr) = +1.$$

If X is a nonsingular projective variety, $Y \subset X$ and Y is nonsingular then there exists a nonsingular variety Z of complementary dimension that intersects it transversally in a nonempty set of points. We can take Z to be the intersection of X with a suitable linear subspace. In this case (7.4) shows that

$$[Y]\cdot[Z] = \deg Y, \tag{7.5}$$

where $\deg Y$ is the degree of Y in the ambient projective space (Example 4.6 of Section 1.4, Chapter 4). It follows of course from this that $[Y] \neq 0$. The proposition is proved. □

We will give a generalisation of this result in Proposition of Section 4.4, Chapter 8. The proof there is based on somewhat different principles, and even for the case of projective varieties gives another proof of the proposition, not using properties of intersection numbers.

1.4 Exercises to Section 1

1 Prove that (7.5) remains valid in the case that Y is a curve, but possibly singular. [Hint: Consider the normalisation morphism.]

2 Prove that if X is a nonsingular projective curve, $\omega \in \Omega^1[X]$ and $\int_\sigma \omega = 0$ for all $\sigma \in H_1(X, \mathbb{Z})$ then $\omega = 0$. [Hint: Consider the function $\varphi(x) = \int_{x_0}^x \omega$, where x_0 is a fixed point and x any point of X, and prove that the integral does not depend on the path of integration, that φ is continuous on X and is analytic as a function of a local parameter at x. Prove that the existence of a maximum of $|\varphi(x)|$ leads to a contradiction to the maximum modulus principle for an analytic function.]

3 Prove that the assertion of Exercise 2 holds for a nonsingular projective variety of any dimension.

4 Prove that if G is an algebraic group acting on a nonsingular projective variety X and the space $G(\mathbb{C})$ is connected then $g^*(\omega) = \omega$ for any $g \in G$ and $\omega \in \Omega^1[X]$. [Hint: Prove that the cycles σ and $g_*\sigma$ are homologous.]

2 Connectedness

The purpose of this section is to prove that $X(\mathbb{C})$ is connected for an irreducible algebraic variety X over the field of complex numbers $\mathbb{C}$. Connectedness of $X(\mathbb{C})$ is a basic topological question, and we will see that it behaves simply under the coarse algebraic geometric operations such as passing to a Zariski open set, passing to a generically finite cover, treating X as a fibred variety, and so on.

We give two alternative proofs of the main theorem. The first is by induction on dimension, and is ultimately based on the Riemann–Roch inequality for curves. The other starts by reducing the statement to the case of X affine, and from then on is based on Noether normalisation: by Theorem 1.18 of Section 5.4, Chapter 1, there exists a finite map $f\colon X \to \mathbb{A}^n$ to affine space. Using the terminology of Chapter 5, we speak of finite morphisms from now on.

Section 2.1 deduces some simple topological properties of algebraic varieties, and Sections 2.2 and 2.3 give the two proofs of the main theorem on the connectedness of $X(\mathbb{C})$. The second proof uses some simple properties of analytic functions of several complex variables, which are proved in Section 2.4.

2.1 Preliminary Lemmas

Lemma 7.1 *If X is an irreducible algebraic variety and $Y \subsetneqq X$ a proper subvariety then the set $X(\mathbb{C}) \setminus Y(\mathbb{C})$ is everywhere dense in $X(\mathbb{C})$.*

Proof Consider first the case that X is an algebraic curve. Then Y consists of a finite set of closed points. Let $\nu\colon X^\nu \to X$ be the normalisation morphism, and

$Y' = \nu^{-1}(Y)$. Since X^ν is nonsingular, every point $y' \in Y'$ has a neighbourhood U in X^ν homeomorphic to the disc $|z| < 1$ in the plane of one complex variable z. Obviously $U \setminus y'$ is everywhere dense in U, and hence also $X^\nu(\mathbb{C}) \setminus Y'$ is everywhere dense in $X^\nu(\mathbb{C})$. Since ν is surjective, it follows that $X(\mathbb{C}) \setminus Y$ is everywhere dense in $X(\mathbb{C})$.

The general case reduces to the 1-dimensional case just considered by a simple induction on $n = \dim X$. Suppose that $n > 1$. For any point $y \in Y(\mathbb{C})$, there exists an irreducible codimension 1 subvariety $X' \subset X$ containing y and not containing any irreducible component of Y passing through y. Indeed, in an affine neighbourhood U of y, we choose one point $y_i \neq y$ on each irreducible component of Y passing through y, and consider a section of U by a hyperplane L of the ambient affine space such that $y \in L$ but $y_i \notin L$ for all of the chosen points y_i. We can take X' to be the closure in X of any irreducible component of the intersection $L \cap X$ passing through y. Set $Y' = X' \cap Y$. By induction, $X'(\mathbb{C}) \setminus Y'(\mathbb{C})$ is everywhere dense in $X'(\mathbb{C})$. In particular, y is in the closure of $X'(\mathbb{C}) \setminus Y'(\mathbb{C})$. Hence it is a fortiori in the closure of $X(\mathbb{C}) \setminus Y(\mathbb{C})$. Since we can take y to be any point of $Y(\mathbb{C})$, this proves the lemma. □

Corollary *If X is an irreducible algebraic variety and $Y \subsetneqq X$ an algebraic subvariety, and the open subset $X(\mathbb{C}) \setminus Y(\mathbb{C})$ of $X(\mathbb{C})$ is connected then so is $X(\mathbb{C})$.*

Proof Indeed, if $X(\mathbb{C}) = M_1 \sqcup M_2$ is a decomposition of $X(\mathbb{C})$ as a union of two disjoint closed sets, then $X(\mathbb{C}) \setminus Y(\mathbb{C})$ breaks up into its intersection with M_1 and with M_2. Since $X(\mathbb{C}) \setminus Y(\mathbb{C})$ is connected, it must be equal to one of these intersections, and hence is contained in M_1 or M_2. But then also its closure is contained in M_1 or M_2. By Lemma 7.1 this closure is the whole of $X(\mathbb{C})$, which means that one of M_1 or M_2 is the empty set. □

Lemma 7.2 *If $V \subset \mathbb{A}^n$ is an open subset in the Zariski topology then $V(\mathbb{C})$ is connected.*

Proof Write $\mathbb{A}^n \setminus V = Y$, and let $x_1, x_2 \in V(\mathbb{C})$. Pass a line L through x_1 and x_2. Then L is not contained in any irreducible component of Y, so that $L \cap Y$ is a finite set $\{y_1, \dots, y_m\}$. Then $L(\mathbb{C})$ is homeomorphic to $\mathbb{C}$ and $L(\mathbb{C}) \cap V(\mathbb{C})$ to $\mathbb{C} \setminus \{y_1, \dots, y_m\}$. It follows that $L(\mathbb{C}) \cap V(\mathbb{C})$ is connected, and hence x_1 and x_2 are contained in the same connected component of $V(\mathbb{C})$. Since x_1 and x_2 were arbitrary points, $V(\mathbb{C})$ is connected. □

2.2 *The First Proof of the Main Theorem*

The first proof that $X(\mathbb{C})$ is connected is by induction on $\dim X$.

Lemma 7.3 *$X(\mathbb{C})$ is connected if X is an irreducible curve.*

Proof Because the connectedness of $X(\mathbb{C})$ is not affected by adding or deleting a finite number of points, we can restrict ourselves at once to the case that X is a nonsingular projective curve.

Suppose that $X(\mathbb{C}) = M_1 \sqcup M_2$ is a decomposition of $X(\mathbb{C})$ as a union of two disjoint closed sets. Let $x_0 \in X(\mathbb{C})$ be any point, with $x_0 \in M_1$, say. By the Riemann inequality proved as Corollary of Section 7.2, Chapter 3, $\ell(rx_0) > 1$ for some large enough r; in other words, there exists a nonconstant function $f \in \mathcal{L}(rx_0)$. Then since f has $x_0 \in X$ as its only pole, it restricts to a continuous function $M_2 \to \mathbb{C}$; now M_2 is a compact set, so that the absolute value $|f|$ achieves a maximum at some point $m \in M_2$. Let U be a neighbourhood of $m \in X$ with local parameter t, so that f is an analytic function of t. By the maximum modulus principle, the modulus of a nonconstant analytic function does not have a maximum at an interior point of its domain, and hence $f(t)$ is constant on U, say $f(U) = \alpha$. Then $f - \alpha \in k(X)$ has infinitely many zeros in X, which is a contradiction. The lemma is proved. □

Lemma 7.4 *For an irreducible nonsingular n-dimensional variety X, there exists an open set U, an irreducible $(n-1)$-dimensional variety V, and a surjective morphism $f\colon U \to V$ with the following properties*:

(1) *Every fibre of f is* 1*-dimensional.*
(2) *Every fibre of f is irreducible.*
(3) *For every point $x \in U$, the differential* $\mathrm{d}_x f\colon \Theta_{U,x} \to \Theta_{V,f(x)}$ *is surjective.*

Proof We will choose V as a model of a subfield $K \subset \mathbb{C}(X)$ of transcendence degree $n-1$ to which we can apply the Bertini theorems, Theorem 2.26 of Section 6.1, Chapter 2 and Theorem 2.27 of Section 6.2, Chapter 2. The only difficulty arises in connection with the first of these: to apply it, we need to know that K is algebraically closed in $\mathbb{C}(X)$.

To achieve this, we choose a transcendence basis $t_1, \dots, t_n$ of $\mathbb{C}(X)$ and take K to be the algebraic closure of $\mathbb{C}(t_1, \dots, t_{n-1})$ in K, that is, the subfield of elements of $\mathbb{C}(X)$ algebraic over $\mathbb{C}(t_1, \dots, t_{n-1})$. Then $\mathbb{C}(t_1, \dots, t_{n-1}) \subset K$ is a finite extension. This follows from the fact that any element $\xi \in K$ has bounded degree over $\mathbb{C}(t_1, \dots, t_{n-1})$; indeed, its degree over $\mathbb{C}(t_1, \dots, t_{n-1})$ is at most its degree over $\mathbb{C}(t_1, \dots, t_n)$, which is bounded by $[K : \mathbb{C}(t_1, \dots, t_n)]$.

Now choose a variety V such that $K = \mathbb{C}(V)$. The inclusion $\mathbb{C}(V) = K \subset \mathbb{C}(X)$ defines a rational map $f\colon X \to V$ with dense image in V. It remains only to choose the open set $U \subset X$ such that f is a morphism, and then shrink it down to satisfy the conclusions (1)–(3) of the lemma. This is possible by the theorem on the dimension of fibres, Theorem 1.25 of Section 6.3, Chapter 1, and the Bertini theorems already referred to. The lemma is proved. □

Theorem 7.1 *If X is an irreducible algebraic variety over $\mathbb{C}$, then $X(\mathbb{C})$ is connected.*

Proof By induction on $n = \dim X$. Consider the map $f\colon U \to V$ whose existence was proved in Lemma 7.4. Suppose that $X(\mathbb{C}) = M_1 \sqcup M_2$ is a decomposition into disjoint closed sets. Because f has connected fibres, every fibre is contained entirely in M_1 or M_2. Therefore $f(M_1)$ and $f(M_2)$ are disjoint. Since f is onto, it follows that $V(\mathbb{C}) = f(M_1) \sqcup f(M_2)$. Now the surjectivity of the differential $d_x f: \Theta_{U,x} \to \Theta_{V,f(x)}$ together with the implicit function theorem implies that $f(M_1)$ and $f(M_2)$ are both open, hence both closed. By induction we can assume that $V(\mathbb{C})$ is connected. It follows from this that either M_1 or M_2 is empty. Thus $U(\mathbb{C})$ is connected, so that $X(\mathbb{C})$ is also by Lemma 7.1 and Corollary of Section 2.1. The theorem is proved. □

2.3 The Second Proof

The second proof of Theorem 7.1 is based on the following result, which reduces the problem to a simpler case.

Lemma *For any irreducible variety X, there exists a Zariski open set $U \subset X$ and a finite morphism $f: U \to V$ onto a Zariski open subset of affine space $V \subset \mathbb{A}^n$ such that the following conditions hold*:

(1) *U is isomorphic to a hypersurface $V(F) \subset V \times \mathbb{A}^1$, defined by $F = 0$, where $F(T) \in k[\mathbb{A}^n][T] \subset k[V \times \mathbb{A}^1]$ is a polynomial that is irreducible over $k[\mathbb{A}^n]$ and has leading coefficient* 1, *and $f: U \to V$ is induced by the projection $V \times \mathbb{A}^1 \to V$.*
(2) *The continuous map $f: U(\mathbb{C}) \to V(\mathbb{C})$ is an unramified cover.*

Proof Both assertions follow at once from Theorem 2.30 of Section 6.3, Chapter 2. □

Second Proof of Theorem 7.1 Let U be the set whose existence is guaranteed by the lemma. By Lemma 7.1 and Corollary of Section 2.1, it is enough to prove that $U(\mathbb{C})$ is connected. Suppose that $U(\mathbb{C}) = M_1 \sqcup M_2$ is a decomposition of $U(\mathbb{C})$ as a union of two disjoint closed subsets. The map $f: U(\mathbb{C}) \to V(\mathbb{C})$ constructed in the lemma takes open sets to open sets and closed sets to closed sets. Since M_1 and M_2 are open and closed in $U(\mathbb{C})$, their images $f(M_1)$ and $f(M_2)$ are open and closed in $V(\mathbb{C})$. By Lemma 7.2, $V(\mathbb{C})$ is connected, so that $f(M_1) = f(M_2) = V(\mathbb{C})$.

Obviously the restriction of f to M_1 defines an unramified cover $f_1: M_1 \to V(\mathbb{C})$. It follows easily from the connectedness of $V(\mathbb{C})$ that the number of inverse images in M_1 of points $v \in V(\mathbb{C})$ is constant for all points v. We write r for this number, and call it the *topological degree* of the cover $f_1: M_1 \to V(\mathbb{C})$. Since also $f(M_2) = V(\mathbb{C})$, we have $r < m$, where $m = \deg f$.

For a point $v \in V(\mathbb{C})$, we choose a neighbourhood V_v of v for which $f^{-1}(V_v) = U_1 \cup \cdots \cup U_r$ with $U_i \cap U_j = \emptyset$ for $i \neq j$, and such that the restriction of f to U_i is a homeomorphism $f_i: U_i \to V_v$ for $i = 1, \dots, r$.

Now let $\theta \in \mathbb{C}[U]$ be a function $\theta \in \mathbb{C}[U]$ that is integral over $\mathbb{C}[\mathbb{A}^n]$, and is a primitive element of the field extension $\mathbb{C}(V) \subset \mathbb{C}(U)$. Consider the restrictions

$\theta_1, \dots, \theta_r$ of θ to the sets $U_1, \dots, U_r$, and write $g_1, \dots, g_r$ for the elementary symmetric functions in $\theta_1, \dots, \theta_r$. The idea of the following argument is to prove that there exist polynomials $p_1, \dots, p_r \in \mathbb{C}[\mathbb{A}^n]$ whose restrictions to V_v at any point $v \in f(M_1)$ equal $g_1, \dots, g_r$. It will follow from this that θ satisfies the relation

$$\theta^r - f^*(p_1)\theta^{r-1} + \dots + (-1)^r f^*(p_r) = 0 \tag{7.6}$$

at all points $x \in M_1$.

Then since a similar relation (with another value $r' < m$) holds at points $x \in M_2$, there exist polynomials $P_1, P_2 \in \mathbb{C}[\mathbb{A}^n][T]$ of degrees $<m$ such that $P_i(\theta) = 0$ on M_i for $i = 1, 2$. Therefore $P_1(\theta)P_2(\theta) = 0$ in $\mathbb{C}[U]$, and since $\mathbb{C}[U]$ has no zerodivisors, we get that $\theta \in \mathbb{C}[U]$ satisfies an equation over $\mathbb{C}[\mathbb{A}^n]$ of degree $<m$. This contradicts the fact that by definition $m = [\mathbb{C}(U) : \mathbb{C}(\mathbb{A}^n)]$.

We proceed to carry out this plan, leaving two technical lemmas on analytic functions to Section 2.4. Note first that the functions $(f_i^{-1})^*(\theta)$ are analytic functions on V_v in the coordinates $z_1, \dots, z_n$ of $\mathbb{A}^n(\mathbb{C})$. Indeed, according to the preceding lemma, together with the implicit function theorem, local parameters at a point $u_i = f^{-1}(v)$ can be expressed as analytic functions in $f^*(z_1), \dots, f^*(z_n)$, and in a sufficiently small neighbourhood, θ is an analytic function of the local parameters at v. Thus $g_1, \dots, g_r$ are also analytic functions of $z_1, \dots, z_n$ in a neighbourhood V_v. Therefore each of the functions g_i is an analytic function on the whole set $V(\mathbb{C})$. Recall that $V(\mathbb{C})$ is obtained from the whole of $\mathbb{A}^n(\mathbb{C})$ by deleting the points of an algebraic variety $S \neq \mathbb{A}^n$. Consider the behaviour of the g_i in a neighbourhood of a point $s \in S(\mathbb{C})$. By the choice of θ, it satisfies an equation

$$\theta^m + f^*(a_1)\theta^{m-1} + \dots + f^*(a_m) = 0, \quad \text{with } a_i \in \mathbb{C}\bigl[\mathbb{A}^n\bigr]. \tag{7.7}$$

The values of $(f_i^{-1})^*(\theta)$ are roots of this equation. Hence the g_i are bounded in any compact neighbourhood of s. It follows from this, using Lemma 7.5, that they extend to analytic functions on the whole of $\mathbb{A}^n(\mathbb{C})$.

Let us prove that the analytic functions g_i on $\mathbb{A}^n(\mathbb{C})$ we have constructed are polynomials in the coordinates $z_1, \dots, z_n$. For this, we bound their order of growth as a function of the growth of $\max |z_i|$. For a point $z = (z_1, \dots, z_n) \in \mathbb{A}^n(\mathbb{C})$ we set $|z| = \max |z_i|$. Applying to (7.7) the well-known bound for the modulus of a root of an algebraic equation f in terms of its coefficients a_i, we get that

$$\bigl|\theta(x)\bigr| \le 1 + \max_i \bigl|a_i\bigl(f(x)\bigr)\bigr| \quad \text{for all } x \in M_1.$$

The a_i are polynomials in $z_1, \dots, z_n$ by assumption. If the maximum of their degrees is k then for any $\varepsilon > 0$ there exists a constant C such that

$$\bigl|\theta(x)\bigr| < C|z|^k \quad \text{for } |z| > \varepsilon.$$

It follows that $(f_i^{-1})^*(\theta)$ satisfies the same inequality for any $i = 1, \dots, r$, and hence

$$\bigl|g_i(z)\bigr| \le C|z|^{ik} \quad \text{for } i = 1, \dots, r.$$

We see that $g_i(z)$ are analytic functions on the whole of $\mathbb{A}^n(\mathbb{C}) = \mathbb{C}^n$ having polynomial growth. By Lemma 7.6, it follows that they are polynomials in $z_1, \dots, z_n$.

This proves (7.6), and thus completes the proof of the theorem. □

2.4 Analytic Lemmas

We prove here the two lemmas used in Section 2.3 on analytic functions of complex variables.

Lemma 7.5 *Let $S \subsetneqq \mathbb{A}^n$ be an algebraic subvariety and g an analytic function on the complement $\mathbb{A}^n(\mathbb{C}) \setminus S(\mathbb{C})$, which is bounded in a neighbourhood of any point $s \in S(\mathbb{C})$. Then g can be extended to a analytic function on the whole of $\mathbb{A}^n(\mathbb{C}) = \mathbb{C}^n$, and the extension is unique.*

Proof The uniqueness of the extension follows at once from the uniqueness of analytic continuation. It is obviously enough to find a neighbourhood U of any $s \in S(\mathbb{C})$ such that g extends as an analytic function from $U \setminus (U \cap S(\mathbb{C}))$ to U. Then by uniqueness of the extension we get a global extension.

To prove that the extension exists, note that we can replace S by a bigger algebraic subvariety, and can therefore assume that S is defined by one equation $f(z_1, \dots, z_n) = 0$. Making a linear coordinate change $\widetilde{z}_n = z_n$ and $\widetilde{z}_i = z_i + c_i z_n$ for $i = 1, \dots, n-1$ for suitable choice of $c_1, \dots, c_{n-1}$, we can arrange that f has leading coefficient 1 as a polynomial in z_n, that is,

$$f(z_1, \dots, z_n) = z_n^k + h_1(z')z_n^{k-1} + \dots + h_k(z'),$$

where $z' = (z_1, \dots, z_{n-1})$. Set $|z'| = \max_{i=1}^{n-1} |z_i|$.

Suppose that s is the origin. Then the restriction of f to the z_n-axis $z' = 0$ factorises as

$$f(0, \dots, 0, z_n) = z_n^m (z_n - \lambda_1) \cdots (z_n - \lambda_{k-m}).$$

By the theorem on the continuity of the roots of an algebraic equation, the roots of $f(z'; z_n) = 0$ tend either to 0 or to $\lambda_1, \dots, \lambda_{k-m}$ as $z' \to 0$. Hence there exists some real number $r > 0$ and $\varepsilon > 0$ such that for $|z'| < \varepsilon$ the equation $f(z'; z_n) = 0$ does not have any roots with $|z_n| = r$.

We now set

$$G(z_1, \dots, z_n) = \frac{1}{2\pi i} \oint_{|w|=r} \frac{g(z'; w)}{w - z_n} dw,$$

and prove that G is an analytic function for $|z'| < \varepsilon$, $|z_n| < r$, and is a continuation of g to this domain.

That G is analytic is verified by a direct integration. By assumption g is analytic at any point $(\alpha_1, \dots, \alpha_{n-1}, w)$ with $|\alpha_i| < \varepsilon$ and $|w| = r$. Therefore for any w on the contour $|w| = r$, the function $g(z_1, \dots, z_{n-1}; w)/(w - z_n)$ is analytic in a neighbourhood of the point $z_1 = \alpha_1, \dots, z_{n-1} = \alpha_{n-1}$, $z_n = \beta$ with $|\beta| < r$. Writing out its Taylor series expansion

$$\frac{g(z'; w)}{w - z_n} = \sum c_{k_1 \dots k_n}(w)(z_1 - \alpha_1)^{k_1} \cdots (z_{n-1} - \alpha_{n-1})^{k_{n-1}}(z_n - \beta)^{k_n} \qquad (7.8)$$

about this point and integrating around the contour $|w| = r$ gives the Taylor series expansion for G.

We now prove that $G = g$ wherever g is defined. For this, we set $z_i = \alpha_i$ with $|\alpha_i| < \varepsilon$ for $i = 1, \dots, n-1$, and consider the functions of one complex variable $g(\alpha; z_n)$ and $G(\alpha; z_n)$. The preceding argument shows that $G(\alpha; z_n)$ is analytic for $|z_n| \le r' < r$, and by assumption $g(\alpha; z_n)$ is analytic at all points z_n with $|z_n| \le r'$ except possibly for the finitely many roots of $f(\alpha; z_n) = 0$; but at these points it is bounded. Hence $g(\alpha; z_n)$ has no poles for $|z_n| \le r'$, and (7.8) shows that $g(\alpha; z_n) = G(\alpha; z_n)$ by the Cauchy integral formula. The lemma is proved. □

Lemma 7.6 *Suppose that $f(z_1, \dots, z_n)$ is an analytic function on the whole of $\mathbb{C}^n$, and that there exists a constant C such that*

$$\bigl|f(z)\bigr| < C|z|^k \quad \text{for } z = (z_1, \dots, z_n), \text{ where } |z| = \max |z_i|. \qquad (7.9)$$

Then f is a polynomial of degree $\le k$.

Proof Suppose that the homogeneous component F_l of the Taylor series expansion

$$f = F_0 + F_1 + \cdots$$

of f about 0 is not identically 0 for some $l > k$. There exist $\alpha_1, \dots, \alpha_n$ with $F_l(\alpha_1, \dots, \alpha_n) \ne 0$. Then $g(w) = f(\alpha_1 w, \dots, \alpha_n w)$ is a function of one variable for which a bound of type (7.9) holds and the coefficient a_l of w^l in the Taylor series is nonzero. Subtracting the first k terms of the Taylor series we get a new function g_1 with Taylor series

$$g_1(w) = a_k w^k + \cdots,$$

for which a bound of type (7.9) holds and $a_l \ne 0$. Now by assumption g_1/w^k is bounded on the whole plane, and is hence constant. This contradicts $a_l \ne 0$. The lemma is proved. □

2.5 Connectedness of Fibres

Theorem 7.2 *Let X and Y be nonsingular irreducible varieties and $f : X \to Y$ a proper morphism such that $f(X) \subset Y$ is dense and X remains irreducible in the algebraic closure of $\mathbb{C}(Y)$ (compare Section 6.1, Chapter 2). Then every fibre of f is connected in the complex topology.*

Proof By the Bertini theorems, Theorem 2.26 of Section 6.1, Chapter 2 and Theorem 2.27 of Section 6.2, Chapter 2, there exists a subvariety $S \subsetneqq Y$ such that for every point $y \notin S$ the fibre $f^{-1}(y)$ is nonsingular and irreducible. We need to consider the fibres $f^{-1}(y_0)$ for $y_0 \in S$. We can find a nonsingular curve C passing through y_0 but not contained in S, and thus reduce to the case that $Y = C$ is a curve. Suppose that $f^{-1}(y_0)$ decomposes into two disjoint closed components Z_1 and Z_2. Consider disjoint neighbourhoods U_1 and U_2 of these. Since f is proper the set $f(X \setminus (U_1 \cup U_2))$ is closed. It does not contain y_0, hence does not intersect some neighbourhood of y_0 in C, which we can take to be a disc V. In V, the sets $f(U_1)$ and $f(U_2)$ can only intersect at y_0. For if $y \neq y_0$ is a point with $y \in f(U_1) \cap f(U_2) \cap V$ then the fibre $f^{-1}(y)$ is contained in $U_1 \cup U_2$ and intersects both U_1 and U_2. This means that it is not connected, which contradicts Theorem 7.1 and the fact that the fibre $f^{-1}(y)$ is irreducible for all $y \in V$ with $y \neq y_0$. Therefore $V \setminus y_0$ breaks up as the union of the two disjoint open sets $(V \setminus y_0) \cap f(U_1)$ and $(V \setminus y_0) \cap f(U_2)$, and so is not connected. But $V \setminus y_0$ is just a punctured disc, and this contradiction proves the theorem. □

Analysing the proof just given, we see easily that we have used the nonsingularity of Y only in a very weak form, namely in the statement that $V \setminus y_0$ is connected. The proof goes through if we impose the following rather weak condition on the singularities of Y: every singular point $y_0 \in Y$ has an arbitrarily small neighbourhood U in the complex topology such that the set of nonsingular points of U is connected. It is a hard theorem of Zariski that a normal singular point has this property (see Mumford's [61, Section 9, Chapter III]). Thus the theorem also holds for a normal variety Y.

2.6 *Exercises to Section 2*

1 Prove that for a quasiprojective variety X, if $X(\mathbb{C})$ is compact then X is projective.

2 Prove that for a variety X, if $X(\mathbb{C})$ is compact then X is complete. [Hint: Use Exercise 7 of Section 2.5, Chapter 6.]

3 Let X be a reduced irreducible scheme of finite type over $\mathbb{C}$ and $X(\mathbb{C})$ its set of closed points with the topology defined in Section 1.1. Prove that if $X(\mathbb{C})$ is Hausdorff then X is separated.

4 Prove that the group of automorphisms of a nonhyperelliptic nonsingular projective curve of genus >1 is finite. [Hint: Prove that if $\varphi\colon X \to \mathbb{P}^{g-1}$ is the embedding corresponding to the canonical class (compare Section 7.1 of Chapter 3), then automorphisms of $\varphi(X)$ are induced by projective transformations of $\mathbb{P}^{g-1}$, and hence form an algebraic group G, which has only finitely many connected components.

Now apply Exercise 4 of Section 1.4 to deduce that the identity component of G acts trivially on differentials.]

5 Extend the result of Exercise 4 to hyperelliptic curves.

6 Let X and Y be topological spaces and $f: X \to Y$ a surjective continuous map. Prove that if Y and every fibre $f^{-1}(y)$ are connected (for $y \in Y$), then so is X.

3 The Topology of Algebraic Curves

Constructing the associated topological space $X(\mathbb{C})$ of an algebraic variety X leads to two types of questions. First of all, it would be interesting to determine what topological spaces arise in this way, and, if possible, to achieve a topological classification of these. Secondly, which of the invariants of the topological space $X(\mathbb{C})$ have an algebraic meaning? In other words, the question is to construct invariants of an algebraic variety X, defined over an arbitrary field k which, when $k = \mathbb{C}$, become the given topological invariants of $X(\mathbb{C})$.

In this section we discuss the answers to both types of questions in the simplest case, when X is a nonsingular projective curve. At the same time, this is practically the only case in which the topological classification of the spaces $X(\mathbb{C})$ is completely known and the algebraic meaning of the topological invariants understood.

3.1 Local Structure of Morphisms

Let X be a nonsingular algebraic curve. Any point $x \in X(\mathbb{C})$ has a (complex) neighbourhood U homeomorphic to a neighbourhood of the origin in the complex plane $\mathbb{C}$. This homeomorphism is defined by any local parameter t at x,

$$t: U \xrightarrow{\sim} t(U) \subset \mathbb{C}. \tag{7.10}$$

We will assume from now on that U is chosen so that $t(U)$ is the domain $|z| < \varepsilon$ in $\mathbb{C}$. The map (7.10) allows us to take t as a coordinate in U, so that $x \in U$ is uniquely determined by the number $t(x)$.

Let $f: X \to Y$ be a morphism of nonsingular curves with $f(X) \subset Y$ dense, and for $x \in X$ set $y = f(x) \in Y$. We now show that x and y have neighbourhoods U and V with coordinates u and v in terms of which the map f has a very simple description.

We choose local parameters t at x and v at $y = f(x)$ and neighbourhoods $U \ni x$ and $V \ni y$ so that t and v define open embeddings (7.10)

$$t: U \to \mathbb{C} \quad \text{and} \quad v: V \to \mathbb{C},$$

and such that $f(U) \subset V$. Considering t as a coordinate on U, and v as a coordinate on V, we can say that f is determined in the neighbourhood U by giving $v(f(x'))$

as a function of $t(x')$ for $x' \in U$. In other words, f is determined by giving $f^*(v)$ as a function of t.

Since $f(X) \subset Y$ is dense, $f^*(v)$ is a function on X that is not identically 0, is regular at x and has $v(x) = 0$. We set

$$f^*(v) = t^k \varphi \quad \text{with } \varphi \in \mathcal{O}_x \text{ and } \varphi(x) \neq 0. \tag{7.11}$$

Now φ corresponds to a formal power series $\Phi(T)$ with a positive radius of convergence and $\Phi(0) \neq 0$. Thus, passing to a smaller neighbourhood U of x if necessary, we can assume that

$$\varphi(x') = \Phi(t(x')) \quad \text{for } x' \in U.$$

Since $\Phi(0) \neq 0$, there exists a power series $\Psi(T) = \Phi(T)^{1/k}$ also having a positive radius of convergence. Hence $u(x') = \Psi(t(x'))t(x')$ defines a function for points x' of some sufficiently small neighbourhood of x. We again denote this neighbourhood by U. We have thus constructed a map

$$u : U \to \mathbb{C}, \quad \text{for which } t^k \varphi = u^k \text{ in } U.$$

The analytic function $u = t\Psi(t)$ is no longer a rational function on X, and is only defined in a sufficiently small complex neighbourhood U of x. However, in this neighbourhood it is obviously continuous. Just as t, it defines a homeomorphism of some neighbourhood of x to an open set in $\mathbb{C}$. Indeed, by the implicit function theorem, the analytic function $z\Psi(z)$ has an inverse in some neighbourhood of 0 and defines a homeomorphism onto some neighbourhood of 0 in $\mathbb{C}$.

By construction $f^*(v) = u^k$ in the open set U. Thus we conclude our analysis, obtaining the following simple local description of f.

Theorem 7.3 *For every morphism $f : X \to Y$ of nonsingular curves with $f(X)$ dense in Y, and every $x \in X$, there exist neighbourhoods $U \ni x$ and $V \ni f(x)$, and homeomorphisms $u : U \to \mathbb{C}$ and $v : V \to \mathbb{C}$ onto neighbourhoods of 0 in $\mathbb{C}$ such that the diagram*

$$\begin{array}{ccc} U & \xrightarrow{u} & \mathbb{C} \\ {\scriptstyle f}\downarrow & & \downarrow{\scriptstyle \rho_k} \\ V & \xrightarrow[v]{} & v \end{array}$$

is commutative. Here $\rho_k(z) = z^k$, where k is defined as the order of zero of the function $f^*(t)$ at x for t a local parameter at y.

If we interpret u and v as coordinates in open sets U and V, then Theorem 1 asserts that f, restricted to these neighbourhoods and expressed in terms of these coordinates has the very simple form

$$v = u^k \tag{7.12}$$

(see Figure 29). The open sets U and V can obviously be chosen in such a way that $u(U)$ and $v(V)$ are both equal to the interior of the disc $|z| < 1$.

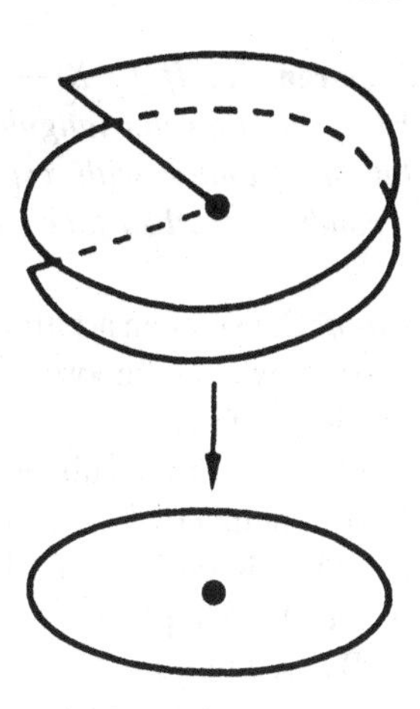

Figure 29 The branch point $z \mapsto z^k$

Definition Neighbourhoods of this form will be called *normal.* The number k in the representations (7.11) and (7.12) is called the *ramification degree* of f at $x \in X$. If for any point $x \in f^{-1}(y)$ the ramification degree of f at x is >1 then y is called a *branch point* or *ramification point* of f.

Obviously, the ramification degree of f at $x \in X$ equals the multiplicity of x in the divisor $f^*(y)$. If X and Y are projective in addition to nonsingular then Theorem 3.5 of Section 2.1, Chapter 3 shows that y is not a branch point if and only if the number of inverse images $f^{-1}(y)$ equals the degree $\deg f$ of f. In other words, the above definition agrees with the definition of branch point given in Section 6.3, Chapter 2. It follows from Theorem 2.29 of Section 6.3, Chapter 2 that a morphism of nonsingular projective curves over $\mathbb{C}$ has only finitely many branch points. (Compare also Exercises 2–4 of Section 8.1, Chapter 3; note that the "ramification multiplicity" e_x defined there is one less than the ramification degree, that is, $e_x = k - 1$.)

3.2 Triangulation of Curves

In this section, we prove that if X is a nonsingular projective algebraic curve, then the topological space $X(\mathbb{C})$ can be triangulated. For the convenience of the reader, we treat the definition and basic facts on the classification of triangulated 2-manifolds in Section 3.4.

The triangulation of $X(\mathbb{C})$ is obtained as a corollary of a more general fact. To state it, we introduce the following definition. If X and Y are topological spaces and $f: X \to Y$ a continuous map, we say that a triangulation Φ of X is *compatible* with a triangulation Ψ of Y with respect to f if for every simplex $E \in \Psi$ we have $f^{-1}(E) = \bigcup E_i$ for $E_i \in \Phi$, and the map $f: E_i \to E$ is a homeomorphism.

Theorem 7.4 *If $f\colon X \to Y$ is a morphism of nonsingular projective curves, and $Y(\mathbb{C})$ has a given triangulation Ψ_0, then $X(\mathbb{C})$ and $Y(\mathbb{C})$ have compatible triangulations Φ and Ψ with respect to f, with Ψ a subdivision of Ψ_0 on $Y(\mathbb{C})$. If $Y(\mathbb{C})$ is a combinatorial surface (see Section 3.4) then so is $X(\mathbb{C})$.*

Proof Consider an arbitrary point $y \in Y(\mathbb{C})$ and set $f^{-1}(y) = \{x_1, \dots, x_k\}$. By Theorem 1, we can choose a neighbourhood V_y of y and disjoint normal neighbourhoods U_i of x_i.

Choose a finite subcover of the cover $\{V_y \mid y \in Y\}$ of $Y(\mathbb{C})$. We get a finite set $\{y_\alpha\}$ of points of $Y(\mathbb{C})$, for each point y_α a neighbourhood V_α, and disjoint normal neighbourhood $U_{\alpha,i}$ of all $x_{\alpha,i} \in f^{-1}(y_\alpha)$. Obviously if $y \in V_\alpha$ and $y \neq y_\alpha$ then y is not a branch point.

By assumption, $Y(\mathbb{C})$ has a given triangulation Ψ_0. By Proposition of Section 3.4, there exists a finer triangulation Ψ having all the branch points of f among its vertexes, and such that each simplex is contained in a neighbourhood V_α.

Let E be an arbitrary simplex of the triangulation Ψ. It is contained in some open chart V_α by construction. If y_α is not a branch point then $f_i : U_{\alpha,i} \to V_\alpha$ is a homeomorphism for any $x_i \in f^{-1}(y_\alpha)$. Write $E_i = f^{-1}(E) \subset U_{\alpha,i}$ for the inverse image. If $t\colon E \to \sigma$ is the map in the definition of triangulation Ψ, set $t_i = t \circ f_i : E_i \to \sigma$. We put the set of all E_i and homeomorphisms t_i into the triangulation Φ.

Now suppose that $E \subset V_\alpha$ and y_α is a branch point. We consider two cases.

Case $y_\alpha \notin E$ In suitable coordinates, the map $f_i : U_{i,\alpha} \to V_\alpha$ is of the form $v = f^*(u) = u^k$. Because $y_\alpha \notin E$ and E is simply connected, v is nowhere zero on E and any branch of $\sqrt[k]{v}$ defines a single valued function there. It follows from this that $f_i^{-1}(E)$ breaks up into k connected components $E_1, \dots, E_k$ such that the map $f_i : E_i \to E$ is a homeomorphism. In this case we add the set of all E_i and homeomorphisms $t_i = t \circ f_i : E_i \to \sigma$ to the triangulation Φ.

Case $y_\alpha \in E$ In this case, we can apply the same arguments to the set $E \setminus y_\alpha$. We get that $f^{-1}(E \setminus y_\alpha)$ breaks up into k connected components $\widetilde{E}_1, \dots, \widetilde{E}_k$. We set

$$E_j = \widetilde{E}_j \cup x_i \quad \text{where } x_i \in f^{-1}(y_\alpha) \cap U_{i,\alpha};$$

then $f_j\colon E_j \to E$ is a homeomorphism. We again add the set of all E_j and homeomorphisms $t_j = t \circ f_j\colon E_j \to \sigma$ to the triangulation Φ.

A simple verification, which we leave to the reader, shows that the sets and maps we have constructed determine a triangulation Φ of $X(\mathbb{C})$, and that Φ is compatible with Ψ with respect to f and satisfies the conditions in the definition of combinatorial surface. The theorem is proved. □

Theorem 7.5 *If X is a nonsingular projective curve then the space $X(\mathbb{C})$ can be triangulated, and is a combinatorial surface.*

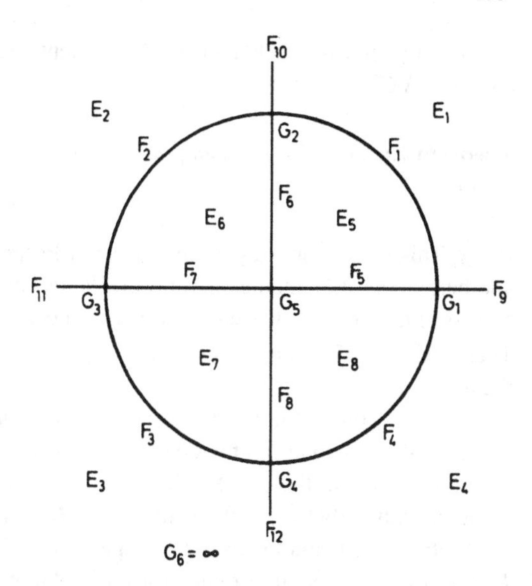

Figure 30 A triangulation of $\mathbb{P}^1(\mathbb{C})$

Proof We prove first that $\mathbb{P}^1_{\mathbb{C}}$ has a triangulation. We indicate a triangulation which, although not the most economic, is useful for subsequent applications.

The decomposition of the surface of an octahedron into its faces, edges and vertexes gives a triangulation. Suppose that the octahedron is inscribed in the 2-sphere $S^2 \subset \mathbb{R}^3$. Projecting this triangulation outwards from an interior point of the octahedron provides a triangulation of S^2. Since $\mathbb{P}^1(\mathbb{C})$ is homeomorphic to S^2, we get a triangulation of $\mathbb{P}^1(\mathbb{C})$.

We identify $\mathbb{P}^1(\mathbb{C})$ with the plane of one complex variable together with a point at infinity. The triangulation constructed in terms of the octahedron is the decomposition realised by the real and imaginary axes and the circle $|z| = 1$ (see Figure 30). It has

$$\begin{cases} 8 \text{ faces } E_i, \text{ with } d(E_i) = 2 \text{ for } i = 1, \dots, 8, \\ 12 \text{ edges } F_i, \text{ with } d(F_i) = 1 \text{ for } i = 1, \dots, 12, \\ 6 \text{ vertexes } G_i, \text{ with } d(G_i) = 0 \text{ for } i = 1, \dots, 6. \end{cases} \tag{7.13}$$

Now consider a nonconstant rational function f on X. It defines a morphism $f : X \to \mathbb{P}^1$, and it remains only to apply Theorem 2. The theorem is proved. □

3.3 Topological Classification of Curves

We apply the topological classification of surfaces, recalled in Section 3.4, to the surface $X(\mathbb{C})$, where X is a nonsingular projective curve. For this, by Theorem A, we need to determine whether $X(\mathbb{C})$ is orientable, and compute its Euler[3] characteristic $e(X(\mathbb{C}))$.

Theorem 7.6 *If X is a nonsingular projective curve then $X(\mathbb{C})$ is an orientable surface.*

Proof This is of course a particular case of Proposition of Section 1.2, but we give another, much more elementary, proof. We use an arbitrary morphism $f: X \to \mathbb{P}^1$, and consider compatible triangulations Φ and Ψ of $X(\mathbb{C})$ and $\mathbb{P}^1_{\mathbb{C}}$ with respect to f. These exist by Theorem 2. Since the 2-sphere S^2 is orientable, the same holds for the triangulation Ψ.

From the compatibility of the triangulations Φ and Ψ it follows that the simplexes of Φ are exhausted by the components of the inverse images of simplexes of Ψ, as described in the proof of Theorem 2. Hence any simplex $E \in \Phi$ is mapped homeomorphically by f onto a simplex $f(E) \in \Psi$.

We fix an orientation of the triangulation Ψ (see Section 3.4). For a triangle $E \in \Phi$ we consider the orientation on it obtained from that of $f(E) \in \Psi$ under the homeomorphism $f: E \to f(E)$. It remains to check that in this way we get a orientation of the whole of Φ. This is very easy. Suppose that two triangles $E', E'' \in \Phi$ have a common side E with vertexes b, c. Write a, b, c for the vertexes of E' and b, c, d for those of E'', and set $f(a) = a'$, $f(b) = b'$, $f(c) = c'$, $f(d) = d'$. Then a', b', c' and b', c', d' are the vertexes of $f(E'), f(E'') \in \Psi$. Suppose that the chosen orientation of Ψ is given in $f(E')$ by ordering its vertexes (a', b', c'). Then by the definition of orientation, the vertexes of $f(E'')$ are ordered by (c', b', d'). By assumption, the order of the vertexes of E' and E'' is the (a, b, c) and (c, b, d), from which one sees that they define opposite orientations on the side E. The theorem is proved. □

We pass to the second question, the determination of the Euler characteristic of $X(\mathbb{C})$ for an irreducible nonsingular projective curve X.

Theorem 7.7 *The Euler characteristic of $X(\mathbb{C})$ is $2 - 2g$, where g is the genus of X.*

Proof We again use a regular map $f: X \to \mathbb{P}^1$ and compatible triangulations Φ and Ψ of $X(\mathbb{C})$ and $\mathbb{P}^1(\mathbb{C})$. We write c_0, c_1, c_2 for the numbers appearing in the definition of the Euler characteristic (7.19) of Φ, and c_0', c_1', c_2' for the same numbers for Ψ. Then

$$e(\mathbb{P}^1(\mathbb{C})) = c_0' - c_1' + c_2' \quad \text{and} \quad e(X(\mathbb{C})) = c_0 - c_1 + c_2.$$

[3]The Euler characteristic is also traditionally denoted by $\chi(X)$, but Hirzebruch's notation $e(X)$ neatly avoids ambiguity with $\chi(\mathcal{O}_X)$ in sheaf cohomology.

Let us see how these numbers are related. By definition of compatible triangulations, for every simplex $E \in \Psi$ we have

$$f^{-1}(E) = \bigcup E_i, \tag{7.14}$$

where $f \colon E_i \to E$ is a homeomorphism. In the proof of Theorem 7.6, we saw that running through all the simplexes $E \in \Psi$, we get in this way among the E_i all the simplexes of Φ.

How many simplexes E_i are there in (7.14)? If $\deg f = n$ and $d(E) > 0$ then there are n. Indeed, there cannot be more than n of them, since every point $y \in \mathbb{P}^1$ has $\leq n$ inverse images. But it also cannot be less than n, since only the finitely many branch points have $<n$ inverse images. Let $d(E) = 0$, so that E is a point $y \in \mathbb{P}^1(\mathbb{C})$. We write $\overline{y}$ for the divisor consisting of one point y with multiplicity 1. We set

$$f^*(\overline{y}) = \sum_{i=1}^{r} k_i \overline{x}_i. \tag{7.15}$$

The number of inverse images of points y equals r, but since $\sum_{i=1}^{r} k_i = n$, we have $r = n - \sum(k_i - 1)$. Hence we get

$$\begin{cases} c_2 = nc_2', \\ c_1 = nc_1', \\ c_0 = nc_0' - \sum(k_i - 1), \end{cases}$$

where the final sum contains the ramification degree of f at all points $x \in X$. In conclusion, we get that

$$e\bigl(X(\mathbb{C})\bigr) = e\bigl(\mathbb{P}^1(\mathbb{C})\bigr)n - \sum(k_i - 1).$$

On the other hand, for example from the triangulation of Figure 30 and (7.13) we see that $e(\mathbb{P}^1(\mathbb{C})) = 8 - 12 + 6 = 2$. Hence, finally,

$$e\bigl(X(\mathbb{C})\bigr) = 2n - \sum(k_i - 1). \tag{7.16}$$

We now consider an arbitrary rational differential form $\omega \neq 0$ on $\mathbb{P}^1$ and compute the divisor $\operatorname{div}(f^*(\omega))$ of the pullback of ω on X. Suppose that $v_y(\omega) = m$ at a point $y \in \mathbb{P}^1$; that is,

$$\omega = t^m g dt,$$

where t is a local parameter at y and $g \in \mathcal{O}_y$ with $g(y) \neq 0$. If the numbers k_i are the same as in (7.16), we get $v_{x_i}(f^*(t)) = k_i$, that is, $f^*(t) = \tau^{k_i} h_i$, where τ is a local parameter at x_i and $h_i \in \mathcal{O}_{x_i}$ with $h_i(x_i) \neq 0$. It follows from this that

$$v_{x_i}\bigl(f^*(\omega)\bigr) = mk_i + k_i - 1.$$

In other words,

$$\operatorname{div}\big(f^*(\omega)\big) = f^*(\operatorname{div}\omega) + \sum (k_i - 1)\overline{x}_i, \tag{7.17}$$

where, as in (7.16), the final sum is taken over all points $x_i \in X$ at which the ramification degree of f is >1.

Since $f^*(\omega)$ is a differential form on X, we have $\deg(\operatorname{div}(f^*(\omega))) = 2g - 2$. In exactly the same way, $\deg(\operatorname{div}(\omega)) = -2$. Finally, for every divisor D on $\mathbb{P}^1$ we have the equality $\deg(f^*(D)) = n \deg D$, so that this holds for the divisor consisting of just one point by Theorem 3.5 of Section 2.1, Chapter 3. Considering the degrees of the divisors on either side of (7.17) we thus get

$$2g - 2 = -2n + \sum (k_i - 1).$$

(See also Exercise 4 of Section 8.1, Chapter 3.)

Comparing this formula with (7.16), we get Theorem 7.7. □

Theorems 7.6 and 7.7, in combination with the topological result Theorem A give us a complete topological classification of nonsingular projective curves. They show that for two such curves the spaces $X(\mathbb{C})$ are homeomorphic if and only if the curves have the same genus.

There is no analogous result known for varieties of dimension >1. We discuss one of the simplest results on the relation between the topological and algebraic properties of complete nonsingular varieties that generalises Theorem 5. Because

$$e\big(X(\mathbb{C})\big) = b_0 - b_1 + b_2$$

(in the notation of Theorem 5), where b_1 is the first Betti number of $X(\mathbb{C})$ and $b_0 = b_2 = 1$, Theorem 5 can be expressed as the equality

$$b_1 = 2g.$$

If X is an arbitrary nonsingular projective variety then a similar result holds:

$$b_1 = 2h^1,$$

where b_1 is the first Betti number of $X(\mathbb{C})$ and $h^1 = \dim_{\mathbb{C}} \Omega^1[X]$ (compare Section 4.5, Chapter 8). Using more delicate constructions, we can also express the other Betti numbers of $X(\mathbb{C})$ in terms of algebraic invariants of X.

In conclusion, we note that the topological classification of nonsingular projective curves can be obtained by other methods, in the framework of the theory of differentiable manifolds. For this one has to use somewhat less elementary topological apparatus, but the treatment is more intrinsic. We note only the general direction of this treatment, omitting all details.

In Section 1.2 we saw how to prove the orientability of $X(\mathbb{C})$ for any nonsingular variety X using notions of the theory of differentiable manifolds. The combinatorial classification of surfaces (Theorem A) must be replaced by its smooth analogue, the theorem that any connected compact oriented surface can be obtained by glueing a finite number of handles onto a sphere. The proof of this theorem follows easily from Morse theory (see for example Wallace [78]). It remains to prove Theorem 7.7. For this we have to consider the tangent bundle Θ of the 2-dimensional manifold $X(\mathbb{C})$. Its first Chern class $c_1(\Theta)$ is an element of $H^2(X(\mathbb{C}), \mathbb{Z})$. This group has a canonical generator, the cohomology class φ for which $\varphi(\omega_X) = 1$, where ω_X is the orientation class of X. Hence $c_1(\Theta) = \nu\varphi$ for some $\nu \in \mathbb{Z}$, and thus determines an integer ν. In our case c_1 is the Euler class, and hence $\nu = e(X(\mathbb{C}))$; that is $c_1(\Theta) = e(X(\mathbb{C}))\varphi$, where $e(X(\mathbb{C}))$ is the Euler characteristic of the surface $X(\mathbb{C})$ (see Husemoller [41], Theorem 7.2).

On the other hand, Θ has rank 1 as an algebraic vector bundle over X, and corresponds to the divisor class $-K$, where K is the canonical class of the curve X. Since $\deg(-K) = 2 - 2g$, the relation

$$e\bigl(X(\mathbb{C})\bigr) = 2 - 2g$$

is a consequence of the following general result.

Proposition *If E is a rank* 1 *vector bundle on a nonsingular projective curve X and D its characteristic class, and $c_1(E)$ the Chern class of the corresponding vector bundle on $X(\mathbb{C})$, then*

$$c_1(E) = (\deg D)\varphi.$$

The proof for the case of a divisor D on a variety X of any dimension is given in Chern [22] (Russian translator's footnote No. 39), or Springer [74, 5–9].

3.4 Combinatorial Classification of Surfaces

We recall here for the reader's convenience a number of elementary topological notions and results.

Let V be an n-dimensional affine space over the real number field $\mathbb{R}$. Then any two points $P, Q \in V$ define a vector $\overrightarrow{PQ}$ belonging to the n-dimensional vector space $\mathbb{R}^n$. And any vector $x \in \mathbb{R}^n$ together with a point $P \in V$ define a point $Q \in V$ such that $\overrightarrow{PQ} = x$; this can also be expressed as $P + x = Q$. For any points $P_1, \dots, P_m \in V$ and any set of numbers $\lambda_1, \dots, \lambda_m \in \mathbb{R}$ such that $\sum \lambda_i = 1$, the point $Q + \sum \lambda_i \overrightarrow{QP_i}$ is independent of the choice of the auxiliary point Q, and can be written in the form $\sum \lambda_i P_i$.

If $P_0, \dots, P_r \in V$ are not contained in any affine space of dimension less than r then the representation of a point R in the form $R = \sum_{i=0}^r \lambda_i P_i$ with $\sum \lambda_i = 1$ is unique. We say that $P_0, \dots, P_r$ are independent points. In this case, the set of points

$R \in V$ that can be represented in the form

$$R = \sum_{i=0}^{r} \lambda_i P_i \quad \text{with } \sum \lambda_i = 1 \text{ and } \lambda_i \geq 0 \tag{7.18}$$

is called an r-*simplex* or r-dimensional simplex. The points P_i are called its *vertexes*.

If $P_{i_1}, \dots, P_{i_{r-s}}$ are $r-s$ of the vertexes of a simplex σ then the points $R \in \sigma$ which have $\lambda_{i_1} = \dots = \lambda_{i_{r-s}} = 0$ in (7.18) themselves form an s-dimensional simplex with vertexes P_j for $j \neq i_1, \dots, i_{r-s}$. A simplex of this form is called a *face* of σ.

Let X be a Hausdorff topological space. We give here a definition of *triangulation* of X (more precisely, finite triangulation). By this, we mean the following data: (a) a finite family Φ of closed subsets E_i of X; (b) a map sending each $E_i \in \Phi$ to an integer $d(E_i) \geq 0$; and (c) for each i, a homeomorphism $t_i \colon E_i \to \sigma_i$ where σ_i is a simplex of dimension $d(E_i)$. Here the following conditions should hold:

(1) $X = \bigcup E_i$;
(2) if $E_i, E_j \in \Phi$ then either $E_i \cap E_j \in \Phi$ or $E_i \cap E_j = \emptyset$;
(3) if $E_k \subset E_i$ then $t_i(E_k)$ is a face of σ_i, and all the faces of the simplex arise in this way.

It follows from the definition that if $d(E_i) = 0$ then E_i is a point $x \in X$. All such points are called the *vertexes* of the triangulation. The subsets $E_i \subset X$ are called the simplexes of the triangulation, and the vertexes of the triangulation contained in a given simplex F_i are called the vertexes of this simplex. It is easy to show that if we know the set $K = \{x_1, \dots, x_N\}$ of vertexes of a triangulation, and which subsets $S \subset K$ are the sets of vertexes of a simplex of the triangulation, then we can recover the topological space X from this information. Thus a triangulation of a space allows us to specify it as a purely combinatorial construction. A topological space that admits at least one triangulation is said to be *triangulable*.

In connection with triangulations of $X(\mathbb{C})$, where X is a nonsingular projective curve, we need triangulations with the following properties:

(a) all the simplexes of Φ have dimension ≤ 2;
(b) every simplex of dimension <2 is the face of some 2-simplex;
(c) every 1-simplex is the face of exactly two 2-simplexes.

A topological space having a triangulation with these properties is a *combinatorial surface*.

In what follows, we use the operation of refining or subdividing a given triangulation. We give a simplified description of the operation of subdivision which is valid for triangulations with $d(E_i) \leq 2$ for all $E_i \in \Phi$.

Suppose that X is a topological space and Φ a triangulation such that $d(E_i) \leq 2$ for all $E_i \in \Phi$, and let $E_r \in \Phi$ be a 1-simplex. Choose any interior point ξ of the interval $t_r(E_r)$ and denote the two intervals into which ξ divides $t_r(E_r)$ by Γ' and Γ''. Set $x = t_r^{-1}(\xi)$. Let E_i for $i \in I$ be the simplexes of the triangulation Φ with

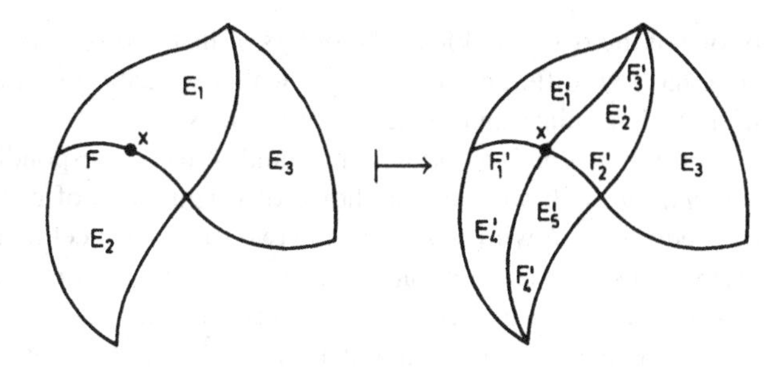

Figure 31 Subdivision of a triangulation

$d(E_i) = 2$ and $E_i \supset E_r$. For each $i \in I$, we subdivide the triangle $t_i(E_i)$ into two triangles T_i' and T_i'' by drawing an interval Γ joining the point $t_i(x)$ to the vertex opposite the side $t_i(E_r)$. (See Figure 31.)

Consider the family Φ' consisting of the following closed subsets of X:

$d = 0$: take $E_j \in \Phi$ with $d(E_j) = 0$, together with the new point x;
$d = 1$: take $E_j \in \Phi$ with $d(E_j) = 1$ and $j \neq r$, together with four new intervals $t_r^{-1}(\Gamma')$, $t_r^{-1}(\Gamma'')$ and $t_i^{-1}(\Gamma)$ for each $i \in I$;
$d = 2$: take $E_j \in \Phi$ with $d(E_j) = 2$ and $j \notin I$, together with four new triangles $t_i^{-1}(T_i')$ and $t_i^{-1}(T_i'')$ for each $i \in I$.

Thus each simplex of the set Φ' is either a simplex of Φ, or part of one. We define t' as the corresponding maps in the triangulation Φ or their restrictions. It is easy to check that conditions (1), (2) and (3) hold, so that Φ' is again a triangulation. It is called a *subdivision* of Φ. The following result follows at once from the definition of subdivision.

Proposition *Let $S \subset X$ be a finite set, $X = \bigcup U_i$ a finite open cover and Φ a triangulation of a space X. Then there exists another triangulation obtained by successive subdivisions of Φ such that the points of S are vertexes and every simplex is contained in one of the U_i.*

Such a triangulation is said to be *subordinate* to the cover $X = \bigcup U_i$.

Proof If a point $s \in S$ is contained in a simplex E_i with $d(E_i) = 1$ then a single subdivision makes it into a vertex of the triangulation. If $s \in E_i$ with $d(E_i) = 2$ then we first choose an edge $E_j \subset E_i$ with $d(E_j) = 1$ and the vertex P of the triangle $t_i(E_i)$ opposite the side $t_i(E_j)$. Take ξ to be the point of intersection of the side $t_i(E_j)$ and the line Γ joining $t_i(s)$ and P. Carrying out the subdivision in this point, we come to the case already considered.

To make the triangulation subordinate to an open cover, it is enough to do this for each triangle. The argument here is very simple and we leave it to the reader. The proposition is proved. □

We recall the notion of *orientation* of a triangulation of a surface. An orientation of an interval is an ordering of its endpoints; an orientation of a triangle is a choice of one of the two possible cyclic orders of its three vertexes. Thus each triangle and edge has two different orientations, which we call *opposite*. Each orientation of a triangle determines an orientation of its sides.

Let X be a combinatorial surface and Φ the corresponding triangulation. By an *orientation* of Φ we mean a choice of orientations of each triangle such that for each edge $E \in \Phi$ with $d(E) = 1$, the two triangles meeting along E define opposite orientations. A triangulation that admits an orientation is said to be *orientable*. If X is connected then a triangulation φ admits either no orientation, or exactly 2. It is easy to check that the triangulation of S^2 or $\mathbb{P}^1(\mathbb{C})$ constructed in Section 3.3 is orientable.

The property that a surface is orientable is independent of its triangulation. In other words, if Φ and Ψ are two different triangulations of the same surface then they are either both orientable, or both nonorientable. The condition for a surface to be orientable can be written in intrinsic terms as $H_2(X, \mathbb{Z}) \neq 0$.

The final topological notion that we require is that of *Euler characteristic* of a surface. If a triangulation Φ has c_0 vertexes, c_1 edges and c_2 triangles, then the Euler characteristic is defined as

$$e_\Phi(X) = c_0 - c_1 + c_2. \tag{7.19}$$

As with orientability, the Euler characteristic of a surface is independent of the triangulation, and so is denoted by $e(X)$. Its invariant definition is

$$e(X) = \dim_K H_0(X, K) - \dim_K H_1(X, K) + \dim_K H_2(X, K),$$

where K is any field of characteristic 0. It is easy to check that the Euler characteristic of S^2 or $\mathbb{P}^1(\mathbb{C})$ is 2.

The main result of the topology of surfaces is that the topological invariants we have introduced, that is, orientability and the Euler characteristic, form a complete system of topological invariants of connected triangulable surfaces.

Theorem A *Two connected triangulable surfaces are homeomorphic if and only if they are orientable or nonorientable together and have equal Euler characteristic.*

For the proof, see Aleksandrov [4, III.7.2], Seifert and Threlfall [68, Chapter 6, Section 39], Hauptsatz or Springer [74, 5–5 and 5–9]. The assumption of triangulability is redundant here; it can be shown that any surface is triangulable (compare Theorem 7.5), although we do not need this.

3.5 The Topology of Singularities of Plane Curves

Let X be a algebraic plane curve and $O \in X$ a singular point. In this case, already arbitrarily small neighbourhoods (or punctured neighbourhoods) of $O \in X \subset \mathbb{C}^2$ have nontrivial topological invariants, and the question that arises is to describe

Figure 32 A torus knot

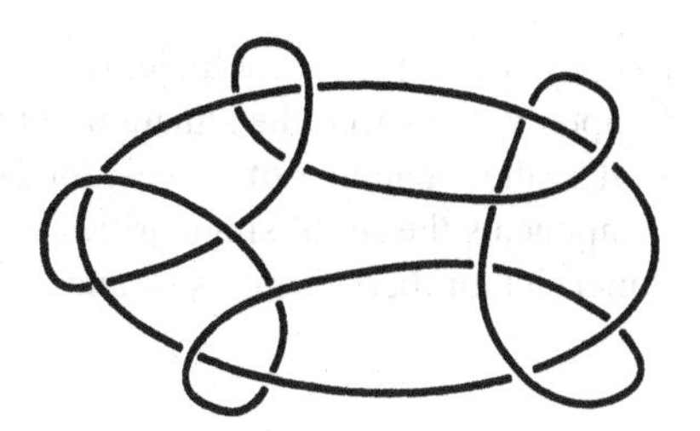

these. For this, suppose that $O = (0,0)$ is the origin of the plane $\mathbb{C}^2$, and surround it by a sufficiently small 3-sphere S_ε with equation $|z_1|^2 + |z_2|^2 = \varepsilon^2$. It can be proved that, for ε sufficiently small, the intersection $X(\mathbb{C}) \cap S_\varepsilon$ is a smooth 1-dimensional manifold, that is, a union of a finite number of smooth curves homeomorphic to the circle S^1. Such a system of curves in S^3 is called a *link*, and if it is a connected curve, a *knot*. The topology of the singular point $O \in X \subset \mathbb{C}^2$ is determined by the link we have constructed, since a sufficiently small neighbourhood of O is the cone over this link.

If X has r branches at O then the corresponding link breaks up as a union of r knots (see Section 5.3, Chapter 2 for a discussion of branches; in what follows, we are mainly concerned with the unibranch case $r = 1$). Each knot is uniquely determined by the characteristic pairs of the corresponding branch; it is a so-called *iterated torus knot*. To describe this, we first describe a *torus knot*, that can be drawn on the surface of a torus. If we view the torus as the quotient $\mathbb{R}^2/\mathbb{Z}^2$, where $\mathbb{Z}^2 \subset \mathbb{R}^2$ is the lattice of points with integer coordinates, then a torus knot of type (p,q), where p and q are coprime natural numbers, is the image of the line $y = (p/q)x$ in $\mathbb{R}^2$. The corresponding curve goes p times round the torus in the direct of one basic cycle, and q times round it in the other.

Suppose that $(a_1, b_1), \dots, (a_m, b_m)$ are the characteristic pairs of our branch (see (2.39) of Section 5.3, Chapter 2). Consider an unknotted circle l_0 in our sphere, and the boundary of a tubular neighbourhood of l_0, that is, the torus described by a small circle with centre a point of l_0 contained in a normal plane to l_0 as the centre moves around l_0. On this torus we construct a torus knot l_1 of type (a_1, b_1) (see Figure 32). Then consider the boundary of a tubular neighbourhood of l_1 and consider a torus knot of type (a_2, b_2) on it, and so on. Repeating this process m times using the specified sequence of characteristic pairs, we get the knot of the corresponding branch. It can be proved that the set of characteristic pairs is a topological invariant of the iterated torus knot.

We do not give proofs of these assertions, although they are quite elementary; see for example Kähler [44] or Milnor [59].

The picture that emerges is reminiscent of the theory of nonsingular curves. There we had one integer invariant, the genus, that uniquely determines the topology of the curve $X(\mathbb{C})$. For singularities, the analogous role is played by the set of characteristic pairs, that uniquely determines the knot. The set of all curves of genus g up to isomorphism depends on a finite number of parameters ($3g - 3$ if $g > 1$). In the same way, the set of singularities with a given set of characteristic pairs (up to formal or local analytic isomorphism) depends on a finite number of parameters:

roughly speaking, the coefficients of the Puiseux expansion of (2.39) of Section 5.3, Chapter 2. However, the situation in the case of singularities seems to be much less well studied than that of curves; for example, it is not known how many connected components the set of singularities with given topology (that is, link) has, what the dimension of these families is, and so on.

3.6 Exercises to Section 3

1 Let k be an algebraically closed field of characteristic 0 and $k((t))$ the field of fractions of the ring of formal power series in one variable. Prove that this field has a unique extension of given degree n, obtained by adjoining $\sqrt[n]{t}$. [Hint: Use the arguments of the proof of Theorem 7.3.]

2 Let X and Y be nonsingular projective curves and $f\colon X \to Y$ a morphism with $f(X) = Y$. Deduce a formula expressing the genus of X in terms of that of Y and the ramification degrees of f at points of X. [Hint: Consider compatible triangulations of X and Y with respect to f. See also Exercise 2–4 of Section 8.1, Chapter 3.]

3 Let X be a projective nonsingular model of the curve with equation

$$y^2 = (x-a)(x-b)(x-c)(x-d),$$

and $f\colon X(\mathbb{C}) \to \mathbb{P}^1(\mathbb{C})$ the continuous map defined by the function x. Let α and β denote disjoint intervals in the sphere $\mathbb{P}^1(\mathbb{C})$ joining ab and cd. Prove that $f^{-1}(\mathbb{P}^1(\mathbb{C}) \setminus (\alpha \cup \beta))$ breaks up into two connected components X_i, each of which maps homeomorphically to $\mathbb{P}^1(\mathbb{C}) \setminus (\alpha \cup \beta)$ under f, where $\mathbb{P}^1(\mathbb{C}) \setminus (\alpha \cup \beta)$ is homeomorphic to a sphere with two discs deleted.

4 In the notation of Exercise 3, prove that for $i = 1, 2$ the boundary of the closure $\overline{X}_i$ is homeomorphic to $\alpha \cup \beta$, and that X is obtained by identifying these boundary components, and is thus homeomorphic to the torus, in agreement with Theorem 7.7.

5 Prove that the knot corresponding to the singular point $(0,0)$ of the curve $y^p = x^q$, where p and q are coprime integers, is the torus knot of type (p,q). [Hint: The case $(p,q) = (2,3)$ is treated in Mumford [60, Section 1B].]

4 Real Algebraic Curves

By a *real algebraic curve*, we mean a scheme X defined over $\mathbb{R}$ such that $X \otimes_{\mathbb{R}} \mathbb{C}$ is an algebraic curve. We assume from now on that $X \otimes_{\mathbb{R}} \mathbb{C}$ is a nonsingular irreducible projective curve. As before, we write $X(\mathbb{R})$ for the set of closed points

$x \in X$ for which $k(x) = \mathbb{R}$. In more simple-minded terms, X is a nonsingular irreducible projective curve defined by equations with real coefficients, and $X(\mathbb{R})$ is the set of points of X with real coordinates.

The set $X(\mathbb{R})$ is a compact 1-dimensional manifold. However, it is not necessarily connected, so that the analogue of Theorem 7.1 is false here. An example of a disconnected variety $X(\mathbb{R})$ has already appeared in Figure 8.

A connected compact 1-dimensional manifold is homeomorphic to the circle. This is not hard to prove directly, and for the connected components of $X(\mathbb{R})$ it follows at once from the fact that they are triangulable, which we prove presently. Thus $X(\mathbb{R})$ is homeomorphic to some number of disjoint circles, so that the unique topological invariant of this space is its number of connected components.

We prove in this chapter the main result relating this topological invariant of the topological space $X(\mathbb{R})$ to algebraic properties of the algebraic curve X.

Harnack's Theorem *If X is a nonsingular projective curve of genus g defined over $\mathbb{R}$ then the number of connected components of $X(\mathbb{R})$ is $\leq g+1$.*

There are several proofs of this theorem. One of these takes place entirely in the real domain; this can be found in Lang [55, Section 7, Chapter X]. We give here another proof which is interesting in that the stated property of $X(\mathbb{R})$ is deduced from its embedding in the space $X(\mathbb{C})$.

4.1 Complex Conjugation

The main role in the proof of Harnack's Theorem given in Section 4.2 is played by the complex conjugation map $\tau\colon \mathbb{P}^n(\mathbb{C}) \to \mathbb{P}^n(\mathbb{C})$, that sends any point $x \in \mathbb{P}^n(\mathbb{C})$ to the point with complex conjugate coordinates $\tau(x)$. Obviously, τ defines a homeomorphism of the topological space $\mathbb{P}^n(\mathbb{C})$ (but *not* an automorphism of the algebraic variety $\mathbb{P}^n$!). Since X is defined by equations with real coefficients, $\tau(X(\mathbb{C})) = X(\mathbb{C})$, and τ defines a homeomorphism of $X(\mathbb{C})$. In other words, τ is the automorphism of the scheme $X \otimes_{\mathbb{R}} \mathbb{C}$ induced by the complex conjugation map $\alpha \mapsto \tau(\alpha)$ on $\mathbb{C}$.

As before, we use a triangulation of $X(\mathbb{C})$, but now it is convenient to choose it invariant under τ, that is, so that together with a simplex E, it also contains the simplex $\tau(E)$. We prove that such a triangulation exists.

For this, we must repeat the whole process of constructing the triangulation of the surface $X(\mathbb{C})$. We start from a triangulation of $\mathbb{P}^1(\mathbb{C})$ invariant under τ and such that $\mathbb{P}^1(\mathbb{R})$ is a union of simplexes. The triangulation indicated in Figure 30 has these properties. Next, one can easily check that we can add extra precision to Proposition of Section 3.4 on subdividing a triangulation: if Φ is invariant under τ then the subdivision Ψ we construct is also invariant under τ. For this, one need only, when subdividing a simplex E into E' and E'', simultaneously subdivide $\tau(E)$ into $\tau(E')$ and $\tau(E'')$.

Finally, we choose a nonconstant function $f \in \mathbb{R}(X)$, that is, a rational function of the coordinates with real coefficients. The corresponding map $f: X(\mathbb{C}) \to \mathbb{P}^1(\mathbb{C})$ will obviously have the property that

$$f(\tau(x)) = \tau(f(x)).$$

It is easy to check that the process of constructing a triangulation Ψ of $X(\mathbb{C})$ compatible with the triangulation Φ of $\mathbb{P}^1(\mathbb{C})$ with respect to f described in the proof of Theorem 7.6 leads to a τ-invariant triangulation Ψ, provided that Φ was τ-invariant. Thus we have the following proposition.

Proposition 7.1 *$X(\mathbb{C})$ has a triangulation Φ invariant under the homeomorphism τ. Then $X(\mathbb{R})$ is obviously made up of simplexes of the triangulation.*

Proposition 7.2 *Let E be a 1-simplex of the triangulation Φ of $X(\mathbb{C})$ contained in $X(\mathbb{R})$, and E' and E'' the two 2-simplexes meeting along E. Then $\tau(E') = E''$.*

Proof Since E' and E'' are the only two simplexes of the triangulation Φ with E as boundary, Φ is τ-invariant and $\tau(E) = E$, it follows that either $\tau(E') = E'$ and $\tau(E'') = E''$, or $\tau(E') = E''$ and $\tau(E'') = E'$.

Let x be an interior point of E, and choose a local parameter $t \in \mathbb{R}(X)$, for example the equation of a hyperplane with real coefficients passing through x and transversal to X. Let $x \in U$ and

$$t: U \to \mathbb{C}$$

be a homeomorphism of U to the interior W of the unit disc $|z| < 1$ in $\mathbb{C}$. Choose U to be so small that it only intersects E, E' and E'' of all the simplexes of Φ.

Since $t \in \mathbb{R}(X)$,

$$t(\tau(x)) = \tau(t(x)), \tag{7.20}$$

and hence $t(E \cap U)$ equals the real diameter of the unit disc $|z| < 1$. We see that $U \setminus (U \cap E)$ breaks up into two connected components $U \cap (E' \setminus E)$ and $U \cap (E'' \setminus E)$. In the same way, $t(U) \setminus t(E)$ breaks up into two components, the upper and lower half-discs. Clearly the two different components of $U \setminus (U \cap E)$ map onto the two different components of the image. But the two half-discs are complex conjugate, so that it follows from (7.20) that

$$\tau(U \cap (E' \setminus E)) = U \cap (E'' \setminus E).$$

Therefore $\tau(E') \cap E'' \neq \emptyset$, and hence $\tau(E') = E''$. The proposition is proved. □

4.2 Proof of Harnack's Theorem

We use chain and homology groups with coefficients in $\mathbb{Z}/2\mathbb{Z}$. We work with coefficients $\mathbb{Z}/2\mathbb{Z}$, so that a chain $S = \sum \varepsilon_i E_i$ is a sum of simplexes E_i with coefficients

$\varepsilon_i = 0$ or 1; we say that E_i *appears* in S if $\varepsilon_i = 1$. For an arbitrary surface F, we write $H_1(F)$ for the group $H_1(F, \mathbb{Z}/2\mathbb{Z})$.

Let $T_1, \dots, T_k$ be all the connected components of $X(\mathbb{R})$. In the triangulation $\Phi(\mathbb{R})$, they are made up of 1- and 0-simplexes. They are obviously cycles in $H_1(X(\mathbb{C}))$, and we denote them by the same symbol.

Proposition *The cycles $T_1, \dots, T_k$ are either linearly independent elements of $H_1(X(\mathbb{C}))$, or related by the single equation*

$$T_1 + \cdots + T_k = 0.$$

Proof If the proposition is false then (after renumbering the T_i if necessary), there exists a relation

$$T_1 + \cdots + T_r = 0 \in H_1\bigl(X(\mathbb{C})\bigr) \quad \text{with } r < k.$$

In other words,

$$T_1 + \cdots + T_r = \partial S,$$

where S is a 2-chain of the triangulation Φ. The 2-simplexes not appearing in S form a chain $\overline{S}$, and since $\partial(S + \overline{S}) = 0$, we get

$$\partial\overline{S} = \partial S = T_1 + \cdots + T_r. \tag{7.21}$$

Thus every 1-simplex appearing in any of the T_i (for $1 \le i \le r$) is a face of exactly one 2-simplex appearing with coefficient 1 in S, and exactly one in $\overline{S}$.

Now note that $\tau(S)$ is also a chain of the triangulation Φ, and

$$\partial\bigl(\tau(S)\bigr) = \tau(\partial S) = \partial S \tag{7.22}$$

in view of (7.21), and because $\tau(T_i) = T_i$. Since $H_1(X(\mathbb{C}))$ is a $\mathbb{Z}/2\mathbb{Z}$-module, (7.22) shows that

$$\partial\bigl(S + \tau(S)\bigr) = 0. \tag{7.23}$$

Because $H_2(X(\mathbb{C})) = \mathbb{Z}/2\mathbb{Z}$ has only two elements, it follows from (7.23) that

$$\text{either} \quad S + \tau(S) = S + \overline{S}, \quad \text{so that } \overline{S} = \tau(S), \tag{7.24}$$

$$\text{or} \quad S + \tau(S) = 0, \quad \text{so that } S = \tau(S), \tag{7.25}$$

Consider an arbitrary 1-simplex E_1 of the triangulation Φ contained in T_i for some $i = 1, \dots, r$. Let E' and E'' be the 2-simplexes that meet along E_1. Then we can assume that E' appears in S, and E'' in $\overline{S}$. By Proposition 7.2 we see that (7.25) is impossible, so that (7.24) must hold.

Now consider the set T_{r+1} (recall we are assuming that $r < k$) and choose any point $t \in T_{r+1}$. Obviously $t \in S$ or $t \in \overline{S}$, but $t \notin S \cap \overline{S}$, since this intersection is $T_1 + \cdots + T_r$, which is disjoint from T_{r+1}. If for example $t \in S$ then $\tau(t) \in \tau(S)$.

But $\tau(t)=t$, since $t \in T_{r+1} \subset X(\mathbb{R})$, and $\tau(S)=\overline{S}$, and hence $t \in S \cap \overline{S}$, which, as we have seen, is false. The proposition is proved. □

To complete the proof of Harnack's theorem, we appeal to yet another topological argument. If F is an arbitrary oriented surface then $H_1(F, \mathbb{Z}/2\mathbb{Z})$ is a module of finite rank m over $\mathbb{Z}/2\mathbb{Z}$. The intersection pairing sends two elements $\alpha, \beta \in H_1(F)$ to an element of $\mathbb{Z}/2\mathbb{Z}$ denoted by (α, β). The function (α, β) is linear in each argument and skewsymmetric, that is, $(\alpha, \alpha)=0$ for any $\alpha \in H_1(F)$. By Poincaré duality it follows that it is nondegenerate, that is, if $\xi_1, \ldots, \xi_m$ is a basis of $H_1(F)$ then

$$\det\left|(\xi_i, \xi_j)\right| \neq 0.$$

It follows from this that any $n > m/2$ elements $\alpha_1, \ldots, \alpha_n \in H_1(F)$ such that $(\alpha_i, \alpha_j)=0$ for $i, j=1, \ldots, n$ are linearly dependent.

We apply this remark to $H_1(X(\mathbb{C}))$. As proved in Section 3, it has rank $2g$. The cycles $T_1, \ldots, T_k$, the connected components of $X(\mathbb{R})$, are disjoint by definition. Hence $(T_i, T_j)=0$ for $i, j=1, \ldots, k$, and hence any $g+1$ of them are linearly dependent in $H_1(X(\mathbb{C}))$. But if the number of components k is $>g+1$, we would get a contradiction to the proposition. Harnack's theorem is proved.

4.3 Ovals of Real Curves

In connection with Harnack's theorem, Hilbert, in his famous 1900 lecture on the problems of math, raised the question of the relative position of connected components of a real nonsingular plane curve $X \subset \mathbb{P}^2$ (see [38, Problem 16]). These components are called *ovals* of X. We only discuss the exact statement of the question in case of curves of even degree. In this case, we can prove that any oval of X is homologous to 0 in $\mathbb{P}^2(\mathbb{R})$, and divides it into two connected components, one homeomorphic to the disc and the other to the Möbius strip. The first component is called the *interior* of the oval. Hence it makes sense to speak of one oval containing another or being contained in it. The problem then consists of determining the possible relative positions of ovals (in the sense of which ovals contain one another) for all real nonsingular plane curves of a given degree. An analogous statement of the question in case of curves of odd degree is possible, but it requires more detailed considerations, and we do not discuss it.

At the present time, the answer to this question is known for curves of degree ≤ 7. We describe the answer for curves of even degree under the assumption that the number of ovals is the maximal allowed by Harnack's theorem. Using the formula for the genus of a nonsingular plane curve of degree $2n$, we see that this maximal number is $2n^2-3n+2$.

A curve of degree 2 can only have 1 oval. A curve of degree 4 has at most four ovals. Here only one relative position is possible, with all four ovals outside one another. A curve of degree 6 has at most 11 oval.

In this case 3 different relative positions are possible: one of the ovals contains 1, 5 or 9 ovals, with none contained in another, and outside it there are respectively 9, 5 or 1 ovals, with again none contained in another.

As the degree increases the complexity grows very rapidly. Thus for curves of degree 7 there are 121 types of relative position of ovals, of which 14 have the maximal number of ovals.

Certain general equalities and congruences are also known, as restriction on the relative position of ovals. For example, from a general result of I.G. Petrovskii it follows that the number of ovals of a curve of degree $2n$ not containing one another is $\leq (3/2)n(n-1)+1$, from which it follows in particular that a curve of degree 6 cannot decompose as 11 ovals with none contained in another. The analysis of all possible types of ovals is not known at present, and it is completely unclear in what terms one could search for such an analysis.

A natural generalisation of Harnack's theorem to higher dimensional varieties is Thom's theorem, according to which, for any nonsingular real algebraic variety X, we have

$$b_*(X(\mathbb{R})) \leq b_*(X(\mathbb{C})).$$

Here $b_*(Z)$ is the total Betti number mod 2, that is $\sum_k \dim H_k(\mathbb{Z}, \mathbb{Z}/2\mathbb{Z})$.

In the same lecture, and in connection with the same problem, Hilbert points out an analogy between questions of ovals of real algebraic curves and limit cycles of differential equations

$$\frac{\mathrm{d}y}{\mathrm{d}x} = \frac{f(x,y)}{g(x,y)},$$

where f and g are polynomials. In this question, there is not even an analogue of Harnack's theorem known, that is, no bound is known for the number of limit cycles in terms of the maximum degrees of f and g. A bound is not even known when this maximum is 2.

Only recently has it been proved that the number of limit cycles is finite for each individual equation, but it is not known if there exists a general bound on the number of limit cycles for all equations with given $N = \max(\deg f, \deg g)$, and a fortiori a value for this bound, even for $N = 2$. In this case, equations with 4 limit cycles have been constructed.

4.4 Exercises to Section 4

1 Prove that if a real curve of degree 4 decomposes as ≥ 3 ovals then none is contained in another. [Hint: Otherwise there would exist a line intersecting the curve in 6 points.]

2 Prove that the normalisation of a real curve is again real.

3 Consider the normalisation of the projective closure of the curve

$$y^2 = -(x - e_1) \cdots (x - e_{2g+2}) \quad \text{with } e_i \in \mathbb{R} \text{ and } e_i \neq e_j.$$

Prove that its number of ovals equals the bound of Harnack's theorem.

4 Verify Thom's theorem for a real quadric in $\mathbb{P}^3$.

5 Prove that a real cubic surface, not a cone, is unirational over $\mathbb{R}$. [Hint: Consider the intersection of the surface with the tangent plane at a nonsingular point. Carry out the same operation at every point of this singular cubic.]

6 Prove that the cubic surface $t(z^2 + y^2) = x^3 - xt^2$ in $\mathbb{P}^3(\mathbb{R})$ consists of two components. Deduce that it is not rational over $\mathbb{R}$. This is a counterexample to the Lüroth problem over $\mathbb{R}$!

Chapter 8
Complex Manifolds

1 Definitions and Examples

1.1 Definition

In the preceding chapter we studied the topological space $X(\mathbb{C})$ associated with an arbitrary algebraic variety defined over the complex numbers $\mathbb{C}$. The example of nonsingular projective curves already gives a feeling for the extent to which $X(\mathbb{C})$ characterises the variety X. We proved that in this case the genus g of X is the unique invariant of the topological space $X(\mathbb{C})$. Thus we can say that the genus is the unique topological invariant of a nonsingular projective curve. The genus is undoubtedly an extremely important invariant of an algebraic curve, but it is very far from determining it. We saw at the end of Section 7.1, Chapter 3 (see also Exercise 8 of Section 2.6, Chapter 3) that there are very many nonisomorphic curves of the same genus. The relation between a variety X and the topological space $X(\mathbb{C})$ is similar in nature for higher dimensional varieties.

By looking more carefully at how the topology of the set $X(\mathbb{C})$ was defined in Section 1.1, Chapter 7, we observe that the same method allows us to associate with X another object, which reflects many more properties of the variety X. We do this here under the assumption that X is a nonsingular variety; the general case is considered in Section 1.5.

We begin as in the preceding chapter: consider a point $x \in X(\mathbb{C})$ and some system of local parameters $t_1, \dots, t_n$ at x; these define a homeomorphism

$$\varphi: U \xrightarrow{\sim} V \subset \mathbb{C}^n \tag{8.1}$$

of some neighbourhood U of x with a neighbourhood $V \subset \mathbb{C}^n$ of the origin. This homeomorphism was used to give $X(\mathbb{C})$ the structure of a $2n$-dimensional topological manifold. An essential point here was the compatibility of the different maps (8.1) defined in various neighbourhoods U by various systems of local parameters. This follows from the fact that if $f \in \mathbb{C}(X)$ is a regular function at x then $g = f \circ \varphi^{-1}$ is complex analytic in a neighbourhood of the origin as a function of

I.R. Shafarevich, *Basic Algebraic Geometry 2*, DOI 10.1007/978-3-642-38010-5_4,

the n complex variables $z_1, \dots, z_n$ of $\mathbb{C}^n$. We have so far made very little use of this property—only the continuity of g, from which it followed that any other local parameters $u_1, \dots, u_n$ are continuous functions of $t_1, \dots, t_n$.

The essence of this argument is that there is an invariant way of defining a continuous complex valued (or real valued) function in a neighbourhood U of a point $x \in X(\mathbb{C})$. Thus it is natural to say that a function $h: U \to \mathbb{C}$ is *continuous* if the function $h \circ \varphi^{-1}$ is continuous on $V \subset \mathbb{C}^n$, and this property is independent of the choice of φ. Recall now that if f is a regular function at x, the function $g = f \circ \varphi^{-1}$ is not just continuous, but also analytic. It follows that if $u_1, \dots, u_n$ is another system of local parameters at x then $u_1, \dots, u_n$ are analytic functions of $t_1, \dots, t_n$ in some neighbourhood $x \in U' \subset U$. Hence if $h: U \to \mathbb{C}$ is a continuous function and $h \circ \varphi^{-1}$ is analytic near 0 then the same holds for any map (8.1) given by another system of local parameters at x.

Thus the following notion is well defined: we say that a complex valued function h on a neighbourhood of $x \in X(\mathbb{C})$ is *analytic* at x if the function $g(z_1, \dots, z_n) = h \circ \varphi^{-1}$ defined by means of a map (8.1) is analytic in a neighbourhood of $0 \in \mathbb{C}^n$ as a function of the variables $z_1, \dots, z_n$.

The functions $h: U \to \mathbb{C}$ that are analytic at all points of an open set U form a ring, that we denote by $\mathcal{O}_{\mathrm{an}}(U)$. Since the definition of analytic is local in nature, the map $U \to \mathcal{O}_{\mathrm{an}}(U)$ defines a sheaf $\mathcal{O}_{\mathrm{an}}$ called the sheaf of analytic functions. $\mathcal{O}_{\mathrm{an}}$ is obviously a subsheaf of the sheaf of continuous functions on $X(\mathbb{C})$, and in turn the sheaf of regular functions $\mathcal{O}$ is a subsheaf of $\mathcal{O}_{\mathrm{an}}$.

In previous chapters of the book we defined an algebraic variety in terms of its underlying topological space (in the Zariski topology) and its sheaf of regular functions. In the same way, a topological space together with a specified sheaf of analytic functions leads to the new notion that we wish to define.

We consider first a domain W in the space $\mathbb{C}^n$ of n complex variables. For any open set $U \subset W$ the set of all functions that are analytic at all points of U forms an algebra $\mathcal{O}_{\mathrm{an}}(U)$ over $\mathbb{C}$, and assigning $U \to \mathcal{O}_{\mathrm{an}}(U)$ defines a sheaf of $\mathbb{C}$-algebras on W, which is a subsheaf of the sheaf of continuous functions on W. This sheaf $\mathcal{O}_{\mathrm{an}}$ is called the *sheaf of analytic functions* on W. Ringed spaces of the form $W, \mathcal{O}_{\mathrm{an}}$ play the role of the simplest objects in our theory, and are analogous to affine schemes in the definition of the general notion of scheme.

Definition A ringed space $X, \mathcal{O}_X$, consisting of a Hausdorff topological space X together with a specified sheaf of $\mathbb{C}$-algebras $\mathcal{O}_X$ which is a subsheaf of the sheaf of continuous functions on X, is a *complex manifold* or *complex analytic manifold* if it satisfies the following condition: for every $x \in X$ there exists a neighbourhood U such that the ringed space obtained by restricting $\mathcal{O}_X$ to U is isomorphic to $W, \mathcal{O}_{\mathrm{an}}$, where W is a domain in $\mathbb{C}^n$ and $\mathcal{O}_{\mathrm{an}}$ the sheaf of analytic functions on W. This isomorphism allows us to introduce coordinates $z_1, \dots, z_n$ on U, that from now on we call *local analytic coordinates*. An *analytic function* on an open set $U \subset X$ is a continuous function on U that is a section of $\mathcal{O}_X$ over U.

A map $f: X \to Y$ of two complex manifolds is *holomorphic* if it is continuous and defines a morphism of ringed spaces; this is equivalent to requiring that the pullback $f^*(h)$ of an analytic function h is analytic.

We say that a closed subset $Y \subset X$ is a *complex submanifold* if there exists a local analytic coordinate system $z_1, \dots, z_n$ on X in a neighbourhood of every $y \in Y$ such that Y is given in this neighbourhood by the equations $z_1 = \cdots = z_m = 0$. This defines a natural structure of complex manifold on Y and the embedding $Y \hookrightarrow X$ is a holomorphic map.

If x is a point of a complex manifold, U a neighbourhood of x and $f\colon U \to W \subset \mathbb{C}^n$ an isomorphism of U with an open set in $\mathbb{C}^n$ then the number n is called the *complex dimension* of X at x. It follows from the definition that the topological space $X(\mathbb{C})$ is a $2n$-dimensional topological manifold at x. Thus the complex dimension is the same for all points of a connected component of a complex manifold. If X is connected then this number n is its *complex dimension*.

We have just constructed a sheaf $\mathcal{O}_{\text{an}}$ on every nonsingular algebraic variety X over $\mathbb{C}$. Obviously $X(\mathbb{C}), \mathcal{O}_{\text{an}}$ is a complex manifold, and we denote it by X_{an}; any morphism $f\colon X \to Y$ of algebraic varieties defines a holomorphic map $X_{\text{an}} \to Y_{\text{an}}$ denoted by f_{an}. The remainder of this chapter is devoted mainly to studying the relation between algebraic varieties and the corresponding complex manifolds. For example, the following questions arise:

(i) Is every complex manifold of the form X_{an} where X is some algebraic variety?
(ii) Is every holomorphic map $X_{\text{an}} \to Y_{\text{an}}$ of the form f_{an}, where $f\colon X \to Y$ is some regular map of algebraic varieties?
(iii) If X_{an} and Y_{an} are isomorphic as complex manifolds then does it follow that X and Y are isomorphic as algebraic varieties?

The answer to all these questions is negative, and counterexamples are not hard to construct: for (i) and (ii) see Exercises 1–3 of Section 1.6; for (iii) see Section 3.2. But if we restrict attention to compact complex manifolds, the same questions become much deeper, and the answers less trivial. We consider these in the following sections.

A number of arguments of the preceding Chapter 7 only used the fact that $X(\mathbb{C})$ is a complex manifold; this applies, for example, to the introduction of an orientation on $X(\mathbb{C})$.

Some Notation and Conventions As mentioned in the preface to this volume, this chapter assumes known the theory of homology and cohomology on manifolds, and we use freely the de Rham treatment of cohomology in terms of differential forms. We gather here some conventions relating to complex manifolds. For more details see, for example, Cartan [20, II.2.3].

An n-dimensional complex manifold is a $2n$-dimensional differentiable manifold. If $z_1, \dots, z_n$ are local complex coordinates and $z_j = x_j + iy_j$ then we can take $x_1, y_1, \dots, x_n, y_n$ as local real coordinates. In particular any differential form on X can be written in terms of the $\mathrm{d}x_j$ and $\mathrm{d}y_j$. But it is usually convenient to use coordinates z_j and $\overline{z}_j$, setting

$$x_j = \frac{1}{2}(z_j + \overline{z_j}), \qquad y_j = \frac{1}{2i}(z_j - \overline{z_j}). \tag{8.2}$$

For example, one uses the definition

$$\frac{\partial}{\partial z} = \frac{1}{2}\left(\frac{\partial}{\partial x} - i\frac{\partial}{\partial y}\right) \quad \text{and} \quad \frac{\partial}{\partial \overline{z}} = \frac{1}{2}\left(\frac{\partial}{\partial x} + i\frac{\partial}{\partial y}\right), \tag{8.3}$$

under which the Cauchy–Riemann equations read "f holomorphic if and only if $\partial f/\partial \overline{z} = 0$." We write the differential d acting on forms as $\mathrm{d} = \mathrm{d}' + \mathrm{d}''$, where d' is the differential with respect to the variables z_j and d'' that with respect to the $\overline{z}_j$. For example, if φ is a function then $\mathrm{d}\varphi = \mathrm{d}'\varphi + \mathrm{d}''\varphi$, where

$$\mathrm{d}'\varphi = \sum \frac{\partial \varphi}{\partial z_j}\mathrm{d}z_j \quad \text{and} \quad \mathrm{d}''\varphi = \sum \frac{\partial \varphi}{\partial \overline{z}_j}\mathrm{d}\overline{z}_j. \tag{8.4}$$

Obviously $(\mathrm{d}')^2 = (\mathrm{d}'')^2 = 0$ and $\mathrm{d}'\mathrm{d}'' = -\mathrm{d}''\mathrm{d}'$.

In writing out a differential m-form η, we can group together all the terms

$$f_{j_1 \cdots j_m} \mathrm{d}z_{j_1} \wedge \cdots \wedge \mathrm{d}z_{j_p} \wedge \mathrm{d}\overline{z}_{j_{p+1}} \wedge \cdots \wedge \mathrm{d}\overline{z}_{j_m} \tag{8.5}$$

involving a given number p of the $\mathrm{d}z_j$, and therefore $q = (m - p)$ of the $\mathrm{d}\overline{z}_j$. By doing this we get a unique decomposition

$$\eta = \sum_{p+q=m} \eta^{(p,q)}, \tag{8.6}$$

where $\eta^{(p,q)}$ is a form of *type* (p, q) or a (p, q)-form.

1.2 Quotient Spaces

We now describe a new method of constructing complex manifolds. Our first application of this is a series of examples that answer some of the questions discussed at the end of Section 1.1.

Let X be a topological space and G a group consisting of homeomorphisms of X. We say that G acts *freely and discretely* on X if the following two conditions hold:

(1) Every point $x \in X$ has a neighbourhood U such that

$$g(U) \cap U = \emptyset \quad \text{for every } g \in G \text{ with } g \neq e. \tag{8.7}$$

(2) Any two points $x, y \in X$ such that $x \neq g(y)$ for any $g \in G$ have neighbourhoods $U \ni x$ and $V \ni y$ such that

$$g(U) \cap V = \emptyset \quad \text{for every } g \in G.$$

We write X/G for the set of equivalence classes of points, where two points x_1, x_2 are equivalent if there exists $g \in G$ such that $g(x_1) = x_2$. Sending each point to its equivalence class defines a map $\pi : X \to X/G$. We introduce a topology on the set X/G, saying that a subset $U \subset X/G$ is open if $\pi^{-1}(U)$ is open in X. Condition (2) guarantees that if X is a Hausdorff space then so is X/G. If $x \in X$, $y = \pi(x)$, and U is a neighbourhood of x satisfying (1) then $V = \pi(U)$ is a neighbourhood of y. Moreover by (8.7) we have

$$\pi^{-1}(V) = \bigcup_{g \in G} g(U), \quad \text{where } g_1(U) \cap g_2(U) = \emptyset \text{ for } g_1 \neq g_2, \tag{8.8}$$

and the map $\pi : U \to V$ is a homeomorphism. We say that a map of topological spaces $\pi : X \to Y$ with this property is a *covering space* or *unramified cover*.

Suppose now that $X, \mathcal{O}_X$ is a complex manifold, and G a group of automorphisms of X acting freely and discretely on X. In this case we construct a sheaf $\mathcal{O}_{X/G}$ on X/G by defining $\mathcal{O}_{X/G}(V)$ to be the set of all continuous functions f on V such that $\pi^*(f)$ is analytic on $\pi^{-1}(V)$, that is, $\pi^*(f) \in \mathcal{O}_X(\pi^{-1}(V))$.

We now prove that $X/G, \mathcal{O}_{X/G}$ is a complex manifold. For this, consider a neighbourhood U of a point $x \in X$ that satisfies both (8.7) and the condition in the definition of a complex manifold, with $\varphi \colon U \to W$ the isomorphism to a domain of $\mathbb{C}^n$. Set $\pi(U) = V$, and consider a continuous function f on V. By (8.8), $f \in \mathcal{O}_{X/G}(V)$ if and only if $\pi^*(f) \in \mathcal{O}_X(g(U))$ for every $g \in G$. On the other hand, since g is an automorphism of X, it defines an isomorphism of U and $g(U)$, and the restrictions of $\pi^*(f)$ to U and $g(U)$ go into one another under this isomorphism. It follows from this that if $\pi_1 : U \to V$ is the restriction of π_1 to U then

$$f \in \mathcal{O}_{X/G}(V) \iff \pi_1^*(f) \in \mathcal{O}_X(U).$$

Since π_1 is a homeomorphism, it follows from this that $\pi_1 : U \to V$ is an isomorphism of ringed spaces. Hence we have an isomorphism

$$\varphi \circ \pi_1^{-1} \colon V \to W \subset \mathbb{C}^n$$

of ringed spaces, whose existence proves that X/G is a complex manifold.

Example 8.1 (Quotients of $\mathbb{C}^n$ by a lattice) We view the n-dimensional complex vector space $\mathbb{C}^n$ as a $2n$-dimensional real vector space, and choose m linearly independent vectors $a_1, \dots, a_m$. Write Ω for the set of vectors of the form

$$a = k_1 a_1 + \cdots + k_m a_m \quad \text{with } k_i \in \mathbb{Z},$$

which is a lattice of rank m. For each $a \in \Omega$ the translation $g_a \colon \mathbb{C}^n \to \mathbb{C}^n$ given by $g_a(z) = z + a$ for $z \in \mathbb{C}^n$ is an automorphism of the complex manifold $\mathbb{C}^n$. Since

$$g_{a+b} = g_a \circ g_b,$$

the translations g_a with $a \in \Omega$ form a group G. Obviously G acts freely and discretely on $\mathbb{C}^n$. Indeed, extend $a_1, \dots, a_m$ to a basis $a_1, \dots, a_{2n}$ of $\mathbb{C}^n$ and write U for the open set consisting of vectors $a = x_1 a_1 + \cdots + x_{2n} a_{2n}$ with $x_i \in$

$(-1/2, 1/2) \subset \mathbb{R}$ for $i = 1, \ldots, m$. Then the set $z + U$ consisting of vectors $z + u$ with $u \in U$ is a neighbourhood of z, and

$$g_a(z + U) \cap (z + U) = \emptyset \quad \text{for } a \in \Omega \text{ and } a \neq 0,$$

that is, condition (1) in the definition of free and discrete action holds. Condition (2) is equally obvious.

Thus $\mathbb{C}^n/G$ is a complex manifold; it is easy to see that it is compact if and only if $m = 2n$.

Suppose that $m = 2n$. In this case the manifold $\mathbb{C}^n/G$ has a very simple topological structure. Since

$$\mathbb{C}^n = \mathbb{R}a_1 + \cdots + \mathbb{R}a_{2n} \quad \text{and} \quad \Omega = \mathbb{Z}a_1 + \cdots + \mathbb{Z}a_{2n},$$

$\mathbb{C}^n/G$ is homeomorphic to the product of $2n$ copies of $\mathbb{R}/\Gamma$, where Γ is the group of translations $t \mapsto t + n$ with $t \in \mathbb{R}$ and $n \in \mathbb{Z}$. Obviously $\mathbb{R}/\Gamma$ is homeomorphic to the circle, and $\mathbb{C}^n/G$ to the $2n$-dimensional torus. Hence when $m = 2n$ the manifold $\mathbb{C}^n/G$ is called a *complex torus*. We will see later that two complex toruses are not usually isomorphic to one another as complex manifolds.

Example 8.2 (Hopf manifolds) Write $X = \mathbb{C}^n \setminus 0$ and let c be a real number with $c > 1$. Let G be the group of transformations

$$(z_1, \ldots, z_n) \mapsto \left(c^k z_1, \ldots, c^k z_n\right) \quad \text{with } k \in \mathbb{Z}.$$

One can easily check directly that G acts freely and discretely on X; but this becomes completely obvious from the following considerations.

Write any point $z \in X$ in the form

$$z = ru,$$

where r is a positive number and $u = (u_1, \ldots, u_n)$ a vector such that $|u_1|^2 + \cdots + |u_n|^2 = 1$. This representation is obviously unique and defines a homeomorphism

$$X \cong \mathbb{R}_+ \times S^{2n-1},$$

where $\mathbb{R}_+$ is the set of positive real numbers and S^{2n-1} the $(2n-1)$-dimensional sphere. In this representation the transformations of G act trivially on S^{2n-1}, and on $\mathbb{R}_+$ by multiplying by powers of c. If we use the log function to map $\mathbb{R}_+$ to $\mathbb{R}$ then this action becomes the translations by vectors of the lattice $\mathbb{Z}\gamma$, where $\gamma = \log c$. From this we see that G acts freely and discretely, and that X/G is homeomorphic to $(\mathbb{R}/\mathbb{Z}) \times S^{2n-1}$, that is, to $S^1 \times S^{2n-1}$. The compact complex manifold we have just constructed is called the *Hopf manifold*.

1.3 Commutative Algebraic Groups as Quotient Spaces

We return to the quotient spaces $\mathbb{C}^n/G$ discussed in Example 8.1, where G is a group consisting of translations by vectors of some lattice Ω. This lattice is a subgroup of $\mathbb{C}^n$, and the quotient space $\mathbb{C}^n/G$ is homeomorphic to the quotient group $\mathbb{C}^n/\Omega$, and is hence a group. It is easy to check that the map

$$m\colon (\mathbb{C}^n/\Omega) \times (\mathbb{C}^n/\Omega) \to \mathbb{C}^n/\Omega$$

defined by the group law is holomorphic. Thus $\mathbb{C}^n/\Omega$ is a commutative complex Lie group.

Suppose that the manifold $\mathbb{C}^n/\Omega$ arises from some algebraic variety X, that is, is of the form X_{an}. In this case one can prove that $m = \mu_{\text{an}}$, where

$$\mu\colon X \times X \to X$$

is a morphism that defines an algebraic group structure on X. In the most interesting case when X is compact this will follow from Theorem 8.5. In this case, X is thus an Abelian variety (see Section 4.3, Chapter 3).

We now show that, conversely, any commutative algebraic group over the complex number field can be represented in the form $\mathbb{C}^n/\Omega$ for some lattice Ω. For this we need an auxiliary result.

Lemma *An invariant differential* 1*-form* $\varphi \in \Omega^1[G]$ *(Sections* 5.1 *and* 6.2*, Chapter* 3*) on a commutative algebraic group is closed.*

Proof Let φ be an invariant differential form on a group G. It is easy to check that then $d\varphi$ is also invariant. Hence it is enough to prove that $d\varphi(e) = 0$, from which it follows that also $d\varphi = 0$.

We write φ in the form $\varphi = \sum \psi_k du_k$, then use (3.72) of Section 6.2, Chapter 3:

$$\sum_l c_{kl}\psi_l = \psi_k(e) \in \mathbb{C}.$$

It follows from this that

$$\sum_l \frac{\partial \psi_l}{\partial u_i} c_{kl} + \sum_l \psi_l \frac{\partial c_{kl}}{\partial u_i} = 0 \quad \text{for } i = 1, \dots, n.$$

Consider this equality at the point e. Since $c_{kl}(e) = \delta_{kl}$, it follows that

$$\frac{\partial \psi_k}{\partial u_i}(e) + \sum_l \psi_l(e) \frac{\partial c_{kl}}{\partial u_i}(e) = 0.$$

To prove the equality

$$\frac{\partial \psi_k}{\partial u_i}(e) = \frac{\partial \psi_i}{\partial u_k}(e) \tag{8.9}$$

expressing that φ is a closed form, it is enough to check that

$$\frac{\partial c_{kl}}{\partial u_i}(e) = \frac{\partial c_{il}}{\partial u_k}(e).$$

It is at this point that we use the commutativity of G. We use (3.71) of Section 6.2, Chapter 3:

$$c_{kl} = \sum_j v_{lj}\big(g^{-1}\big)\frac{\partial w_{lj}}{\partial u_k}(e).$$

Because the group is commutative,

$$\mu^*(u_k)(g_1, g_2) = \sum_j v_{kj}(g_1)w_{kj}(g_2) = \sum_j w_{kj}(g_1)v_{kj}(g_2).$$

Hence

$$c_{kl}(g) = \sum_j v_{lj}\big(g^{-1}\big)\frac{\partial w_{lj}}{\partial u_k}(e) = \sum_j w_{lj}\big(g^{-1}\big)\frac{\partial v_{lj}}{\partial u_k}(e),$$

and therefore

$$\frac{\partial c_{kl}}{\partial u_i}(e) = \sum_j \frac{\partial v_{lj}}{\partial u_i}(e)\frac{\partial w_{lj}}{\partial u_k}(e) = \sum_j \frac{\partial w_{lj}}{\partial u_i}(e)\frac{\partial v_{lj}}{\partial u_k}(e) = \frac{\partial c_{il}}{\partial u_k}(e).$$

This proves (8.9) and the lemma. □

Now consider an arbitrary n-dimensional connected commutative algebraic group A defined over $\mathbb{C}$. By Proposition of Section 6.2, Chapter 3, the space of invariant differential 1-forms on A is n-dimensional. Write $\omega_1, \dots, \omega_n$ for a basis. The differential forms ω_i are closed by the lemma, so that there exist holomorphic functions $f_1, \dots, f_n$ defined in some complex neighbourhood U of the zero element $0 = e \in A$ such that

$$\omega_i = \mathrm{d}f_i \quad \text{and} \quad f_i(e) = 0.$$

This is a simple local fact that can easily be checked directly; a more general assertion is proved in Lemma of Section 4.3. Since the ω_i are invariant, it follows that

$$\mathrm{d}\big(t_g^*(f_i)\big) = \mathrm{d}f_i$$

in the domain $U \cap t_g^{-1}(U)$. Thus

$$t_g^*(f_i) = f_i + \alpha_i \quad \text{with } \alpha_i \in \mathbb{C}. \tag{8.10}$$

But $(t_g^*(f_i))(g_1) = f_i(g + g_1)$, where we write the group law on A additively. Hence (8.10) means that $f_i(g + g_1) = f_i(g) + \alpha_i$ if g, $g + g_1 \in U$. In particular, setting

$g = 0$ we get that

$$f_i(g + g_1) = f_i(g) + f_i(g_1) \quad \text{for } g, \text{ and } g + g_1 \in U. \tag{8.11}$$

Thus $f_1, \ldots, f_n$ defines a "local group homomorphism" of a neighbourhood of $e \in A$ to a neighbourhood of $0 \in \mathbb{C}^n$. We see by Section 6.2, Chapter 3 that $\mathrm{d}_e f_i$ form a basis of Θ_e^*, and therefore the Jacobian $\det|\partial f_i/\partial t_j(e)| \neq 0$ for any system of local parameters $t_1, \ldots, t_n$ at 0. Hence the map φ defined by

$$\varphi(g) = \big(f_1(g), \ldots, f_n(g)\big)$$

is an analytic isomorphism of a neighbourhood U of e in A to a neighbourhood V of 0 in $\mathbb{C}^n$, and by (8.11) is a "local group isomorphism".

We now construct a homomorphism $\psi\colon \mathbb{C}^n \to A$, setting

$$\psi(z) = k\varphi^{-1}(z/k) \quad \text{for } z \in \mathbb{C}^n,$$

where k is a sufficiently large integer such that $z/k \in V$. The fact that this is well defined (independent of k) follows at once from (8.11). We have thus constructed a homomorphism $\psi\colon \mathbb{C}^n \to A$ that is equal to φ^{-1} on $V \subset \mathbb{C}^n$. From the fact that ψ is isomorphic on V and A is connected it follows easily that $\psi(\mathbb{C}^n) = A$. Write Ω for the kernel of ψ. Then $\Omega \cap V = 0$, that is, Ω is a discrete subgroup of $\mathbb{C}^n$. It follows easily from this that Ω is a lattice, and therefore $A \cong \mathbb{C}^n/\Omega$ where Ω is a lattice.

We have thus proved the following result.

Theorem *Let A be an n-dimensional connected commutative algebraic group defined over $\mathbb{C}$. Then A_{an} is isomorphic to $\mathbb{C}^n/\Omega$, where Ω is a lattice.*

1.4 Examples of Compact Complex Manifolds not Isomorphic to Algebraic Varieties

We now take up the first of the questions posed at the end of Section 1.2 in connection with the definition of complex manifold: does every complex manifold arise from some algebraic variety, that is, is it of the form X_{an}? Since the question is only really interesting if we restrict ourselves to compact complex manifolds, we consider it in this setting.

That the question is much more delicate under this restriction is apparent if only because it has a positive answer for 1-dimensional complex manifolds: every compact 1-dimensional manifold is isomorphic to X_{an}, where X is a nonsingular projective curve. This assertion is called the *Riemann mapping theorem.* We will not give a proof; in any case, it requires some arguments of an analytic nature. A proof can be found, for example, in Springer [74, Chapter 8] or Forster [28].

The fact that our question has a negative answer in dimension >1 is therefore all the more interesting. We meet once again a familiar phenomenon of algebraic geometry: many of the difficulties do not yet arise in dimension 1. We now give some example of nonalgebraic compact complex manifolds, restricting ourselves for simplicity to 2-dimensional manifolds.

Since this is a question that relates to notions at the very heart of complex analytic geometry, we will work out two principles for constructing such examples: in Example 8.3 we use almost exclusively algebraic considerations, in Examples 8.4–8.5 more geometric ones.

Example 8.3 Our manifold is a complex torus, that is, is of the form $\mathbb{C}^2/\Omega$, where $\Omega \subset \mathbb{C}^2$ is a lattice of rank 4. If it were algebraic, it would be an Abelian variety, as we proved in Section 1.3. We now prove that Abelian varieties have a property which turns out not to be satisfied by $\mathbb{C}^2/\Omega$ for some choice of the lattice Ω. This property is the *Poincaré complete irreducibility theorem*; it consists of the following assertion.

Proposition *If A and B are Abelian varieties and $\varphi\colon A \to B$ a surjective homomorphism then there exists an Abelian subvariety $C \subset A$ such that* $\dim C = \dim B$ *and $\varphi\colon C \to B$ is surjective.*

Proof We assume for brevity that $\dim A = 2$, $\dim B = 1$, and that the ground field has characteristic 0; we will only make use of the proposition under these assumptions.

Consider a point $a \in A$ and the fibre $Y = \varphi^{-1}(\varphi(a))$ of φ through a. By the theorem on dimension of fibres (Theorem 1.25 of Section 6.3, Chapter 1) Y is a curve. There exists another irreducible curve X on A through a and not contained in Y. This follows from the fact that A is algebraic. It is enough to consider an affine neighbourhood of a and take X to be a suitable hyperplane section, or a component of this. Let $\psi : X \to B$ be the restriction of φ; this is obviously a morphism of finite degree, and has finite fibres $\psi^{-1}(b)$.

For a divisor $D = \sum_{i=1}^r k_i x_i$ on X, write $S(D)$ for the point $k_1 x_1 \oplus \cdots \oplus k_r x_r$ where $\oplus$ is the group law on A, and set

$$f(b) = S\bigl(\psi^*(b)\bigr) \quad \text{for } b \in B.$$

We assert that f is a morphism from B to A. We first check that it is a rational map. Let θ be a primitive element of the field extension $k(X)/k(B)$, so that

$$k(X) = k(B)(\theta).$$

The affine coordinates t_j of some point $x \in X$ are thus of the form $F_j(\theta)$ with $F_j \in k(B)(\theta)$. Let $\theta_1, \dots, \theta_n$ be all the conjugates of θ over the field $k(B)$. Then since θ and θ_i are conjugate, the points x_i with coordinates $F_j(\theta_i)$ belong to X and have the same image in B:

$$\psi(x_i) = \psi(x) = b.$$

The coordinates of the points of $f(b)$ can obviously be expressed as symmetric rational functions in the coordinates of x_i, that is, they are symmetric functions in the θ_i, and are hence elements of $k(B)$. This proves that f is a rational map.

Since A is complete and B is a nonsingular curve, it follows that f is a morphism, and from Theorem 3.16 of Section 4.3, Chapter 3 that it is a homomorphism. Set $C = f(B)$. To prove the proposition it is enough to check that $\varphi(C) \neq 0$. But by definition

$$\varphi(C) = \varphi\big(f(B)\big) = \nu(B),$$

where $\nu\colon B \to B$ is the endomorphism of multiplying by n:

$$\nu(b) = b \oplus \cdots \oplus b.$$

Since the ground field has characteristic 0, $\ker \nu$ is finite (Exercise 2 of Section 4.5, Chapter 3), hence $\nu(B) \neq 0$. The proposition is proved. □

Now we can complete the construction of the example. The idea is to construct a complex torus for which the proposition does not hold. Consider the lattice $\Omega \subset \mathbb{C}^2$ with basis consisting of the 4 vectors

$$(1,0), \quad (i,0), \quad (0,1), \quad (\alpha,\beta).$$

It is easy to see that these are linearly independent over $\mathbb{R}$ provided that β is not real. Set $A = \mathbb{C}^2/\Omega$ and $B = \mathbb{C}^1/\Omega'$, where Ω' is the lattice with basis 1 and β. The map $(z_1, z_2) \mapsto z_2$ induces a holomorphic map $\varphi\colon A \to B$ which is a group homomorphism. Suppose that A is an Abelian variety. It follows from the Riemann existence theorem that B is an algebraic curve (we will verify this directly in Theorem 9.3). As we will see a little later (Theorem 8.5), it follows from this that φ is a morphism. We can apply the Poincaré complete irreducibility theorem to deduce that there exists a 1-dimensional Abelian subvariety $C \subset A$ such that $\varphi(C) = B$.

Write $\Lambda \subset \mathbb{C}^2$ for the inverse image of C. It is a closed subgroup of $\mathbb{C}^2$, and it is easy to determine all the closed subgroups of any $\mathbb{R}^n$: a simple argument (see Pontryagin [66, Section 19, Ex. 33, p. 110]) shows that they are of the form $\mathbb{Z}e_1 + \cdots + \mathbb{Z}e_s + \mathbb{R}e_{s+1} + \cdots + \mathbb{R}e_{s+r}$, where $e_1, \ldots, e_{s+r}$ are linearly independent over $\mathbb{R}$. In our case $\Lambda \supset \Omega$, and hence it contains 4 linearly independent vectors over $\mathbb{R}$. Write Λ_0 for the connected component of 0 in Λ. This is an $\mathbb{R}$-vector subspace. Since $C \subset A$ is locally defined by one equation and is nonsingular, $\Lambda_0 \subset \mathbb{C}^2$ is locally defined by an equation $f(z_1, z_2) = 0$ with nonzero linear part. Let

$$f = f_1 + f_2 + \cdots$$

be the Taylor series of f written out in homogeneous components. Then for sufficiently small $\alpha \in \mathbb{R}$

$$f(\alpha z_1, \alpha z_2) = \alpha f_1 + \alpha^2 f_2 + \cdots,$$

and since Λ_0 is an $\mathbb{R}$-vector subspace, it follows that $f_2 = \cdots = 0$ and $f = f_1$. We see that Λ_0 is defined by a linear equation $f_1 = 0$. We deduce that

$$\Lambda = \mathbb{Z}e_1 + \mathbb{Z}e_2 + \mathbb{R}e_3 + \mathbb{R}e_4 = \mathbb{Z}e_1 + \mathbb{Z}e_2 + \Lambda_0,$$

where $\Lambda_0 = \mathbb{R}e_3 + \mathbb{R}e_4$ is a $\mathbb{C}$-vector subspace of $\mathbb{C}^2$; in other words, $e_4 = \lambda e_3$ with $\lambda \in \mathbb{C}$. To conclude, recall that $\Lambda \supset \Omega$. It follows from this that the projection of Ω to $\mathbb{Z}e_1 + \mathbb{Z}e_2$ maps a rank 2 sublattice of Ω to 0. Hence $\Lambda_0 \cap \Omega$ is a rank 2 sublattice of Λ_0. We thus arrive at the conclusion that, in our case, the Poincaré complete reducibility theorem means simply that there exists a complex line

$$\Lambda_0 = \mathbb{R}e_3 + \mathbb{R}e_4 \quad \text{with } e_4 = \lambda e_3 \text{ for some } \lambda \in \mathbb{C},$$

such that $\Lambda_0 \cap \Omega$ is a rank 2 sublattice of Λ_0, and which projects to the whole line z_1, that is, is not equal to the line $z_2 = 0$.

In other words, to check that the theorem holds, we have to find a vector e in Ω such that $z_1 \neq 0$, and $\lambda e \in \Omega$ for some value $\lambda \in \mathbb{C} \setminus \mathbb{R}$. Now let's see if this is always possible. Suppose that

$$e = a(1,0) + b(i,0) + c(0,1) + d(\alpha,\beta),$$
$$\lambda e = a(\lambda,0) + b(i\lambda,0) + c(0,\lambda) + d(\lambda\alpha,\lambda\beta),$$

with $a, b, c, d \in \mathbb{Z}$. The z_2-coordinate of every vector of Ω is contained in $\mathbb{Z} + \mathbb{Z}\beta$. In particular $(c + d\beta)\lambda \in \mathbb{Z} + \mathbb{Z}\beta$, and hence λ must be contained in the field $\mathbb{Q}(\beta)$ (recall that $c + d\beta \neq 0$ by assumption). Similarly by considering the z_1-coordinate we get that $\alpha \in \mathbb{Q}(\beta, \lambda, i) = \mathbb{Q}(\beta, i)$. Now this condition is obviously not always satisfied: we need only set $\beta = i$ and $\alpha = \sqrt{2}$.

This completes the construction of the counterexample. It is interesting now to go through our argument once more to understand at what point we have made essential use of the assumption that A is algebraic. It is easy to see that all of the arguments work also for complex manifolds, except one. This single essential argument occurs in the proof of the Poincaré complete reducibility theorem, where we passed a curve X not equal to any of the fibres $\varphi^{-1}(b)$ through a point of A. We conclude that this cannot be done for the torus A constructed in our example. Thus A has a map φ to a curve B such that the only compact complex submanifolds of A are the fibres $\varphi^{-1}(b)$; the next section contains more details about the notion of complex submanifold, but for the moment it can be taken to mean simply the image of a projective curve under a holomorphic map to A. We see that 1-dimensional complex submanifolds are very scarce in A. This is the main respect in which it differs from algebraic surfaces, which are crisscrossed by curves in all directions.

Example 8.4 The second example also relates to a 2-dimensional torus, but we will use some topological arguments. Again let

$$\Omega = \mathbb{Z}e_1 + \mathbb{Z}e_2 + \mathbb{Z}e_3 + \mathbb{Z}e_4;$$

then $A = \mathbb{C}^2/\Omega$ is homeomorphic to the torus $(\mathbb{R}/\mathbb{Z})^4$. Therefore $H_2(A,\mathbb{Z}) = \mathbb{Z}^6$, with generators given by 6 cycles $S_{i,j}$, the images in A of the planes $\mathbb{R}e_i + \mathbb{R}e_j$ with $1 \le i < j \le 4$.

In this example we start with the argument where we left off in Example 8.3. If A were algebraic, we would be able to find an algebraic curve $C \subset A$. If $\nu\colon C^\nu \to C$ is the normalisation map, then triangulating C^ν using Theorem 7.5 makes C^ν into a singular cycle in $H_2(A,\mathbb{Z})$. In particular

$$C \sim \sum_{1\le i<j\le 4} a_{i,j} S_{i,j} \quad \text{with } a_{i,j} \in \mathbb{Z}.$$

We now prove that C is not homologous to 0, so that not all the $a_{i,j}$ are equal to 0. For this note that the differential form

$$\frac{1}{2i}(\mathrm{d}z_1 \wedge \mathrm{d}\bar{z}_1 + \mathrm{d}z_2 \wedge \mathrm{d}\bar{z}_2)$$

on $\mathbb{C}^2$ is invariant under Ω, and hence defines a differential form ω on A. We prove that $\int_C \omega > 0$, from which it follows that C is not homologous to 0. If we consider z_1 and z_2 as functions in a neighbourhood of $x \in C$ then in a neighbourhood of $y = \nu^{-1}(x)$ our form is equal to

$$\frac{1}{2i}\left(\left|\frac{\mathrm{d}\nu^*(z_1)}{\mathrm{d}t}\right|^2 + \left|\frac{\mathrm{d}\nu^*(z_2)}{\mathrm{d}t}\right|^2\right)\mathrm{d}t \wedge \mathrm{d}\bar{t} > 0, \tag{8.12}$$

where t is a local parameter at y.

Now consider in the same way the differential form η on A corresponding to $\mathrm{d}z_1 \wedge \mathrm{d}z_2$. On the one hand by Stokes' theorem

$$\int_C \eta = \sum a_{i,j} \int_{S_{i,j}} \eta,$$

where the integral $\int_{S_{i,j}} \eta$ is easy to compute: we leave the reader to check that if $e_i = (\alpha_i, \beta_j) \in \mathbb{C}^2$ then

$$\int_{S_{i,j}} \eta = \alpha_i\beta_j - \alpha_j\beta_i.$$

On the other hand $\int_C \eta = 0$. Indeed, in the same way as for (8.12), $\nu^*(\eta)$ on C^ν is equal to

$$\left(\frac{\mathrm{d}\nu^*(z_1)}{\mathrm{d}t}\right)\left(\frac{\mathrm{d}\nu^*(z_2)}{\mathrm{d}t}\right)\mathrm{d}t \wedge \mathrm{d}t = 0.$$

Thus under the assumption that A is algebraic, we see that there is a relation

$$\sum a_{i,j}(\alpha_i\beta_j - \alpha_j\beta_i) = 0 \quad \text{with } a_{i,j} \in \mathbb{Z}, \tag{8.13}$$

and not all $a_{i,j}=0$. Of course, it is not difficult to choose $\alpha_1,\dots,\alpha_4$ and $\beta_1,\dots,\beta_4$, in such a way that the numbers $\alpha_i\beta_j$ are linearly independent over $\mathbb{Z}$. Then the corresponding torus A will not be algebraic.

One sees easily that 1-dimensional submanifolds are even more scarce on this torus than in that of Example 8.3: in fact it doesn't have any compact 1-dimensional complex submanifolds at all.

Remark 8.1 Write the coordinates of the basis vectors e_i of Ω as a 2×4 matrix

$$\Omega=\begin{pmatrix}\alpha_1 & \alpha_2 & \alpha_3 & \alpha_4\\ \beta_1 & \beta_2 & \beta_3 & \beta_4\end{pmatrix}$$

and consider the skewsymmetric 4×4 matrix $M=(a_{i,j})$, where $a_{i,j}$ are as in (8.13). Then (8.13) can be written as a

$$\Omega M\Omega'=0, \tag{8.14}$$

where Ω' is the transpose matrix. The existence of a matrix M satisfying this relation is obviously a necessary condition for the torus corresponding to the matrix Ω to be projective, or even algebraic.

Other conditions are provided by inequalities of type (8.12). To get as many of these conditions as possible, consider the 2-form

$$\begin{aligned}\omega&=\frac{1}{2i}(\lambda_1\overline{\lambda}_1 dz_1\wedge d\overline{z}_1+\lambda_1\overline{\lambda}_2 dz_1\wedge d\overline{z}_2+\overline{\lambda}_1\lambda_2 dz_2\wedge d\overline{z}_1+\lambda_2\overline{\lambda}_2 dz_2\wedge d\overline{z}_2)\\ &=(\lambda_1 dz_1+\lambda_2 dz_2)\wedge\overline{(\lambda_1 dz_1+\lambda_2 dz_2)}.\end{aligned}$$

Arguing exactly as in the proof of (8.12) gives that $\int_C\omega\ge 0$. Moreover, it is easy to prove that if the torus $A=\mathbb{C}^2/\Omega$ is projective and C corresponds to a hyperplane section under some embedding then $\int_C\omega=0$ only if $\lambda_1=\lambda_2=0$ (see Exercise 9). This final condition can be expressed in another way. Let Ω^* be the Hermitian conjugate of Ω. Then it is easy to see that the 2×2 matrix $\Omega M\Omega^*$ is Hermitian, that is, it corresponds to a Hermitian form $F(x)$. A simple substitution shows that $\int_C\omega=F(\lambda)$ where $\lambda=(\lambda_1,\lambda_2)$. Thus the relation deduced above means that F is a positive definite form. This can be written

$$\Omega M\Omega^*>0. \tag{8.15}$$

Thus the relations (8.14) and (8.15) are necessary conditions for the torus corresponding to a period matrix Ω to be projective. In exactly the same way, the relations are necessary for an n-dimensional complex torus with $n\times 2n$ period matrix Ω to be projective. They are called the *Frobenius relations*. It can be proved that they are also sufficient for the torus to be projective. A hint of the idea of proof is given in the remark at the end of Section 2.2, Chapter 9.

Remark 8.2 In our treatment of Example 8.4, we could have replaced formula (8.12) with a reference to Section 1.3, Chapter 7. Namely, repeating word-for-word the arguments given there shows that if $[C^\nu]$ is the orientation cycle of the curve C^ν then $\nu_*([C^\nu])$ is not homologous to 0 in $A(\mathbb{C})$. The reference to the triangulation of the curve C^ν can similarly be replaced by integration over this cycle. This is the most convenient way of proceeding in the following example.

Example 8.5 Let X be a Hopf manifold (Example 8.2). Since X is homeomorphic to $S^1 \times S^{n-1}$, it has second Betti number $b_2 = 0$ for $n > 1$. Proposition of Section 1.3, Chapter 7 shows that X is a nonprojective manifold. It is not hard to prove that it is also nonalgebraic.

1.5 Complex Spaces

Complex manifolds are the analogues of nonsingular algebraic varieties. It would be quite inconvenient to have to restrict ourselves to this notion. Indeed, singular varieties arise as subvarieties, or as images under regular maps, even in the study of nonsingular algebraic varieties. Furthermore, the majority of the arguments used in Chapter 5 to demonstrate the necessity of introducing the definition of scheme apply just as well to the complex analytic situation. The complex analytic notion corresponding to scheme is not used in the remainder of this book. However, it would be a shame to make no mention of it at all. We therefore give the definition and discuss without proof a few of the basic properties. The proof of these properties can be found, for example, in Gunning and Rossi [36, Chapters I–V].

We start with one particular case. Let $W \subset \mathbb{C}^n$ be a domain in the space of n complex variables and $f_1, \dots, f_k$ holomorphic functions on W. Denote by Y the set of common zeros of $f_1, \dots, f_k$ in W. We define a sheaf $\mathcal{O}_Y$ on Y as follows: let $\mathcal{O}_W$ be the sheaf of holomorphic functions on W. We then set

$$\mathcal{O}_Y(V) = \mathcal{O}_W(\overline{V})/(f_1, \dots, f_k),$$

for V an open subset of Y, where $\overline{V}$ is an open set in W such that $V = Y \cap \overline{V}$ (every open subset of Y is of this form), and $(f_1, \dots, f_k)$ the ideal of $\mathcal{O}_W(\overline{V})$ generated by $f_1, \dots, f_k$. Since Y is the set of common zeros of $f_1, \dots, f_k$, the right-hand side does not depend on the choice of $\overline{V}$. A topological space with this definition of sheaf will be called a *local model*.

We proceed to the global definitions. We say that a ringed space $X, \mathcal{O}$ such that $\mathcal{O}$ is a sheaf of $\mathbb{C}$-algebras is a *complex ringed space*. Any open set $U \subset X$ is itself a complex ringed space with the restriction of $\mathcal{O}$ to U as structure sheaf.

Definition A *complex space* or *complex space* is a complex ringed space $X, \mathcal{O}$ such that for every point $x \in X$ there exists a neighbourhood $U \ni x$ which is isomorphic as a ringed space to some local model in the sense just explained.

Just as for schemes, the stalks of the structure sheaf of a complex space are local rings. If they have no nilpotent elements then we say that the space X is *reduced*. In this case, $\mathcal{O}$ is a subsheaf of the sheaf of continuous functions on X, and on the local model Y, the stalk $\mathcal{O}_y$ consists of functions induced on Y by holomorphic functions on W in a neighbourhood of y. We then say that a continuous function $f \in \mathcal{O}_y$ is a *holomorphic function* on Y at y. In what follows, we only deal with reduced complex spaces, without further mention of this assumption. In this connection, morphisms of complex spaces are called *holomorphic maps*.

Suppose that a closed subset $X' \subset X$ has the following property: for any $x \in X'$ there exists a neighbourhood $x \in U \subset X$ and holomorphic functions $f_1, \dots, f_k$ such that X' is equal in U to the set of common zeros of $f_1, \dots, f_k$. We give X' the sheaf obtained by restricting holomorphic functions on X to X'. It is easy to check that we get in this way a complex space; we say that X' is a *subspace* of X.

A complex space X is *reducible* if $X = X' \cup X''$ with $X', X'' \subsetneqq X$ two proper subspaces. It is not hard to prove that any complex space X is a union of a set of irreducible subspaces

$$X = \bigcup X_\alpha,$$

with only finitely many X_α passing through each point $x \in X$. In what follows, we consider only irreducible complex spaces.

A point $x \in X$ is *nonsingular* if it has a neighbourhood isomorphic to a complex manifold; otherwise x is *singular*. It can be shown that the set of nonsingular points of an irreducible complex space is connected, and therefore has a well-defined dimension as a complex manifold, which we call the *dimension* of X. Any subspace of X distinct from the whole of X has smaller dimension. In particular, one can prove that the locus of singular points is a complex subspace. Because of this, a complex space X is a union of a finite number of complex manifolds (not closed in X): the set of nonsingular points, the set of nonsingular points of the singular locus, and so on.

Complex spaces are the complex analytic analogues of algebraic varieties, and even of schemes, at least in the sense that every scheme of finite type over the complex number field $\mathbb{C}$ has an associated complex space X_{an} (here we again allow schemes and complex spaces that are not necessarily reduced). We describe the construction of the space X_{an}.

We start by associating with X the topological space of complex points of its reduced subscheme $\widetilde{X} = X_{\mathrm{red}}(\mathbb{C})$, with the complex topology. An affine scheme X of finite type over $\mathbb{C}$ defines a local model, where the domain W is the whole of the ambient affine space $\mathbb{C}^N$ containing X. It is easy to see that this local model does not depend on the embedding $X \hookrightarrow \mathbb{A}^N$. The structure sheaf of the model constructed in this way is denoted by $\mathcal{O}_{\mathrm{an}}$.

If X is an arbitrary scheme of finite type over $\mathbb{C}$ with an affine open cover $X = \bigcup U^{(i)}$, the structure sheaves $\mathcal{O}_{\mathrm{an}}^{(i)}$ just defined on $U^{(i)}$ together determine a sheaf $\mathcal{O}_{\mathrm{an}}$ on the space $\widetilde{X}$. The pair $\widetilde{X}, \mathcal{O}_{\mathrm{an}}$ is the complex space X_{an} associated with the scheme X.

In Section 1.4 we met the question of the connections between the notions of complex manifold and nonsingular algebraic varieties. Similar questions of course arise concerning complex spaces and their connections with arbitrary algebraic varieties. There is an analogue in this set-up of the only positive result stated in Section 1.4, the Riemann existence theorem; namely, any compact reduced 1-dimensional complex space is isomorphic to an algebraic curve. This result can be reduced to the Riemann existence theorem via the process of normalising a complex space, which we now discuss briefly, omitting all proofs.

We say that a reduced complex space is *normal* if the local rings $\mathcal{O}_x$ of the structure sheaf are integrally closed. Following very closely the arguments that we gave for the case of algebraic varieties, one can construct the normalisation $\nu\colon X^\nu \to X$ of any reduced irreducible complex space X; that is, X^ν is a normal complex space and ν a holomorphic map satisfying the conditions of Theorem 2.21 of Section 5.2, Chapter 2. If X is compact then so is X^ν. A detailed treatment of all the arguments is given, for example, in Abhyankar [1, p. 447]. In the case of 1-dimensional complex spaces discussed from now on, the situation is somewhat simpler, and the reader could think through these arguments on his or her own as a (not quite trivial) exercise.

Let X be a compact reduced 1-dimensional complex space with structure sheaf $\mathcal{O}$. By the Riemann existence theorem, X^ν is a projective algebraic curve. We give X the Zariski topology, in which the closed sets are finite sets or the whole of X, and define the sheaf $\widetilde{\mathcal{O}}$ by setting $\widetilde{\mathcal{O}}(U) = \mathcal{O}(U) \cap \mathbb{C}(X^\nu)$. It is not hard to check that this defines an algebraic curve $\widetilde{X}$ with $\widetilde{X}_{\mathrm{an}} = X$.

1.6 Exercises to Section 1

1 Construct an example of a holomorphic map $g\colon \mathbb{C}^1 \to \mathbb{C}^1$ not of the form f_{an} for any morphism $f\colon \mathbb{A}^1 \to \mathbb{A}^1$.

2 Let X be a nonsingular irreducible algebraic curve and f a holomorphic function on X_{an}. Prove that if f is bounded on the set $X(\mathbb{C})$ then $f \in \mathbb{C}$.

3 Prove that the disc $|z| < 1$ in $\mathbb{C}^1$ is not isomorphic to X_{an} for any nonsingular curve X.

4 Let A be an elliptic curve, $e \in A$ the zero of the group law and $m > 0$ an integer; give another proof of the fact that the number of solutions of the equation $mx = e$ with $x \in A$ is equal to m^2 (compare Example 3.4 of Section 3.4, Chapter 3). If A is an n-dimensional Abelian variety, prove that $mx = e$ has m^{2n} solutions.

5 Prove that a 1-dimensional Hopf manifold is isomorphic to a complex torus.

6 Let $X = (\mathbb{C}^2 \setminus 0)/G$ be a 2-dimensional Hopf manifold as in Example 8.2. Prove that the map $\mathbb{C}^2 \setminus 0 \to \mathbb{P}^1$ defined by $(z_1, z_2) \mapsto (z_1 : z_2)$ induces a holomorphic map $X \to \mathbb{P}^1$ whose fibres are 1-dimensional complex toruses.

7 In the notation of Exercise 6, prove that X contains no 1-dimensional complex subspaces other than the fibres of $X \to \mathbb{P}^1$.

8 Let X be the complex space $\mathbb{C}^2$, g the automorphism of X given by $g(z_1, z_2) = (-z_1, -z_2)$ and G the group $\{1, g\}$. Prove that the quotient space X/G (see Exercise 1 of Section 3.6, Chapter 5) is a complex space, and is isomorphic to the cone in $\mathbb{A}^3$ given by $xy = z^2$.

9 Let $X = \mathbb{C}^2/\Omega$ be a complex torus, and $x \in X$ a point. Identify the tangent plane to X at x with $\mathbb{C}^2$ by the quotient map $\mathbb{C}^2 \to X$, and use this to give coordinates on X. As in Remark 8.1, suppose that ω is the 2-form

$$\frac{1}{2i}\big(|\lambda_1|^2 dz_1 \wedge d\overline{z}_1 + \lambda_1\overline{\lambda}_2 dz_1 \wedge d\overline{z}_2 + \overline{\lambda}_1\lambda_2 dz_2 \wedge d\overline{z}_1 + |\lambda_2|^2 dz_2 \wedge d\overline{z}_2\big);$$

let $C \subset X$ be a complex curve. Prove that if $x \in C$ is a nonsingular point and the tangent vector to C at x has coordinates μ_1, μ_2, where $\lambda_1\overline{\mu}_1 + \overline{\lambda}_2\mu_2 \neq 0$, then $\int_C \omega > 0$ (that is, is $\neq 0$). Deduce from this that $\int_C \omega > 0$ if the torus X is projective and C is a hyperplane section.

2 Divisors and Meromorphic Functions

2.1 Divisors

We return now to the theory of complex manifolds, and consider the question of constructing an analogue of the theory of divisors for them. We must begin by treating some simple properties of the stalk $\mathcal{O}_x$ of the structure sheaf of a complex manifold. By definition $\mathcal{O}_x$ is isomorphic to the ring $\mathbb{C}\{z_1, \dots, z_n\}$ of power series in $z_1, \dots, z_n$ that converge in some neighbourhood of x (the neighbourhood depending on the power series). This ring is very similar to the ring of formal power series. In particular, it is a regular local ring, and satisfies the analogue of the Weierstrass preparation theorem, stated in exactly the same way and proved in almost the same way as for formal power series (Lemma 2.1 of Section 3.1, Chapter 2); see Siegel [71, Chapter 5, Section 2, p. 5]. It follows from this theorem, word-for-word as in the case of formal power series, that $\mathbb{C}\{z_1, \dots, z_n\}$ is a UFD. In particular, $\mathbb{C}\{z_1, \dots, z_n\}$ has no zerodivisors.

Let U be a connected complex manifold and $\mathcal{O}(U)$ the ring of functions holomorphic on the whole of U. Then $\mathcal{O}(U)$ has no zerodivisors. Indeed, if $f, g \in \mathcal{O}(U)$ and $fg = 0$ then the set of points where $f \neq 0$ is open, and $g = 0$ on this set. But then $g = 0$ on the whole of U by the uniqueness of analytic continuation. Elements of the field of fractions of $\mathcal{O}(U)$ are called *meromorphic fractions* on U. If $V \subset U$ is a connected open subset then the restriction $\mathcal{O}(U) \to \mathcal{O}(V)$ extends to an isomorphic inclusion of the field of meromorphic fractions on U to that on V. We will often identify two meromorphic fractions that correspond in this way.

Definition A *divisor* on a complex manifold X is specified by a cover $X = \bigcup U_\alpha$ of X by connected open sets and a meromorphic fraction φ_α on each U_α, such that $\varphi_\beta/\varphi_\alpha$ is holomorphic and nowhere 0 on $U_\alpha \cap U_\beta$ for each α, β.

Equality of two divisors and the sums of divisors is defined word-for-word as for locally principal divisors on algebraic varieties. A divisor is *effective* if each meromorphic fraction φ_α is holomorphic on its open set U_α.

Theorem 8.1 *Every divisor is a difference of two effective divisors with no common components.*

Lemma *If f and g are holomorphic functions at a point $x \in \mathbb{C}^n$ and are relatively prime as elements of the ring $\mathcal{O}_x = \mathbb{C}\{z_1, \dots, z_n\}$ then there exists a neighbourhood U of x such that f and g are holomorphic in U and relatively prime as elements of $\mathcal{O}_y$ for every $y \in U$.*

Proof of the Lemma Multiplying f and g by invertible elements of $\mathcal{O}_x$ and using the Weierstrass preparation theorem, we can arrange that f and g are polynomials in z_1 with coefficients in $\mathbb{C}\{z_2, \dots, z_n\}$ and leading coefficients 1. Since they are relatively prime, there exist $u, v \in \mathbb{C}\{z_1, \dots, z_n\}$ such that

$$fu + gv = r, \quad \text{with } r \in \mathbb{C}\{z_2, \dots, z_n\}, \tag{8.16}$$

the equality holding in a neighbourhood U of x. Suppose that f and g have a common factor $h \in \mathcal{O}_y$ for some $y \in U$. Then $h \mid r$, and, again using the Weierstrass preparation theorem, we see that h is an invertible element of $\mathcal{O}_y$ times an element $h_1 \in \mathbb{C}\{z_2, \dots, z_n\}$. But $h_1 \mid f$, and since f has leading coefficient 1 as a polynomial in z_1 over $\mathbb{C}\{z_2, \dots, z_n\}$, it follows that h_1 is invertible in $\mathbb{C}\{z_2, \dots, z_n\}$. The lemma is proved. □

Proof of Theorem 8.1 We can assume that D is given by a cover $\bigcup U_\alpha$ and a collection of meromorphic fractions φ_α with

$$\varphi_\alpha = f_\alpha / g_\alpha \quad \text{in } U_\alpha,$$

where f_α and g_α are holomorphic in U_α and relatively prime at every point $y \in U_\alpha$. Then since $\mathcal{O}_y$ is a UFD, it follows that f_α defines a divisor D' and g_α a divisor D'', with both D' and D'' effective and $D = D' - D''$. This proves the theorem. □

It is obvious that every effective divisor D defines a complex subspace of X, defined by $\varphi_\alpha = 0$ in the open set U_α. This is called the *support* of D, and denoted by $\operatorname{Supp} D$. If $D = D' - D''$ is the representation of D as a difference of effective divisors with no common components, we set $\operatorname{Supp} D = \operatorname{Supp} D' \cup \operatorname{Supp} D''$ by definition. Using the notion of dimension of a complex space introduced at the end of Section 1.5, we state the following result.

Proposition *The support of a divisor has codimension* 1.

Proof First of all, we must define this subspace by the most economic system of equations. For this, at each point $x \in U_\alpha$ we factorise φ_α into irreducible factors in $\mathcal{O}_x$, and write ψ_x for the product of these factors each with power 1. Then ψ_x is holomorphic in some neighbourhood U_x of x, and all these functions define the same subset $\operatorname{Supp} D$ as the functions φ_α (although possibly a different divisor). Thus we can assume from the start that the divisor is defined by functions φ_α with no multiple factors in $\mathcal{O}_x$ for $x \in U_\alpha$.

By the Weierstrass preparation theorem we can assume that for some $x \in U_\alpha$ the function φ_α is of the form

$$\varphi_\alpha = z_1^m + a_1 z_1^{m-1} + \cdots + a_m, \quad \text{with } a_i \in \mathbb{C}\{z_2, \dots, z_n\},$$

where $z_1, \dots, z_n$ are local parameters at x. By the above assumption on the φ_α we can assume that $\partial\varphi_\alpha/\partial z_1$ is relatively prime to φ_α in $\mathcal{O}_x$, and hence $\partial\varphi_\alpha/\partial z_1$ is not identically 0 on $\operatorname{Supp} D$ in a neighbourhood of x. Now divide the points $y \in \operatorname{Supp} D$ into two types: those for which all the $\partial\varphi_\alpha/\partial z_i(y) = 0$ for $i = 1, \dots, n$, and the remaining points. The points of the first type obviously form a complex subspace $S \subset \operatorname{Supp} D$, and as we have just seen, $S \neq \operatorname{Supp} D$.

The proposition is an obvious consequence of the following two assertions: (a) a point of the first kind is a singular point of the subspace $\operatorname{Supp} D$, that is, $\operatorname{Supp} D$ is not isomorphic to a complex manifold in a neighbourhood of these points; and (b) in a neighbourhood of a point of the second type $\operatorname{Supp} D$ is isomorphic to an $(n-1)$-dimensional complex manifold.

Assertion (a) follows from the representation

$$\mathcal{O}_{y,\operatorname{Supp} D} = \mathcal{O}_y/(\varphi_\alpha) \tag{8.17}$$

of the local ring of a point $y \in \operatorname{Supp} D$ (we leave the verification of (8.17) to the reader). If y is a point of the first type then $\varphi_\alpha \in \mathfrak{m}_y^2$, where $\mathfrak{m}_y$ is the maximal ideal of the local ring $\mathcal{O}_y$. It follows at once from this that $\mathcal{O}_{y,\operatorname{Supp} D}$ is not a regular local ring, and hence $\operatorname{Supp} D$ is not a manifold.

Assertion (b) is a direct consequence of the implicit function theorem. If say $\partial\varphi_\alpha/\partial z_1(y) \neq 0$, then z_1 is a holomorphic function of $z_2, \dots, z_n$ on $\operatorname{Supp} D$ in a neighbourhood of y, and hence $z_2, \dots, z_n$ define an isomorphism of this neighbourhood with a domain in $\mathbb{C}^{n-1}$. The proposition is proved. □

We will not develop the theory of divisors on complex manifolds any further. This can be done, leading to results completely analogous to those we obtained for algebraic varieties. Namely, each divisor can be expressed in a unique way as a linear combination of irreducible effective divisors, and irreducible divisors are in one-to-one correspondence with codimension 1 complex subspaces. The proofs of these facts are contained in Weil [79, Sections 5–8, Appendix, pp. 148–158] or Griffiths and Harris [33, Section 1, Chapter 1]; they are quite elementary and do not depend on other sections of these books.

2.2 Meromorphic Functions

Now we consider the functions on a complex manifold that are the analogues of rational functions of an algebraic variety, the meromorphic functions. The main tool is the notion of meromorphic fraction introduced in Section 2.1.

Definition A *meromorphic function* on a complex manifold X is specified by a cover $X = \bigcup U_\alpha$ of X by connected open sets, and a system of meromorphic fractions φ_α on U_α such that the restrictions of φ_α and φ_β to $U_\alpha \cap U_\beta$ are equal for all α, β. Such a system φ_α of meromorphic fractions is said to be *compatible*.

A cover $X = \bigcup U_\alpha$ and compatible system of functions φ_α define the same meromorphic function as another cover $X = \bigcup V_\beta$ and compatible system ψ_β if φ_α and ψ_β are equal when restricted to $U_\alpha \cap U_\beta$ for all α, β.

Let $\varphi_\alpha = f/g$ be a meromorphic fraction on U_α, with f and g holomorphic on U_α; if $g(x) \neq 0$ at some point $x \in U_\alpha$, then φ_α equals the holomorphic function f/g in a neighbourhood of x. This notion extends naturally to meromorphic functions. Thus for any meromorphic function φ on X there exists an open subset $U \subset X$ and a holomorphic function f on U such that the restriction of φ to U is equal to f. We say that φ is *holomorphic* at points of U.

Algebraic operations on meromorphic functions are defined in terms of the corresponding meromorphic fractions; all the meromorphic functions on a complex manifold X obviously form a ring. If X is connected, this ring is a field, the *meromorphic function field* of X. Indeed, suppose that φ is given by a cover $\{U_\alpha\}$ and a compatible system of functions $\{\varphi_\alpha\}$. If $\varphi \neq 0$, then at least one $\varphi_\alpha \neq 0$; but from the compatibility of the functions φ_α it follows that $\varphi_\beta \neq 0$ for all β such that $U_\alpha \cap U_\beta \neq \emptyset$. Now since X is connected, it follows that all the $\varphi_\alpha \neq 0$, and the function φ^{-1} exists, given by the system φ_α^{-1}. In what follows we only consider connected manifolds X. The field of meromorphic functions on X is denoted by $\mathcal{M}(X)$.

If the manifold is of the form X_{an} where X is an irreducible nonsingular algebraic variety, then rational functions on X obviously define meromorphic functions on X_{an}. In other words $\mathbb{C}(X) \subset \mathcal{M}(X_{\text{an}})$. Of course, equality does not hold in general. However, if X is complete then the two fields coincide, as we will prove in Section 3.

By putting together the definitions of meromorphic function and divisor, we see that each meromorphic function φ defines a divisor, that we denote by $\operatorname{div}\varphi$. By definition it follows that $\operatorname{div}\varphi$ if effective if and only if φ is holomorphic on the whole of X. For a compact connected manifold this is only possible if φ is a constant, by analogy with Theorem 1.10 and Corollary 1.1 of Section 5.2, Chapter 1.

Theorem 8.2 *A function φ that is holomorphic at every point of a compact connected manifold X is constant.*

Proof The modulus $|\varphi|$ is obviously a continuous function on X, and it therefore takes its maximum at some point x_0. Consider a neighbourhood U of x_0 isomorphic to an open set $V \subset \mathbb{C}^n$; we can assume that V consists of points $(z_1, \ldots, z_n)$ with

$\sum |z_i|^2 < 1$, and that the isomorphism $f: U \to V$ satisfies $f(x_0) = 0 = (0, \ldots, 0)$. Then $\psi = (f^{-1})^*(\varphi)$ is a holomorphic function on V, and its modulus has a maximum at 0. For any point $(\alpha_1, \ldots, \alpha_n) \in V$, consider a 1-dimensional complex subspace $z_i = \alpha_i t$ for $i = 1, \ldots, n$. On this, ψ defines a holomorphic function of one argument t, which is constant by the maximum modulus principle. It follows from this that ψ is constant on V, and hence φ is constant on U. Since X is connected, φ is constant on the whole of X by the uniqueness of analytic continuation. The theorem is proved. □

Since the divisors of meromorphic functions satisfy the natural identities

$$\operatorname{div}(\varphi\psi) = \operatorname{div}\varphi + \operatorname{div}\psi \quad \text{and} \quad \operatorname{div}(\varphi/\psi) = \operatorname{div}\varphi - \operatorname{div}\psi,$$

the theorem implies the following result.

Corollary *On a compact complex manifold, a meromorphic function is determined uniquely up to a constant factor by its divisor.*

In the light of the definition of meromorphic functions and their divisors, we can look at the examples of nonalgebraic compact complex manifolds worked out in Section 1.4 from another point of view. We start with Example 8.4, an algebraic torus A which is nonalgebraic because it does not contain any algebraic curve. As we said in Section 1.5, 1-dimensional complex subspaces are algebraic curves. Therefore the torus A does not contain any 1-dimensional complex subspace, that is, no nonzero divisor. It follows that the divisor of any meromorphic function on A equals 0, and hence every such function is constant by Theorem 8.2 and Corollary. In other words $\mathcal{M}(A) = \mathbb{C}$. We thus have a new characteristic of the nonalgebraic nature of the torus A: it has far fewer meromorphic functions than an algebraic variety, on which at least all the rational functions are meromorphic.

Now consider Example 8.3. We constructed in that example a 2-dimensional torus A and a holomorphic map $f: A \to B$ to an elliptic curve. The torus A is nonalgebraic because the only irreducible curves it contains are the fibres $f^{-1}(b)$.

By Theorem 8.1, the divisor of an arbitrary meromorphic function φ on A can be expressed as

$$\operatorname{div}\varphi = D' - D'',$$

where D' and D'' are effective divisors. From the proof of this theorem it is easy to see that the set $\operatorname{Supp} D' \cap \operatorname{Supp} D''$ consists only of isolated points, and since distinct fibres $f^{-1}(b)$ do not intersect at all, it follows that

$$\operatorname{Supp} D' = \bigcup f^{-1}(b_i'), \quad \operatorname{Supp} D'' = \bigcup f^{-1}(b_j'') \quad \text{with } b_i' \neq b_j''.$$

Applying this argument to the functions $\varphi - c$ with $c \in \mathbb{C}$, we see that any meromorphic function φ on A is constant on the fibres of f. Choose local parameters z_1, z_2 at a point $a \in A$ such that $z_1 = f^*(t)$, where t is a local parameter on B at

$b = f(a)$. In this coordinate system φ can be written as a meromorphic fraction that does not depend on z_2, that is, φ is locally of the form $f^*(\psi)$ where ψ is a meromorphic fraction on B. It follows that on the whole manifold A we also have equality $\varphi = f^*(\psi)$ with $\psi \in \mathcal{M}(B)$. But B is an algebraic curve, and by the theorem already quoted that we will prove in the next section, $\mathcal{M}(B) = \mathbb{C}(B)$. Thus we have proved that for the torus A of Example 8.3,

$$\mathcal{M}(A) = f^*\bigl(\mathbb{C}(B)\bigr).$$

We see that this equality again reflects the nonalgebraic nature of A. For an algebraic surface X the field $\mathcal{M}(X)$ contains $\mathbb{C}(X)$, and hence has transcendence degree at least 2, but in our case the transcendence degree is equal to 1.

The same arguments apply to Example 8.5 (see Exercises 6–7 of Section 1.6): on a Hopf surface we have $\mathcal{M}(X) = \mathbb{C}(\mathbb{P}^1)$.

2.3 The Structure of the Field $\mathcal{M}(X)$

The examples given at the end of Section 2.2 show that a compact complex manifold X may have "too few" meromorphic functions compared with an algebraic variety of the same dimension: more precisely, the transcendence degree of $\mathcal{M}(X)$ may be smaller than the dimension of X. A whole series of important properties of compact complex manifolds follow from the fact that X cannot have "too many" meromorphic functions. We now prove this.

Theorem 8.3 *The meromorphic function field $\mathcal{M}(X)$ on a compact complex manifold X has transcendence degree $\le \dim X$.*

The proof of this theorem is quite elementary. We precede it with a simple remark.

Schwarz' Lemma *Let $f(z) = f(z_1, \dots, z_n)$ be a holomorphic function in the polydisc defined by $|z_i| \le 1$ for $i = 1, \dots, n$, and suppose that $M = \max_{|z_i|\le 1} |f(z)|$. Write $\mathfrak{m}_0$ for the maximal ideal of the local ring of analytic functions at the origin. If $f \in \mathfrak{m}_0^h$, that is, if f and all its derivatives of degree $\le h-1$ vanish at 0, then*

$$\bigl|f(z)\bigr| \le M \max_i |z_i|^h \tag{8.18}$$

for z in the open polydisc $|z_i| < 1$ for $i = 1, \dots, n$.

Proof For $z = (z_1, \dots, z_n) \in \mathbb{C}^n$ we write $|z| = \max_i |z_i|$. For fixed z with $|z| < 1$, set $g(t) = f(tz)$ for $t \in \mathbb{C}$. Then $g(t)$ is a holomorphic function in the disc $|t| \le |z|^{-1}$ and the first h coefficients of its Taylor series at 0 vanish. Therefore $g(t)/t^h$ is holomorphic for $|t| \le |z|^{-1}$. By the maximum modulus principle, in this disk

$|g(t)/t^h| \le M/|z|^{-h} = M|z|^h$. Setting $t = 1$ gives the inequality (8.18). The lemma is proved. □

Proof of the Theorem In the proof of the theorem, we use the fact that polynomials in ν variables of degree $\le k$ form a vector space whose dimension is given by the binomial coefficient

$$\binom{k+\nu}{\nu} = \frac{(k+1)(k+2)\cdots(k+\nu)}{\nu!}.$$

Note that, as a function of k, this is a polynomial of degree ν.

Let $f_1, \dots, f_{n+1}$ be $n+1$ meromorphic functions on a compact n-dimensional complex manifold X. Our aim is to prove the existence of a polynomial $F(T_1, \dots, T_{n+1})$ such that

$$F(f_1, \dots, f_{n+1}) = 0. \tag{8.19}$$

We choose three neighbourhoods $U_x \supset V_x \supset W_x \ni x$ for each point $x \in X$. The first U_x is chosen so that

$$f_i = \frac{P_{i,x}}{Q_{i,x}} \quad \text{for } i = 1, \dots, n+1, \tag{8.20}$$

where $P_{i,x}$, $Q_{i,x}$ are holomorphic in U_x and relatively prime at each point $y \in U_x$; the existence of U_x follows from Lemma of Section 2.1. The second V_x is chosen so that its closure $\overline{V}_x$ is contained in U_x, and V_x has a local coordinate system $(z_1, \dots, z_n)$ with $|z| < 1$. The third neighbourhood W_x is given by $|z| < 1/2$.

Because for different points x and y the expressions (8.20) both represent the same function f_i, and since $P_{i,x}$ and $Q_{i,x}$ are relatively prime, it follows that

$$Q_{i,x} = Q_{i,y}\varphi_{i,x,y},$$

where $\varphi_{i,x,y}$ are holomorphic and nowhere 0 in $U_x \cap U_y$.

From the system of neighbourhoods W_x, we choose a finite cover:

$$X = \bigcup W_\xi;$$

this is the point where compactness is used. Write r for the number of the sets W_ξ (that is, the number of points $\xi \in X$), and set

$$\varphi_{\xi,\eta} = \prod_{i=1}^{n+1} \varphi_{i,\xi,\eta} \quad \text{and} \quad C = \max_{\xi,\eta} \max_{V_\xi \cap V_\eta} |\varphi_{\xi,\eta}|.$$

Note that $|\varphi_{\xi,\eta}|$ is bounded in $V_\xi \cap V_\eta$, since its closure in contained in $U_\xi \cap U_\eta$, where $\varphi_{\xi,\eta}$ is holomorphic. Moreover, $C \ge 1$, since $\varphi_{\xi,\eta}\varphi_{\eta,\xi} = 1$.

For a polynomial $F(T_1,\dots,T_{n+1})$ (as yet to be determined) of degree k in $T_1,\dots,T_{n+1}$, set

$$F(f_1,\dots,f_{n+1})=\frac{R_x}{Q_x^k}\quad\text{in } V_x,$$

where

$$Q_x=\prod_{i=1}^{n+1}Q_{i,x}.$$

Obviously $R_\xi=\varphi_{\xi,\eta}^k R_\eta$ in $V_\xi\cap V_\eta$.

After introducing this notation, we can proceed with the substance of the proof. As a first approximation to (8.19), we show that for any given h the polynomial F can be chosen so that $F\neq 0$ and

$$R_\xi\in\mathfrak{m}_\xi^h \tag{8.21}$$

for all r of the points ξ. These conditions can be written out as the set of relations

$$\bigl(D^s R_\xi\bigr)(\xi)=0,$$

where D^s is the partial derivative of order $s<h$. Hence they are linear relations on the coefficients of the polynomial F; the number of relations is equal to $r\binom{n+h-1}{n}$. If we choose the degree k of F such that

$$\binom{n+k+1}{n+1}>r\binom{n+h-1}{n}, \tag{8.22}$$

then there exists a nonzero polynomial F for which (8.21) holds.

By Schwarz' lemma, for this choice of F the functions R_ξ will be small in the neighbourhoods W_ξ: if

$$M=\max_\xi\max_{x\in V_\xi}\bigl|R_\xi(X)\bigr|,$$

then

$$\bigl|R_\xi(x)\bigr|\le\frac{M}{2^h}\quad\text{for } x\in W_\xi. \tag{8.23}$$

Now this circumstance will imply that $M=0$, that is, that (2) holds for sufficiently large k and h. Indeed, suppose that the maximum value M is taken at a point $x_0\in V_\eta$. Then $x_0\in W_\xi$ for some point ξ. Hence

$$M=\bigl|R_\eta(x_0)\bigr|=\bigl|R_\xi(x_0)\bigr|\bigl|\varphi_{\xi,\eta}(x_0)\bigr|^k.$$

If k and h are such that (8.22) holds, then also (8.23) holds, and hence

$$M\le\frac{M}{2^h}C^k.$$

Now it remains to choose k and h so that, in addition to (8.22), we also have

$$C^k/2^h < 1,$$

and we get $M = 0$. It is possible to make this choice: if $C = 2^\lambda$ then $\lambda \geq 0$, since $C \geq 1$, and we need only take $h \geq \lambda k$ and h, k satisfy (8.22). For example, for $k = h/m$, where m is any integer with $m > \lambda$, on the left-hand side of (8.22) we get a polynomial in h of bigger degree than on the right-hand side, and hence for sufficiently large h divisible by m, the left-hand side will indeed be greater than the right-hand side. The theorem is proved. □

Using similar arguments it can be proved that if the transcendence degree of the field $\mathcal{M}(X)$ equals k and $f_1, \dots, f_k$ are algebraically independent meromorphic functions on X, then the degree of the irreducible relation

$$F(f, f_1, \dots, f_k) = 0$$

satisfied by an arbitrary meromorphic function f is bounded from above. Therefore the field $\mathcal{M}(X)$ is not only of finite transcendence degree, but also finitely generated.

2.4 Exercises to Section 2

1 We define a complex analytic vector bundle by analogy with the way it was done in Section 1.2, Chapter 6, with the difference that E and X are complex manifolds and $p: E \to X$ is a holomorphic map. Prove that the correspondence between vector bundles and transition matrixes established in Section 1.2, Chapter 6 holds also for complex analytic vector bundles.

2 Prove that the correspondence $D \mapsto \mathcal{O}(D)$ described in Section 1.4, Chapter 6 between divisors and line bundles extends to complex analytic divisors and line bundles. For this one has to formulate a definition of this correspondence in terms of transition matrixes (6.13). Prove that linearly equivalent divisors also define isomorphic line bundles in the complex analytic category.

3 Let X be a complex manifold, $U_\alpha \subset X$ an open set isomorphic to an open set in $\mathbb{C}^n$, and $z_{\alpha,1}, \dots, z_{\alpha,n}$ the inverse image of the coordinates in $\mathbb{C}^n$ under this isomorphism. If U_β is another such open set, then in $U_\alpha \cap U_\beta$ set

$$\varphi_{\alpha\beta} = \det\left|\frac{\partial(z_{\alpha,1}, \dots, z_{\alpha,n})}{\partial(z_{\beta,1}, \dots, z_{\beta,n})}\right|.$$

Prove that the $\varphi_{\alpha\beta}$ are transition functions for some line bundle $\mathcal{K}$. Prove that if $X = Y_{\mathrm{an}}$ with Y an algebraic variety then $\mathcal{K} = K_{\mathrm{an}}$, where K is the line bundle corresponding to the canonical class of Y. In the general case, $\mathcal{K}$ is called the *canonical line bundle* of X.

4 Let X be a complex manifold, and $X = \bigcup U_\alpha$ an open cover such that there exist isomorphisms $\varphi_\alpha \colon U_\alpha \to V \subset \mathbb{C}^n$ to open sets of $\mathbb{C}^n$. Suppose given holomorphic functions f_α on $\varphi_\alpha(U_\alpha)$ such that under the isomorphisms $\varphi_\beta \circ \varphi_\alpha^{-1} \colon \varphi_\alpha(U_\alpha \cap U_\beta) \to \varphi_\beta(U_\alpha \cap U_\beta)$, the forms $f_\alpha \mathrm{d}z_1 \wedge \cdots \wedge \mathrm{d}z_n$ and $f_\beta \mathrm{d}z_1 \wedge \cdots \wedge \mathrm{d}z_n$ go into one another. By definition, such a collection of functions defines a holomorphic n-form ω on X. Prove that the functions $\varphi_\alpha^*(f_\alpha)$ define a divisor on X; it is called the *divisor of the n-form ω*, and denoted by $\operatorname{div}\omega$. Prove that the divisors of any two holomorphic n-forms are linearly equivalent. Prove that if X has a holomorphic differential form, the line bundle defined by its divisor is isomorphic to the canonical line bundle.

5 Prove that the canonical line bundle of a complex torus is trivial.

6 Let $\Omega \subset \mathbb{C}^n$ be a lattice of rank $2n$ and $X = \mathbb{C}^n/\Omega$ the n-dimensional torus (Example 8.1). Suppose that $\chi \colon \Omega \to \mathbb{C}^*$ is a group homomorphism of Ω to the multiplicative group of nonzero complex numbers. Define an action of Ω on $\mathbb{C}^n \times \mathbb{C}^1$ by

$$a(x,z) = \big(x + a, \chi(a)z\big) \quad \text{for } x \in \mathbb{C}^n,\ z \in \mathbb{C}^1 \text{ and } a \in \Omega.$$

Prove that Ω acts freely and discretely on $\mathbb{C}^n \times \mathbb{C}^1$. The projection $\mathbb{C}^n \times \mathbb{C}^1 \to \mathbb{C}^n$ commutes with the action of Ω and defines a map $p \colon E_\chi = (\mathbb{C}^n \times \mathbb{C}^1)/\Omega \to \mathbb{C}^n/\Omega = X$. Prove that p is holomorphic and makes E_χ into a line bundle over X (compare Exercise 1).

7 In the notation of Exercise 6, prove that two line bundles E_χ and $E_{\chi'}$ are isomorphic if and only if there exists a nowhere vanishing holomorphic function g on $\mathbb{C}^n$ such that $g(x+a)g(x)^{-1} = \chi'(a)\chi(a)^{-1}$ for all $x \in \mathbb{C}^n$ and $a \in \Omega$.

8 In the notation of Exercises 6–7, suppose in addition that $|\chi(a)| = |\chi'(a)| = 1$ for every $a \in \Omega$. Prove that the bundles E_χ and $E_{\chi'}$ are isomorphic only if $\chi = \chi'$.

9 Prove that the analogue of Theorem 6.3 does not hold in the theory of complex analytic line bundles; that is, not every line bundle is defined by some divisor.

3 Algebraic Varieties and Complex Manifolds

3.1 Comparison Theorems

We are now in a position to prove some basic facts showing that for a complete (or projective) algebraic variety X over $\mathbb{C}$, many of the properties of the corresponding complex manifolds X_{an} can be reduced to algebraic properties of X.

Theorem 8.4 *If X is a complete algebraic variety over $\mathbb{C}$ then a meromorphic function on the complex manifold X_{an} is a rational function on X.*

Proof Suppose that f is a meromorphic function on X_{an}. Since X_{an} is compact, f is algebraic over $\mathbb{C}(X)$ by Theorem 8.3. Hence it is enough to prove that a meromorphic function f on X_{an} that is algebraic over $\mathbb{C}(X)$ is a rational function on X. The completeness of X does not play any role in the proof of this fact.

Suppose that f is a root of an irreducible equation

$$F(f) = f^m + a_1 f^{m-1} + \cdots + a_m = 0$$

over $\mathbb{C}(X)$. Discarding the poles of the rational functions a_i from X, we can assume that the a_i are regular functions on X. Then f is also holomorphic on X. This follows from the fact that the ring $\mathcal{O}_{x,\mathrm{an}}$ is a UFD, and is therefore integrally closed in its field of fractions.

Consider in the product $X \times \mathbb{A}^1$ the set X' of points (x, z) satisfying the relation

$$F(z) = z^m + a_1(x)z^{m-1} + \cdots + a_m(x) = 0.$$

Then X' is an irreducible algebraic variety, and $\mathbb{C}(X') = \mathbb{C}(X)(f)$. Write $p\colon X' \to X$ for the natural projection. We again pass to smaller sets X and X' by discarding from X the set of points at which the discriminant of the polynomial $F(T)$ vanishes, and from X' the inverse image under p of this set. We continue to denote these smaller irreducible varieties by X and X'. We have thus achieved that the inverse image $p^{-1}(x)$ of any point $x \in X$ consists of m distinct points (x, z), and that $F'_T(x, z) \neq 0$ for any such point.

It follows that if $z_1, \dots, z_n$ are local parameters at $x \in X$ then $p^*(z_1), \dots, p^*(z_n)$ are local parameters at any point of $p^{-1}(x)$. Hence there exists a sufficiently small complex neighbourhood U of x in X such that $p^{-1}(U)$ breaks up into m disjoint sets $U_1, \dots, U_m$, and the projection $p\colon U_i \to U$ is an isomorphism of the corresponding complex manifolds. We only need that p is a homeomorphism. The assertion means that $X'(\mathbb{C})$ is an unramified cover of $X(\mathbb{C})$. The function f defines a continuous map $X(\mathbb{C}) \to X'(\mathbb{C})$ given by $\varphi(x) = (x, f(x))$, which is a section of this unramified cover, that is, $p \circ \varphi = 1$.

The information we have obtained is already enough to prove that $m = 1$, and hence $f \in \mathbb{C}(X)$. Indeed, if $m > 1$ then $\varphi(X) \neq X'$, because $\varphi(x)$ is a single point, and $p^{-1}(x)$ consists of m points. We show that $\varphi(X(\mathbb{C}))$ and $X'(\mathbb{C}) \setminus \varphi(X(\mathbb{C}))$ are closed and disjoint, from which it follows that $X'(\mathbb{C})$ is disconnected. This contradicts Theorem 7.1 because X' is an irreducible algebraic variety.

All the assertions remaining to check are local in nature, that is, it is enough to check them for the sets U and $p^{-1}(U)$ instead of X and X', where U is any neighbourhood of a point $x \in X$. In particular, we can choose U to be connected and such that

$$p^{-1}(U) = U_1 \cup \cdots \cup U_m \quad \text{with } U_i \cap U_j = \emptyset \text{ for } i \neq j.$$

Then $\varphi(U)$ must coincide with one of the U_i, from which everything we need follows obviously. The theorem is proved. □

Theorem 8.5 *If X and Y are complete algebraic varieties then any holomorphic map $f: X_{\text{an}} \to Y_{\text{an}}$ is of the form $f = g_{\text{an}}$ where $g: X \to Y$ is a morphism.*

Proof Choose a point $x \in X$ and set $y = f(x)$; let U be an affine neighbourhood of y. Suppose that $U \subset \mathbb{A}^N$, and write $t_1, \dots, t_N$ for the coordinates in $\mathbb{A}^N$. By Theorem 8.4, the holomorphic functions $f^*(t_i)$ are rational functions on X. If we prove that these are regular at x then we get $f = g_{\text{an}}$ in some neighbourhood of x, where g is the morphism defined by $f^*(t_1), \dots, f^*(t_N)$. Thus we construct a system of morphisms $g_\alpha: V_\alpha \to Y$ on open sets V_α that cover X. Obviously these define a single morphism $g: X \to Y$ for which $f = g_{\text{an}}$.

Thus everything reduces to the following local assertion: □

Lemma *If a rational function g is holomorphic at a point x then it is regular at x.*

Proof Write $\mathcal{O}_{x,\text{an}}$ for the ring of functions holomorphic in some neighbourhood of x, and $\widehat{\mathcal{O}}_x$ for the ring of formal power series. Taking a holomorphic function to its power series defines an inclusion $\mathcal{O}_{x,\text{an}} \subset \widehat{\mathcal{O}}_x$, so that

$$\mathcal{O}_x \subset \mathcal{O}_{x,\text{an}} \subset \widehat{\mathcal{O}}_x.$$

Set $g = u/v$ with $u, v \in \mathcal{O}_x$. The fact that g is holomorphic at x means that $v \mid u$ in $\mathcal{O}_{x,\text{an}}$. Then a fortiori $v \mid u$ in $\widehat{\mathcal{O}}_x$. But according to Lemma of Section 7, Appendix, this implies that $v \mid u$ in $\mathcal{O}_x$, and therefore g is regular at x. This proves the lemma, and with it Theorem 8.5. □

Corollary *Under the assumptions of Theorem* 8.5, *X_{an} and Y_{an} are isomorphic as complex manifolds if and only if X and X are isomorphic as algebraic varieties.*

The local version of this result also holds: if two singular points of algebraic varieties are formally analytically equivalent then some complex neighbourhoods of these points are isomorphic as complex spaces.

Theorem 8.6 *If X is a projective variety then any complex submanifold V of the complex manifold X_{an} is of the form Y_{an}, where Y is a closed algebraic subvariety of X.*

Proof Since X is contained in a projective space, it is enough to prove the theorem in the case $X = \mathbb{P}^N$. Moreover, it is enough to prove the assertion for connected subvarieties $V \subset \mathbb{P}^N_{\text{an}}$, since from the fact that V is compact it follows that it has only a finite number of connected components. Hence we assume in what follows that V is connected.

Write Y for the closure of V in the Zariski topology of $\mathbb{P}^N$; that is, Y is the intersection of all algebraic subvarieties of $\mathbb{P}^N$ containing V. We prove that Y is an irreducible projective variety. For this, it is enough to prove that its homogeneous ideal $\mathfrak{A}_Y$ is prime, that is, that if P and Q are homogeneous polynomials such that

$PQ = 0$ on V then $P = 0$ or $Q = 0$ on V. If P does not vanish on the whole of V then the set $U \subset V$ of points where $P \neq 0$ is open in V. Suppose that the homogeneous coordinate x_0 is not 0 on the whole of V. If $\deg Q = k$ then Q/x_0^k is a meromorphic function on V and vanishes on U. Hence by the uniqueness of analytic continuation, $Q/x_0^k = 0$ on the whole of the connected component of V containing U, that is, on the whole of V. Therefore $Q = 0$ on V.

It follows from the definition of Y that every rational function $\varphi \in \mathbb{C}(Y)$ defines a meromorphic function on V, in other words,

$$\mathbb{C}(Y) \subset \mathcal{M}(V). \tag{8.24}$$

Set $\dim V = n$, $\dim Y = m$. We have $m \geq n$, because $V \subset Y$, and Y is an m-dimensional complex manifold in the neighbourhood of any nonsingular point. But by Theorem 8.3, the transcendence degree of $\mathcal{M}(V)$ is at most n, so that from the inclusion (8.24) we get

$$\dim Y = \dim V = n. \tag{8.25}$$

It is easy to deduce from (8.25) that $Y_{\mathrm{an}} = V$, which is what we have to prove. Indeed, write S for the set of singular points of Y. The algebraic variety $Y \setminus S$ is irreducible, and hence by Theorem 7.1, $(Y \setminus S)_{\mathrm{an}}$ is a connected manifold. The subset $V \setminus (V \cap S)$ is closed in $(Y \setminus S)_{\mathrm{an}}$, since V is closed in Y_{an}. On the other hand, it follows from (8.25) that $V \setminus (V \cap S)$ is open in $(Y \setminus S)_{\mathrm{an}}$. Hence $V \setminus (V \cap S) = (Y \setminus S)_{\mathrm{an}}$, that is, $(Y \setminus S)_{\mathrm{an}} \subset V$. Since V is closed, and $(Y \setminus S)_{\mathrm{an}}$ is everywhere dense in Y by Lemma 7.1, it follows that $Y_{\mathrm{an}} \subset V$, that is $V = Y_{\mathrm{an}}$. The theorem is proved. □

3.2 Example of Nonisomorphic Algebraic Varieties that Are Isomorphic as Complex Manifolds

We now construct two nonisomorphic algebraic varieties X and Y such that the associated complex manifolds X_{an} and Y_{an} are isomorphic. By Theorem 8.5 and Corollary of Section 3.1, if this happens then X and Y cannot be complete.

We first describe the example. Let C be a nonsingular projective plane cubic curve, $Q \in C$ a point, B the noncomplete curve $C \setminus Q$ and $P \in B$ some point. We saw in Section 1.4, Chapter 6 that any divisor on B corresponds to a line bundle $E \to B$. Let X be the line bundle corresponding to P, and Y the product $B \times \mathbb{A}^1$, the trivial line bundle, corresponding to the divisor 0. We need to prove two assertions: (1) the algebraic varieties X and Y are not isomorphic; and (2) the complex manifolds X_{an} and Y_{an} are isomorphic.

Proof of (1) First we observe that X and Y are not isomorphic as line bundles. By Theorem 6.3, for this it is enough to prove that the corresponding divisors are not linearly equivalent, that is, the divisor P on B is not linearly equivalent to 0 on B.

If P were linearly equivalent to 0, there would be a regular function f on B with zero of order 1 at P and no other zero. The divisor of f on C would have to be $P - kQ$. By Theorem 3.5 and Corollary of Section 2.1, Chapter 3, $k = 1$, and by Theorem 3.8 of Section 2.1, Chapter 3 this would contradict the fact that C is not rational.

Now suppose that there exists an isomorphism $\varphi\colon X \to Y$ as algebraic varieties. Write p_X and p_Y for the projections of X and Y to B defined by their structure of line bundles. For any point $b \in B$ the curve $p_X^{-1}(b)$ is isomorphic to $\mathbb{A}^1$, and hence so is $\varphi(p_X^{-1}(b))$. If $p_Y(\varphi(p_X^{-1}(b)))$ were not a point of B then p_Y^* would define an inclusion of $\mathbb{C}(B)$ into $\mathbb{C}(\varphi(p_X^{-1}(b)))$, which would contradict Lüroth's theorem (Section 1.3, Chapter 1), because $\varphi(p_X^{-1}(b))$ is rational and B is not. Thus φ takes fibres of X into fibres of Y. We see that there exists a map $\psi\colon B \to B$ such that the diagram

$$\begin{array}{ccc} X & \xrightarrow{\varphi} & Y \\ {\scriptstyle p_X}\downarrow & & \downarrow{\scriptstyle p_Y} \\ B & \xrightarrow[\psi]{} & B \end{array}$$

commutes. If s_X is the zero section of X then $\psi = p_Y \circ \varphi \circ s_X$, and it follows that ψ is a morphism, and hence an automorphism of B. Write $\psi \times 1$ for the automorphism of the line bundle $Y = B \times \mathbb{A}^1$, that acts as ψ on B and as the identity on $\mathbb{A}^1$. Then $\varphi' = (\psi \times 1)^{-1} \circ \varphi$ is also an isomorphism of X and Y, but now for b in B we have $p_Y(\varphi'(p_X^{-1}(b))) = b$ and $\psi' = p_Y \circ \varphi \circ s_X = 1$.

Set $t = \varphi' \circ s_X\colon B \to Y$. This is a section of the line bundle Y. Recall now that Y is a vector bundle, so that it makes sense to speak of subtracting vectors in its fibres. We set

$$\varphi''(x) = \varphi'(x) - t\bigl(p_X(x)\bigr).$$

This is obviously again an isomorphism of X and Y, but now not only is each fibre taken into itself, but also the zero point of each fibre is preserved. However, the only automorphisms of $\mathbb{A}^1$ preserving 0 are the linear maps $\alpha \to \lambda\alpha$. Therefore φ'' must be an isomorphism of the vector bundles X and Y, but we have already seen that they are not isomorphic.

This completes the proof that X and Y are not isomorphic as algebraic varieties. □

Proof of (2) We will make use of the fact that the correspondence $D \mapsto L_D$ of Section 1.4, Chapter 6 between divisors and line bundles carries over word-for-word to complex manifolds and meromorphic functions (compare Exercises 1–2 of Section 2.4). In particular, if we prove that P is equal to the divisor of some meromorphic function φ on B then we will have proved that the manifolds X_{an} and Y_{an} are isomorphic, and even isomorphic as complex analytic line bundles. Thus our problem reduces to that of constructing a holomorphic function on B having a single zero of order 1 at P and no other zero.

To make everything entirely concrete, suppose that the curve C is given by the equation

$$y^2 = x^3 + ax + b, \tag{8.26}$$

and Q is the point at infinity. Then B is given by (8.26) in the affine plane.

On C consider the 3 rational differential 1-forms

$$\omega_1 = \frac{\mathrm{d}x}{y}, \qquad \omega_2 = x\frac{\mathrm{d}x}{y} \quad \text{and} \quad \omega_3 = \frac{1}{2}\frac{y - y_0}{x - x_0}\frac{\mathrm{d}x}{y},$$

where $P = (x_0, y_0)$. Consider the behaviour of these at $Q \in C$. At this point $t = x/y$ is a local parameter, and

$$x = \frac{u}{t^2}, \quad y = \frac{v}{t^3} \quad \text{with } u, v \in \mathcal{O}_Q \text{ and } u(Q) = v(Q) = 1.$$

It follows from this that ω_1 is regular at Q, ω_2 has a pole of order 2 and ω_3 a pole of order 1. Dividing (8.26) by y^2, we see easily that $xt^2 \equiv 1$ and $yt^3 \equiv 1$ modulo t^4. It follows that

$$\omega_2 = \left(-\frac{2}{t^2} + f\right)\mathrm{d}t \quad \text{and} \quad \omega_3 = \left(-\frac{1}{t} + g\right)\mathrm{d}t \quad \text{with } f, g \in \mathcal{O}_Q. \tag{8.27}$$

Since C has genus 1, by the results of Section 3.3, Chapter 7, the topological space $C(\mathbb{C})$ is homeomorphic to the torus. Let α and β be a basis of its 1-dimensional homology group, for example a parallel and a meridian.

It is easy to see that ω_1 is regular on C, and ω_2 has a single pole at Q of order 2. The integral of ω_1 along a 1-cycle σ depends only on the homology class of σ. If σ is homologous to $a\alpha + b\beta$ then

$$\int_\sigma \omega_1 = a\int_\alpha \omega_1 + b\int_\beta \omega_1.$$

Although ω_2 is not regular at Q, its integral along a small contour around Q equals 0, since there is no term in $1/t$ in its expansion (8.27). Therefore the same formula hold for it:

$$\int_\sigma \omega_2 = a\int_\alpha \omega_2 + b\int_\beta \omega_2,$$

provided that the cycle σ does not pass through Q.

Finally, for the form ω_3 we get a similar expression

$$\int_\sigma \omega_3 = a\int_\alpha \omega_3 + b\int_\beta \omega_3 + 2\pi i n \quad \text{with } n \in \mathbb{Z}, \tag{8.28}$$

since ω_3 has poles of order 1 at P and at Q, and its expansion at P is similar to (8.27): $\omega_3 = (1/u + h)\mathrm{d}u$ with $h \in \mathcal{O}_P$, where u is a local parameter at P.

The vectors $(\int_\alpha \omega_1, \int_\beta \omega_1)$ and $(\int_\alpha \omega_2, \int_\beta \omega_2)$ in $\mathbb{C}^2$ are linearly independent over $\mathbb{C}$. Indeed, if a linear combination of them with coefficients λ, μ were equal to 0 we would get a relation

$$\int_\sigma (\lambda\omega_1 + \mu\omega_2) = 0$$

for any 1-cycle σ. Therefore the integral $\varphi(x) = \int_q^z (\lambda\omega_1 + \mu\omega_2)$ (for some choice of base point q) would be a well-defined meromorphic function on C_{an}. By Theorem 8.4 it would have to be a rational function on C. If $\mu \neq 0$ then it would have a single pole of order 1 at Q, which is impossible, since the curve C is not rational. If $\mu = 0$ then it is regular everywhere, which is also impossible.

Using the linear independence of these vectors, choose λ and μ such that

$$\left(\int_\alpha \omega_3, \int_\beta \omega_3\right) = \lambda\left(\int_\alpha \omega_1, \int_\beta \omega_1\right) + \mu\left(\int_\alpha \omega_2, \int_\beta \omega_2\right).$$

Now set $\eta = \omega_3 - \lambda\omega_1 - \mu\omega_2$. The equality (8.28) shows that $\int_\sigma \eta = 2\pi i n$ with $n \in \mathbb{Z}$ for any cycle σ. Hence $\int_q^z \eta$ is a multivalued function $B_{\mathrm{an}} \setminus P \to \mathbb{C}$, and its exponential

$$\varphi = \exp \int_q^z \eta$$

is well defined, and is holomorphic and nowhere zero on $B_{\mathrm{an}} \setminus P$. In a neighbourhood of P we have $\omega_3 = (1/u + h)du$ with $h \in \mathcal{O}_P$, and η has a similar expansion; hence $\varphi = u\psi$, where ψ is holomorphic and nonzero at P. This proves that the divisor of φ on B_{an} consists of the point P with coefficient 1; in other words, in the complex analytic category, the divisor P is linearly equivalent on B_{an} to 0, so that X_{an} and Y_{an} are isomorphic. (Note that φ has an essential singularity at $Q \in C$, since if it were meromorphic it would be a rational function by Theorem 8.4.) (2) is proved. □

3.3 Example of a Nonalgebraic Compact Complex Manifold with Maximal Number of Independent Meromorphic Functions

The transcendence degree of $\mathcal{M}(X)$, which is finite by Theorem 8.3, is the main invariant by means of which it is natural to try to classify compact complex manifolds. We now discuss what is known in this direction, omitting all proofs.

From this point of view, the complex manifolds that are closest to algebraic varieties are those for which the transcendence degree of $\mathcal{M}$ equals the dimension of X. We start by constructing an example of a complex manifold with this property which is not an algebraic variety. The construction is closely related to that used in Section 2.3, Chapter 6 to construct an example of a nonprojective algebraic variety. For it we use the notion of blowup of a nonsingular subvariety in the case that the ambient space is a complex manifold. The reader will easily check that the definitions and elementary properties given in Section 2.2, Chapter 6 carry over word-for-word to this case.

We consider projective 3-space $\mathbb{P}^3$ and a curve $C \subset \mathbb{P}^3$ having a double point x_0 with two distinct tangent directions, for example the curve given by $z=0$, $y^2 = x^2+x^3$ (as in (1.2) of Section 1.2, Chapter 1). There exists a neighbourhood U of x_0 in the complex topology of $\mathbb{P}^3$ such that the complex space $U \cap C$ is reducible, and breaks up into two irreducible 1-dimensional nonsingular submanifolds C' and C'' that intersect transversally at x_0 (the two branches of C at x_0).

We perform first the blowup $\sigma_1 \colon U_1 \to U$ of U with centre in C'. The inverse image $C_1' = \sigma_1^{-1}(C')$ is a nonsingular surface, and $\sigma_1 \colon C_1' \to C'$ has fibres isomorphic to $\mathbb{P}^1$. Set $L_1 = \sigma_1^{-1}(x_0)$. The inverse image $\sigma_1^{-1}(C'')$ of C'' is reducible, consisting of two 1-dimensional components, L_1 together with a nonsingular curve C_1'' that maps isomorphically to C'' under σ_1. Both these components are nonsingular and intersect transversally at the point $x_1 = L_1 \cap C_1''$. Now consider the blowup $\sigma_2 \colon \overline{U} \to U_1$ of U_1 with centre in C_1''. The inverse image $\sigma_2^{-1}(L_1)$ again consists of two 1-dimensional components: $\sigma_2^{-1}(L_1) = \overline{L} \cup \overline{L}_1$, where $\overline{L} = \sigma_2^{-1}(x_1)$, and $\overline{L}_1$ is a curve mapping isomorphically to L_1 under σ_2. We set $\overline{\sigma} = \sigma_2 \circ \sigma_1 \colon \overline{U} \to U$. On the other hand, consider the blowup $\sigma \colon V \to (\mathbb{P}^3 \setminus x_0)$ of $\mathbb{P}^3 \setminus x_0$ with centre in the submanifold $C \setminus x_0$. Since over $U \setminus x_0$ the composite blowup $\overline{\sigma}$ coincides with the blowup of $C - x_0$, the two manifolds and maps $\overline{\sigma} \colon \overline{U} \to U$ and $\sigma \colon V \to \mathbb{P}^3 \setminus x_0$ glue together to give a single map

$$\sigma \colon X \to \mathbb{P}^3.$$

Obviously $\mathbb{C}(\mathbb{P}^3) \subset \mathcal{M}(X)$, so that the transcendence degree of $\mathcal{M}(X)$ equals 3. We now prove that X is not an algebraic variety. For this we suppose that it is an algebraic variety, and use the notion of numerical equivalence of curves on X, introduced in Section 2.3, Chapter 6 in connection with the analogous example. We use the fact that on an algebraic variety, a nonsingular irreducible curve cannot be numerically equivalent to 0. Indeed, as we saw in Section 2.3, Chapter 6, for this it is enough to construct an effective divisor intersecting our curve in a nonempty finite set of points. Let $E \subset X$ be our curve and $W \subset X$ an affine open set (we are assuming that X is an algebraic variety, remember). In W we can find a divisor intersecting $W \cap E$ in a nonempty finite set of points, for example by choosing two distinct points $x, x' \in W \cap E$, and taking a hypersurface F in the ambient space containing x and not x'. The closure $\overline{F}$ of F in X will have the property we require.

Now to get a contradiction to the assumption that X is an algebraic variety, we need only find an irreducible curve in X numerically equivalent to 0. For this, we use the fact that under a blowup with centre a curve, the inverse images of all points of this curve are numerically equivalent. Choose points $x \in C \setminus x_0$, $x' \in C' \setminus x_0$ and $x'' \in C'' \setminus x_0$, and let $L = \sigma^{-1}(x)$, $L' = \sigma^{-1}(x')$ and $L'' = \sigma^{-1}(x'')$. If we consider

L'' as the inverse image of the point $\sigma_1^{-1}(x'')$ under σ_2 we get that

$$L \sim L'' \sim \overline{L}. \tag{8.29}$$

On the other hand, on U_1,

$$\sigma_1^{-1}(x') \sim L,$$

and on $\overline{U}$

$$L' \sim \sigma_2^{-1}(\overline{L}_1) = \overline{L} + \overline{L}_1.$$

Thus

$$L \sim \overline{L} + \overline{L}_1.$$

In conjunction with (8.29) this shows that $\overline{L}_1 \sim 0$.

Note that in all these arguments, instead of numerical equivalence of curves on X we could have used equivalence of the corresponding cycles under homology, using the results of Section 1.3, Chapter 7.

Dimension 3 in our example is the smallest possible case, since it can be proved that a compact complex manifold of dimension 2 with two algebraically independent meromorphic functions is algebraic, and is hence projective, as we have already indicated in Section 2.3, Chapter 6.

Complex manifolds X for which the transcendence degree of $\mathcal{M}(X)$ equals $\dim X$ are very close to algebraic varieties. In this case the field $\mathcal{M}(X)$ is isomorphic to the rational function field $\mathbb{C}(X')$ of an algebraic variety X' with $\dim X' = \dim X$, so that X is *bimeromorphic* to an algebraic variety. This fact can be made more precise, by proving an analogue of Chow's lemma (see Section 2.1, Chapter 6). All this suggests that there is a purely algebraic description of these complex manifolds, and that analogous objects can be defined over an arbitrary field. Such an object, called an *algebraic space* or Moishezon manifold, has indeed been introduced by Artin and Moishezon. For this see Knutson's book [48].

3.4 The Classification of Compact Complex Surfaces

We proceed to the type of complex manifold that comes next in our classification, for which the transcendence degree of $\mathcal{M}(X)$ equals $\dim X - 1$. By the Riemann existence theorem, this case is not possible for $\dim X = 1$, and we should expect to meet it first when $\dim X = 2$, that is, for complex surfaces. We know some examples of these surfaces: they are the complex toruses of Example 8.3 and the Hopf surfaces of Example 8.5. A general description of surfaces in this class is given by the following theorem of Kodaira:

Theorem *A compact complex surface X for which the transcendence degree of $\mathcal{M}$ is 1 has a holomorphic map $p\colon X \to Y$ to an algebraic curve Y such that $\mathcal{M}(X) = p^*(\mathbb{C}(Y))$, and such that all but a finite number of the fibres $p^{-1}(y)$ are elliptic curves.*

An analogous result can also be proved for complex manifolds of arbitrary dimension, but in a weaker form:

Theorem *If X is a compact n-dimensional complex manifold, and the transcendence degree of $\mathcal{M}(X)$ equals $n-1$, then X is bimeromorphic to a manifold X' having a holomorphic map $p\colon X' \to Y$ to an $(n-1)$-dimensional algebraic variety Y, such that $\mathcal{M}(X) = \mathcal{M}(X') = p^*(\mathbb{C}(Y))$ and $p^{-1}(y)$ is an elliptic curve for all points y in an open dense set of the Zariski topology of Y.*

Complex manifolds of other types have been studied almost exclusively in the case of complex surfaces. For these, there remains only one type, when $\mathcal{M}(X) = \mathbb{C}$. We now describe the classification of this type of surfaces, obtained by Kodaira.

We observe first that the notion of -1-curve carries over in a natural way to complex manifolds. One can prove that any complex surface can be obtained by a finite number of blowups of a surface not containing -1-curves. Kodaira proved that for a compact surface X without -1-curves and with no nonconstant meromorphic functions, the first Betti number b_1 can take only one of 3 values, 4, 1 or 0. If $b_1 = 4$ then X is a complex torus. We already know an example of a complex torus on which all meromorphic functions are constant (Example 8.4).

If $b_1 = 0$ then the canonical line bundle of X is trivial. (The canonical line bundle is defined by analogy with the case of algebraic varieties, and is a replacement for the canonical class in cases where we cannot use rational or meromorphic functions; see Exercises 3–4 of Section 2.4.) All surfaces of this type are homeomorphic to one another and to the algebraic K3 surfaces (see Section 6.7, Chapter 3). They are called *complex analytic K3 surfaces* (see Exercises 1–5).

The case $b_1 = 1$ has so far not been investigated so fully. Examples of such surfaces are obtained by generalising the construction of Hopf varieties. Namely, compact complex surfaces of the form $(\mathbb{C}^2 \setminus 0)/G$, where G is a group acting freely and discretely on $\mathbb{C}^2 \setminus 0$, are called *generalised Hopf surfaces*. For example, we could take G to be the cyclic group generated by the automorphism $(z_1, z_2) \mapsto (\alpha_1 z_1, \alpha_2 z_2)$, where $|\alpha_1| < 1$, $|\alpha_2| < 1$. It can be shown that if there do not exist integers n_1, n_2, not both 0, such that $\alpha_1^{n_1} = \alpha_2^{n_2}$, then all meromorphic functions on this surface are constants.

There also exist other classes of surfaces without meromorphic functions with $b_1 = 1$, called the *Inoue–Hirzebruch surfaces*. Some of these surfaces have $b_2 = 0$, and some $b_2 > 0$. This is the class of complex surfaces that is least well studied.

Thus, according to the value of the invariant k, the transcendence degree of $\mathcal{M}(X)$, compact complex surfaces can be classified as follows:

$k = 2$: algebraic surfaces;
$k = 1$: surfaces with a pencil of elliptic curves;
$k = 0$: complex toruses, K3 surfaces, or surfaces with $b_2 = 1$.

One is struck in this classification by the amazing similarity with the classification of algebraic surfaces treated in Section 6.7, Chapter 3. In all probability, this analogy can only be understood in connection with a generalisation of both theories to manifolds of arbitrary dimension. This is one of the most interesting problems in the theory of algebraic varieties and complex manifolds.

3.5 Exercises to Section 3

1 Let $A = \mathbb{C}^2/\Omega$ be a 2-dimensional torus, g the automorphism given by $g(x) = -x$ and G the group $\{1, g\}$. Prove that the ringed space $\widetilde{X} = A/G$ (see Exercise 1 of Section 3.6, Chapter 5) is a complex space, with 16 singular points $z_1, \dots, z_{16}$ corresponding to the points $x \in A$ with $2x = 0$.

2 In the notation of Exercise 1, prove that each of the singular points $z_i \in \widetilde{X}$ has a neighbourhood isomorphic to a neighbourhood of the vertex of the quadratic cone (see Exercise 8 of Section 1.6).

3 In the notation of Exercises 1–2, prove that there exists a complex manifold X and a holomorphic map $\varphi \colon X \to \widetilde{X}$ such that X has 16 mutually disjoint curves $C_1, \dots, C_{16}$, each of which is isomorphic to $\mathbb{P}^1_{\mathrm{an}}$, with $\varphi(C_i) = z_i$, and $\varphi \colon X \setminus \bigcup C_i \to \widetilde{X} \setminus \bigcup z_i$ an isomorphism. [Hint: Use Exercise 10 of Section 4.6, Chapter 2.]

4 We use the notation of the previous exercises; let z_1 and z_2 be coordinates on $\mathbb{C}^2$. Prove that the differential 2-form $dz_1 \wedge dz_2$ on $\mathbb{C}^2$ defines a holomorphic nowhere vanishing 2-form on A (compare Exercises 3–4 of Section 2.4). Prove that it also defines a holomorphic nowhere vanishing 2-form on X. Deduce from this that the canonical line bundle of X is trivial.

5 In the notation of the previous exercises, prove that if on the torus A all meromorphic functions are constant, then the same holds for X. Prove that X is not isomorphic to a complex torus (prove for example that X does not have holomorphic 1-forms). Thus X is an example of a nonalgebraic K3 surface.

6 Prove that for every nonsingular projective variety X of dimension ≥ 3 there exists a nonalgebraic compact complex n-dimensional manifold X' such that $\mathcal{M}(X') = \mathbb{C}(X)$.

4 Kähler Manifolds

We now describe a class of complex manifolds that are close to algebraic varieties. This class is characterised by the existence of a Riemannian metric of a special type, and its theory gives some idea of the powerful metric methods that can be used in the study of complex manifolds, and in particular, of algebraic varieties.

4.1 Kähler Metric

We begin by discussing the question on the level of linear algebra, treating some properties of Hermitian forms that we need in what follows. Let L be a n-dimensional complex vector space and φ a *Hermitian form* on L. Recall that this means that $\varphi(x, y) \in \mathbb{C}$ for $x, y \in L$, and the following conditions hold for all $x_1, x_2, y \in L$ and for all $\alpha_1, \alpha_2 \in \mathbb{C}$:

$$\varphi(y, x) = \overline{\varphi(x, y)},$$
$$\varphi(\alpha_1 x_1 + \alpha_2 x_2, y) = \alpha_1 \varphi(x_1, y) + \alpha_2 \varphi(x_2, y).$$

Introducing coordinates in L, we can write φ in the form $\varphi(x, y) = \sum c_{\alpha\beta} x_\alpha \overline{y}_\beta$, with $c_{\beta\alpha} = \overline{c}_{\alpha\beta}$. Viewing L as a $2n$-dimensional real vector space, and setting $\varphi(x, y) = \alpha(x, y) + i\beta(x, y)$, where $\alpha = \Re\varphi$, $\beta = \Im\varphi$, we get two $\mathbb{R}$-bilinear forms α and β, with α symmetric and β skewsymmetric. The fact that α and β come from a complex Hermitian form implies that

$$\alpha(ix, iy) = \alpha(x, y), \qquad \beta(ix, iy) = \beta(x, y) \quad \text{and} \quad \alpha(x, y) = \beta(ix, y).$$

Conversely, if β is any skewsymmetric $\mathbb{R}$-bilinear form on L, the relation $\alpha(x, y) = \beta(ix, y)$ determines the form α uniquely, and hence also $\varphi = \alpha + i\beta$, which is a Hermitian form provided that $\beta(ix, iy) = \beta(x, y)$.

The form $\omega = -\beta$, where β is the form just constructed, is called the *associated skewsymmetric bilinear form* of the Hermitian form φ, and φ the *associated Hermitian form* of the skewsymmetric form ω. We view ω as an element of the second exterior power $\bigwedge^2 L^*$ of the dual vector space L^* of L. A simple computation shows how ω is constructed from φ in coordinates. Namely, if $\varphi(x, y) = \sum c_{\alpha\beta} x_\alpha \overline{y}_\beta$ with $c_{\beta\alpha} = \overline{c}_{\alpha\beta}$ then

$$\begin{aligned}
\omega(x, y) &= \frac{i}{2} \sum (c_{\alpha\beta} x_\alpha \overline{y}_\beta - \overline{c}_{\alpha\beta} \overline{x}_\alpha y_\beta) \\
&= \frac{i}{2} \sum (c_{\alpha\beta} x_\alpha \overline{y}_\beta - c_{\beta\alpha} y_\beta \overline{x}_\alpha) \\
&= \frac{i}{2} \sum c_{\alpha\beta} (x_\alpha \overline{y}_\beta - y_\alpha \overline{x}_\beta).
\end{aligned}$$

In other words, $\omega = (i/2) \sum c_{\alpha\beta} \xi_\alpha \wedge \overline{\xi}_\beta$, where ξ_α is the basis of L^* dual to the chosen basis of L, that is, such that $\xi_\alpha(x) = x_\alpha$.

Now suppose that the Hermitian form φ is positive definite. Then, in some basis, φ and ω can be written

$$\varphi(x, y) = \sum x_\alpha \overline{y}_\alpha \quad \text{and} \quad \omega = \frac{i}{2} \sum \xi_\alpha \wedge \overline{\xi}_\alpha.$$

If $a_1, \dots, a_n$ are elements of a commutative ring satisfying $a_1^2 = \dots = a_n^2 = 0$ then $(a_1 + \dots + a_n)^n = n! a_1 \cdots a_n$. Applying this to the elements $\xi_\alpha \wedge \overline{\xi}_\alpha$ of the even subalgebra of the exterior algebra, we get

$$\omega^n = \left(\frac{i}{2}\right)^n n! \xi_1 \wedge \overline{\xi}_1 \wedge \cdots \wedge \xi_n \wedge \overline{\xi}_n.$$

Setting $\xi_\alpha = u_\alpha + i v_\alpha$ we see that $\xi_\alpha \wedge \overline{\xi}_\alpha = -2i u_\alpha \wedge v_\alpha$, so that $\omega^n = n!\Omega$, where Ω is the standard volume form $\mathrm{d}x_1 \wedge \mathrm{d}y_1 \wedge \cdots \wedge \mathrm{d}x_n \wedge \mathrm{d}y_n$ on L (as a $2n$-dimensional vector space over $\mathbb{R}$). In a more intrinsic form we can write these relations in the form

$$\omega^n = \left|\det |c_{\alpha\beta}|\right|^2 \mathrm{d}x_1 \wedge \mathrm{d}y_1 \wedge \cdots \wedge \mathrm{d}x_n \wedge \mathrm{d}y_n, \tag{8.30}$$

if $\omega = (i/2)\sum c_{\alpha\beta}\xi_\alpha \wedge \overline{\xi}_\beta$. In particular $\omega^n \neq 0$, and a fortiori $\omega^m \neq 0$ for $m < n$.

Now let X be an n-dimensional complex manifold, with a given positive definite Hermitian form. That is, φ defines a positive definite Hermitian form on the tangent space at each point $x \in X$. In some domain U with local coordinates $z_1, \dots, z_n$ the form φ can be written $\sum c_{\alpha\beta}\mathrm{d}z_\alpha \mathrm{d}\overline{z}_\beta$, where $c_{\alpha\beta}$ are functions on U that we will assume to be complex valued real analytic functions of $z_1, \dots, z_n$, $\overline{z}_1, \dots, \overline{z}_n$. A form of this type defines a Riemannian metric on X, which is called a *Hermitian metric*. Any complex manifold admits many Hermitian metrics, in the same way that any differentiable manifold admits many Riemannian metrics.

We first discuss the local properties of a Hermitian metric φ, in a sufficiently small domain U. The simplest possible question is: can φ be transformed to the flat form $\sum \mathrm{d}z_\alpha \mathrm{d}\overline{z}_\alpha$ by a complex analytic coordinate change, at least in a small neighbourhood of P? In Riemannian geometry, it is well known that the answer is as follows: a metric $\sum g_{\alpha\beta}\mathrm{d}x_\alpha \mathrm{d}x_\beta$ is not distinguished from a flat metric either by the values $g_{\alpha\beta}(P)$ at P of its matrix entries, or by the values $(\partial g_{\alpha\beta}/\partial x_\gamma)(P)$ at P of their partial derivatives. More precisely, there exists a system of coordinates (called *normal* or *geodesic coordinates*) such that $g_{\alpha\beta}(P) = \delta_{\alpha\beta}$ and $(\partial g_{\alpha\beta}/\partial x_\gamma)(P) = 0$. The obstruction to making the metric flat involves the second partial derivatives of the coefficients of the metric; this is the *curvature tensor*.

The situation in the complex analytic case is more delicate. Of course, we can carry out a complex analytic (or even linear) coordinate change so that in the new coordinates $c_{\alpha\beta}(P) = \delta_{\alpha\beta}$. However, there is a very simple obstruction to finding a coordinate system in which $(\partial c_{\alpha\beta}/\partial z_\gamma)(P) = (\partial c_{\alpha\beta}/\partial \overline{z}_\gamma)(P) = 0$; that is to say, already the first derivatives of the matrix entries of the Hermitian metric distinguish it from a flat metric.

Namely, consider on each tangent space at $x \in U$ the skewsymmetric form ω associated with the Hermitian form φ. Together these define a differential 2-form, written $\omega = (i/2)\sum c_{\alpha\beta}\mathrm{d}z_\alpha \wedge \mathrm{d}\overline{z}_\beta$ in local coordinates. The description of the map $\varphi \mapsto \omega$ given at the start of this section shows that it is intrinsic, that is, the construction of the form ω is independent of the choice of the coordinate system $z_1, \dots, z_n$. But then the differential $\mathrm{d}\omega$ of this form is also defined by the Hermitian form φ in an invariant way. In particular, the condition $\mathrm{d}\omega = 0$ is independent of the coordinate system, and is a necessary condition for the metric to become flat after a complex analytic coordinate change. We can write these conditions in the explicit form

$$\frac{\partial c_{\alpha\beta}}{\partial z_\gamma} = \frac{\partial c_{\gamma\beta}}{\partial z_\alpha} \quad \text{and} \quad \frac{\partial c_{\alpha\beta}}{\partial \overline{z}_\gamma} = \frac{\partial c_{\alpha\gamma}}{\partial \overline{z}_\beta}.$$

It is of course easy to write down Hermitian metrics that do not satisfy these relations. For such a metric, already the first derivatives show that it cannot be made flat. It is natural to consider metrics for which this first obstruction vanishes.

Definition A Hermitian metric φ on a complex manifold is a *Kähler metric* if its associated differential 2-form ω is closed, that is, $\mathrm{d}\omega = 0$. A manifold with a given Kähler metric is called a *Kähler manifold.*

4.2 Examples

Example 8.6 (Quotient manifolds) Let X be a complex Kähler manifold with Kähler metric φ and G a group of analytic automorphisms of X acting freely and discretely (see Section 1.2). If each automorphism $g \in G$ preserves the Kähler metric φ then it induces a metric φ^* on the quotient space X/G. To define φ^*, we must take an open set $U \subset X/G$ whose inverse image $\pi^{-1}(U)$ under the natural projection $\pi\colon X \to X/G$ breaks up as a disjoint union of open sets U_α, each of which maps isomorphically to U under π. We first restrict the metric φ to one of the sets U_α, then use the isomorphism π to transfer it to U. Since all the U_α are taken to one another by automorphisms $g \in G$, and φ is invariant under G, the resulting metric on U does not depend on the choice of U_α and is entirely uniquely defined. From this it is easy to deduce that the metrics defined on different neighbourhoods $U \subset X/G$ glue together to give a metric φ^* on the whole of X/G. Obviously this is a Kähler metric, since being Kähler is a local property, and φ^* coincides locally with φ.

A very important special case is $\mathbb{C}^n$ with the flat metric $\varphi = \sum \mathrm{d}z_\alpha \mathrm{d}\overline{z}_\alpha$ and G a group of translations in vectors of a lattice Ω. Obviously translations preserve the metric φ, and the differential 2-form $\omega = (i/2)\sum \mathrm{d}z_\alpha \wedge \mathrm{d}\overline{z}_\alpha$ associated with φ is closed. It follows that any torus $\mathbb{C}^n/G$ is a Kähler manifold.

Example 8.7 (The Fubini–Study metric on $\mathbb{P}^n$) From now on, as discussed at the end of Section 1.1, we write the differential d acting on forms as $\mathrm{d} = \mathrm{d}' + \mathrm{d}''$, where d' and d'' are the differential with respect to z_i and $\overline{z}_i$.

Let $\zeta_0, \dots, \zeta_n$ be homogeneous coordinates on $\mathbb{P}^n$ and ζ an arbitrary linear form. Then $\zeta_\alpha/\zeta = z_\alpha$ are rational functions on $\mathbb{P}^n$. Set

$$H = \log \sum_{\alpha=0}^{n} |z_\alpha|^2 \quad \text{and} \quad \omega = i\mathrm{d}'\mathrm{d}''H. \tag{8.31}$$

Note first that the 2-form ω is independent of the choice of the linear form ζ. For this it is enough to check that if η is another linear form then $\mathrm{d}'\mathrm{d}'' \log|\zeta/\eta|^2 = 0$.

Indeed, wherever $\eta \neq 0$, the function $\zeta/\eta = z$ is one of the local coordinates, and the assertion reduces to an easy exercise in functions of one complex variable:

$$\mathrm{d}'\mathrm{d}'' \log |z|^2 = 0, \quad \text{that is,} \quad \frac{\partial^2}{\partial z \partial \overline{z}} \log |z|^2 = 0;$$

the meaning of this is that $\partial^2/\partial z\partial\overline{z} = \partial^2/\partial^2 x + \partial^2/\partial^2 y$ is the Laplace operator, and $\log(x^2 + y^2)$ is an elementary solution of the Laplace equation.

The 2-form ω is obviously closed, because $\mathrm{d}(\mathrm{d}'\mathrm{d}''h) = 0$ for any function h. We have the following explicit coordinate expression for it:

$$\begin{aligned}
\omega &= \frac{i}{2}\mathrm{d}'\mathrm{d}'' \log \sum |z_\alpha|^2 \\
&= \frac{i}{2}\mathrm{d}'\Big(\sum z_\alpha \mathrm{d}\overline{z}_\alpha \Big/ \sum |z_\alpha|^2\Big) \\
&= \frac{i}{2}\frac{\sum \mathrm{d}z_\alpha \wedge \mathrm{d}\overline{z}_\alpha}{\sum |z_\alpha|^2} - \frac{i}{2}\frac{(\sum \overline{z}_\alpha \mathrm{d}z_\alpha) \wedge (\sum z_\alpha \mathrm{d}\overline{z}_\alpha)}{(\sum |z_\alpha|^2)^2},
\end{aligned} \tag{8.32}$$

where as before $z_\alpha = \zeta_\alpha/\zeta$.

We now show that the Hermitian metric associated with ω is positive definite, so that ω is a Kähler metric on $\mathbb{P}^n$. It is easiest to do this using the homogeneity property of ω. Namely, the unitary group $\mathrm{U}(n+1)$ of $\mathbb{C}^{n+1}$ with the metric $\sum |\zeta_\alpha|^2$ also acts on $\mathbb{P}^n = \mathbb{P}(\mathbb{C}^{n+1})$. The 2-form ω is invariant under this action, because

$$g^*(\omega) = i\mathrm{d}'\mathrm{d}'' \log \frac{\sum |\zeta_\alpha|^2}{|g^*\zeta|^2} = i\mathrm{d}'\mathrm{d}'' \log \frac{\sum |\zeta_\alpha|^2}{|\zeta|^2} = \omega$$

for $g \in \mathrm{U}(n+1)$. Now $\mathrm{U}(n+1)$ acts transitively on $\mathbb{P}^n$, so that it is enough to check that the associated Hermitian form of ω is positive definite at any one point, for example, at $P = (1 : 0 : \cdots : 0)$.

Choose the linear form ζ to be the homogeneous coordinate ζ_0, so that $z_1, \ldots, z_n$ are local coordinates at P. Now since $z_\alpha = \overline{z}_\alpha = 0$ at P and $\mathrm{d}z_0 = 0$, the form (8.32) simplifies to $(i/2)\sum_{\alpha=1}^n \mathrm{d}z_\alpha \wedge \mathrm{d}\overline{z}_\alpha$, and we see that the associated Hermitian form is $\sum \mathrm{d}z_\alpha \mathrm{d}\overline{z}_\alpha$, which is positive definite.

The Kähler metric we have constructed on $\mathbb{P}^n$ is called the *Fubini–Study metric*. We mention without proof another interpretation of it. We view $\mathbb{P}^n$ as the image of the sphere $S^{2n+1} \subset \mathbb{C}^{n+1}$ defined by $\sum_0^n |\zeta_\alpha|^2 = 1$. Each point $P = (\zeta_0 : \cdots : \zeta_n) \in \mathbb{P}^n$ corresponds to an entire great circle of S^{2n+1} consisting of point $(\zeta_0 e^{2\pi i\theta}, \ldots, \zeta_n e^{2\pi i\theta})$. Then the distance in the Fubini–Study metric of $\mathbb{P}^n$ between two points $P, Q \in \mathbb{P}^n$ is equal to the distance in the spherical geometry of S^{2n+1} between the corresponding great circles. For the proof, see Kostrikin and Manin [52, Chapter III, Section 10].

Example 8.8 (The induced Kähler metric on a projective manifold $X \subset \mathbb{P}^n$) Let X be a Kähler manifold and $Y \subset X$ a complex submanifold. The restriction of differential forms from X to Y takes a closed form to a closed form. The restriction of

Hermitian forms takes a positive definite form to a positive definite form. Finally the relation between a Hermitian form and its associated 2-form is preserved, as follows at once from the definition. It follows from all of this that the restriction of a Kähler metric given on X to a complex submanifold $Y \subset X$ defines a Kähler metric on Y.

In particular we see that any projective variety X has a Kähler metric. This metric is defined not by intrinsic properties of X but by its embedding into $\mathbb{P}^n$.

4.3 Other Characterisations of Kähler Metrics

We first show that the formula (8.31) used to write down the Fubini–Study metric was no accident.

Proposition 8.1 *The 2-form ω associated with a Kähler form can be written in the form*

$$\omega = \mathrm{d}'\mathrm{d}'' H$$

in a neighbourhood of any point, where H is a C^∞ function of the real coordinates. A form of type $\mathrm{d}'\mathrm{d}''H$ is obviously always of type $(1,1)$ and closed.

We observe first that the condition for a 2-form ω to be closed is $\mathrm{d}\omega = \mathrm{d}'\omega + \mathrm{d}''\omega = 0$. But ω is of *type* $(1,1)$ (that it, is of degree 1 in both the $\mathrm{d}z_\alpha$ and the $\mathrm{d}\overline{z}_\alpha$), and hence $\mathrm{d}'\omega$ and $\mathrm{d}''\omega$ are of type $(2,1)$ and $(1,2)$ respectively. Hence from $\mathrm{d}'\omega + \mathrm{d}''\omega = 0$ it follows that the two summands separately are zero, $\mathrm{d}'\omega = \mathrm{d}''\omega = 0$.

The proof of the proposition is preceded by a lemma on integrating differential forms; this is an analogue of the Poincaré lemma (closed forms are locally exact), depending on additional parameters $y_1, \dots, y_m$, with respect to which differentiation does not takes place.

Lemma *Consider a differential p-form $\omega = \sum f_{i_1 \dots i_p} \mathrm{d}x_{i_1} \wedge \dots \wedge \mathrm{d}x_{i_p}$, defined in a neighbourhood of the origin of $\mathbb{R}^n$, and with coefficients $f_{i_1 \dots i_p}$ which are real analytic functions of the coordinates $x_1, \dots, x_n$, and also depend analytically on some auxiliary variables $y_1, \dots, y_m$.*

If $p > 0$ and $\mathrm{d}\omega = 0$ then there exists a $(p-1)$-form

$$\eta = \sum g_{i_1 \dots i_{p-1}} \mathrm{d}x_{i_1} \wedge \dots \wedge \mathrm{d}x_{i_{p-1}},$$

defined in a possibly smaller neighbourhood of the origin of $\mathbb{R}^n$, and with coefficients $g_{i_1 \dots i_{p-1}}$ which are real analytic functions, such that $\mathrm{d}\eta = \omega$. This assertion holds both in the case when the coefficients f are real analytic functions, and when they are holomorphic functions.

The statement is clearly false if $p = 0$: then $\omega = g(x_1, \dots, x_n, y_1, \dots, y_m)$ is a function, and $\mathrm{d}\omega = 0$ just means $\partial g / \partial x_i = 0$, that is, $\omega = g(y_1, \dots, y_m)$ is independent of the variables $x_1, \dots, x_n$.

Proof We prove this by induction on the number of variables $x_1, \dots, x_n$, putting x_n among the variables $y_1, \dots, y_m$; we use a tilde $\widetilde{}$ to denote differentials with respect to the remaining variables $x_1, \dots, x_{n-1}$, and also forms involving only differentials of $x_1, \dots, x_{n-1}$. We can write

$$\omega = \widetilde{\omega} + \widetilde{\xi} \wedge \mathrm{d}x_n,$$

where $\widetilde{\xi}$ is a $(p-1)$-form. Then

$$\mathrm{d}\omega = \mathrm{d}\widetilde{\omega} + \mathrm{d}(\widetilde{\xi} \wedge \mathrm{d}x_n) = \widetilde{\mathrm{d}}\widetilde{\omega} + (-1)^p \frac{\partial \widetilde{\omega}}{\partial x_n} \wedge \mathrm{d}x_n + \widetilde{\mathrm{d}}\widetilde{\xi} \wedge \mathrm{d}x_n,$$

where $\partial/\partial x_n$ stands for differentiating all the coefficients of a form. The assumption $\mathrm{d}\omega = 0$ gives

$$\widetilde{\mathrm{d}}\widetilde{\omega} = 0 \quad \text{and} \quad (-1)^p \frac{\partial \widetilde{\omega}}{\partial x_n} + \widetilde{\mathrm{d}}\widetilde{\xi} = 0.$$

By the inductive hypothesis we can write

$$\widetilde{\omega} = \widetilde{\mathrm{d}}\widetilde{\varphi}, \tag{8.33}$$

where moreover

$$\widetilde{\mathrm{d}}\left((-1)^p \frac{\partial \widetilde{\varphi}}{\partial x_n} + \widetilde{\xi}\right) = 0. \tag{8.34}$$

If $p > 1$ then $(-1)^p \partial\widetilde{\varphi}/\partial x_n + \widetilde{\xi}$ is a $(p-1)$-form with $p - 1 > 0$, so that by induction on p we can find a $(p-2)$-form $\widetilde{\psi}$ such that

$$(-1)^p \frac{\partial \widetilde{\varphi}}{\partial x_n} + \widetilde{\xi} = \widetilde{\mathrm{d}}\widetilde{\psi}.$$

Then setting $\eta = \widetilde{\varphi} + \widetilde{\psi} \wedge \mathrm{d}x_n$ gives

$$\mathrm{d}\eta = \widetilde{\mathrm{d}}\widetilde{\varphi} + \left(\widetilde{\mathrm{d}}\widetilde{\psi} + (-1)^{p-1} \frac{\partial \widetilde{\varphi}}{\partial x_n}\right) \wedge \mathrm{d}x_n = \omega,$$

as required.

If $p = 1$ the inductive hypothesis is not applicable. But then it follows from (8.34) that $\widetilde{\xi} = \partial\widetilde{\varphi}/\partial x_n + f(x_n)$. Now recall that in (8.33) we can change the form $\widetilde{\varphi}$ (in the present case a function) by adding a function $g(x_n)$ to it. In particular, we can choose this function $g(x_n)$ such that $\partial g/\partial x_n + f(x_n) = 0$; then $\omega = \mathrm{d}(\widetilde{\varphi} + g)$. The lemma is proved. □

Proof of Proposition 8.1 Let ω be a $(1,1)$-form with $\mathrm{d}'\omega = \mathrm{d}''\omega = 0$. In the equality $\mathrm{d}''\omega = 0$, write ω as $\omega = \sum \eta_\alpha \wedge \mathrm{d}z_\alpha$ where the η_α are $(0,1)$-forms. Then also $\mathrm{d}''\eta_\alpha = 0$ for $\alpha = 1, \dots, n$. We apply the lemma to this equality, viewing η_α as forms in the variables $\overline{z}_1, \dots, \overline{z}_n$ with coefficients depending on $z_1, \dots, z_n$. Then

$\eta_\alpha = \mathrm{d}''\zeta_\alpha$ and hence $\omega = \mathrm{d}''\varphi$ where φ is a $(1,0)$-form. The equality $\mathrm{d}'\omega = 0$ implies $\mathrm{d}''(\mathrm{d}'\varphi) = 0$. The form $\alpha = \mathrm{d}'\varphi$ is a $(2,0)$-form, and $\mathrm{d}''\alpha = 0$ implies that it is holomorphic. On the other hand, from the fact that $\alpha = \mathrm{d}'\varphi$ it follows that $\mathrm{d}'\alpha = 0$. Applying the lemma again, this time in its holomorphic version, we get that there exists a holomorphic 1-form β such that $\alpha = \mathrm{d}'\beta$. Thus $\mathrm{d}'\varphi = \mathrm{d}'\beta$ where β is holomorphic, and therefore $\mathrm{d}'(\varphi - \beta) = 0$. Applying the lemma a third time we get that $\varphi - \beta = \mathrm{d}'H$ where H is a function. Thus in conclusion, $\omega = \mathrm{d}''\varphi = \mathrm{d}''(\mathrm{d}'H + \beta) = \mathrm{d}''\mathrm{d}'H$, since β is holomorphic and hence $\mathrm{d}''\beta = 0$. The proposition is proved. □

Our second characterisation of Kähler metrics is related to the arguments we started from in Section 4.1. We saw there that the Kähler condition is the necessary condition for the existence of a complex analytic coordinate system such that the metric coincides with a flat metric up to terms of degree ≥ 2. We now show that this condition is also sufficient.

Proposition 8.2 *For a Kähler metric, there exists a complex analytic coordinate system at each point P such that the matrix entries $c_{\alpha\beta}$ of the metric satisfy the conditions*

$$c_{\alpha\beta}(P) = \delta_{\alpha\beta}, \quad \textit{and} \quad \frac{\partial c_{\alpha\beta}}{\partial z_\gamma}(P) = \frac{\partial c_{\alpha\beta}}{\partial \overline{z}_\gamma}(P) = 0.$$

Proof We start from the associated form ω of the Kähler metric, and its representation $\omega = \mathrm{d}'\mathrm{d}''H$ established in Proposition 8.1, where H is an analytic function in the coordinates z_α and $\overline{z}_\alpha$. In the Taylor series expansion of H, the terms of degree 0 and 1 have no effect whatsoever on ω, and we can assume that they are 0. On the other hand, the terms of degree ≥ 4 have no effect on the values at P of the matrix entries $c_{\alpha\beta}$ of the Hermitian metric and their first derivatives. Finally, terms involving only monomials in the z_α or $\overline{z}_\alpha$ separately are killed by the operator $\mathrm{d}'\mathrm{d}''$, and we can discard them too.

Consider the terms of degree 2 of the form $c_{\alpha\beta} z_\alpha \overline{z}_\beta$. On applying $\mathrm{d}'\mathrm{d}''$, these give the matrix entries of the Hermitian form on the tangent space at P. Since this form is positive definite, we can assume, at the cost of a linear change of variables, that $c_{\alpha\beta} = \delta_{\alpha\beta}$, that is, that the terms of degree 2 are $\sum z_\alpha \overline{z}_\alpha$.

Finally, among terms of degree 3, we need only consider those of the form $d_{\alpha\beta\gamma} z_\alpha z_\beta \overline{z}_\gamma$ and $e_{\alpha\beta\gamma} z_\alpha \overline{z}_\beta \overline{z}_\gamma$. We write these terms in the form $\sum \varphi_\gamma \overline{z}_\gamma + \sum \overline{\psi}_\gamma z_\gamma$, where φ_γ and ψ_γ are quadratic forms in $z_1, \dots, z_n$. From the conditions $c_{\alpha\beta} = \overline{c}_{\beta\alpha}$ on the matrix entries of the Hermitian metric and the representation $c_{\alpha\beta} = \partial^2 H / \partial z_\alpha \partial \overline{z}_\beta$ it follows that H must be a real valued function, hence $\varphi_\gamma = \psi_\gamma$. Now a transformation of the form $z_\gamma \mapsto z_\gamma + \varphi_\gamma$ kills these terms; this proves the proposition. □

Finally we mention without proof another characterisation of Kähler metrics. A Kähler metric on a manifold induces a Riemannian metric; but a Riemannian metric defines an $\mathbb{R}$-linear map of the tangent space at any point P into the tangent space at any infinitely near point, and by integrating along a curve, a map to the tangent space at any point Q joined to P by a curve. This map is called a *connection* or *parallel transport*. The formulas defining parallel transport depend only on the matrix entries of the Riemann metric and their first derivatives. Since for a Kähler metric the matrix entries and their first derivatives are the same as for a flat metric, also parallel transport will have the same properties as ordinary parallel transport $x \mapsto x + a$ in the metric $\sum \mathrm{d}z_i \mathrm{d}\overline{z}_i$. In particular, this is a $\mathbb{C}$-linear map. It can be shown that Kähler metrics on complex manifolds are exactly the Hermitian metrics for which parallel transport is a $\mathbb{C}$-linear map: the connection is a $\mathbb{C}$-linear map $\Theta_X \to \Theta_X \otimes \Omega^1$.

4.4 Applications of Kähler Metrics

We saw in Section 4.1 that if ω is the associated skewsymmetric form of a Hermitian metric on an n-dimensional complex vector space L then $\omega^n = n!\Omega$, where Ω is the volume form of L with respect to this metric (see (8.30)). It follows that for any n-dimensional complex manifold X with a Hermitian metric $\omega^n = n!\Omega$, where ω is the associated 2-form of the metric and Ω the volume form of this metric. In particular

$$\int_{[X]} \omega^n = n! \int_{[X]} \Omega = n! \operatorname{Vol} X,$$

where $\operatorname{Vol} X$ is the volume of X. This relation can be applied to an m-dimensional subvariety $Y \subset X$. Since we know that restricting to a submanifold preserves the relation between the metric and its associated form, $\int_{[Y]} \omega^m = m! \operatorname{Vol} Y$, where $\operatorname{Vol} Y$ is the volume of Y in the given metric. This relation, expressing the volume of any complex submanifold in terms of integrals of a fixed differential form, is called *Wirtinger's theorem.*

When the metric is Kähler, Wirtinger's theorem gives us much more. In this case, the 2-form ω is closed, and therefore so are all its powers ω^m. It follows that $\int_{[Y]} \omega^m$ depends only on the homology class $[Y]$ of Y. Thus $\operatorname{Vol} Y$ is some invariant of the homology class containing the submanifold Y. The geometric meaning of this invariant is that it is the lower bound for the volume $\operatorname{Vol} Z$ as Z runs through all real submanifolds that are homologous as cycles to $[Y]$. In other words, if we view a complex manifold X with a Kähler metric as a Riemannian manifold then a complex submanifold is a minimal submanifold, realising the minimal volume in its homology class. This is very easy to prove. If ω is the associated 2-form of a Riemannian metric, $\dim Y = m$, and Z is a real $2m$-dimensional submanifold homologous to the cycle $[Y]$ then, as we have seen

$$\operatorname{Vol} Y = \frac{1}{m!} \int_{[Y]} \omega^m = \frac{1}{m!} \int_{[Z]} \omega^m.$$

It remains to prove that $|\int_{[Z]} \omega^m| \le \int_{[Z]} \Omega$, where Ω is the volume form on Z. In local coordinates $x_1, \dots, x_{2m}$ on Z we have $\omega^m = f \mathrm{d}x_1 \wedge \dots \wedge \mathrm{d}x_{2m}$ and $\Omega =$

$g dx_1 \wedge \cdots \wedge dx_{2m}$, and it is enough to prove that $|f| \le |g|$. We only need to verify this inequality point-by-point, and hence everything reduces to an assertion in linear algebra, that we state as the following lemma.

Lemma *Let L be an n-dimensional complex vector space, $\varphi(x, y)$ a Hermitian metric on L, and $\omega(x, y)$ the associated skewsymmetric form of φ. Suppose that $F \subset L$ is a $2m$-dimensional real vector subspace of L, where L is viewed as a $2n$-dimensional vector space over $\mathbb{R}$. Then for any basis $f_1, \dots, f_{2m}$ of F we have*

$$\frac{1}{m!}\left|\omega^m(f_1, \dots, f_{2m})\right| \le \left|\mathrm{Vol}(f_1, \dots, f_{2m})\right|, \tag{8.35}$$

where $\mathrm{Vol}(f_1, \dots, f_{2m})$ *is the volume of the parallelepiped constructed on the vectors $f_1, \dots, f_{2m}$, with respect to the metric φ.* (See Exercise 5 for the converse implication.)

Note that if we partition the basis vectors $f_1, \dots, f_{2m}$ into two subsets $f_1, \dots, f_{2r}$ and $f_{2r+1}, \dots, f_{2m}$ that are orthogonal with respect both to the scalar product $\alpha(x, y)$ and the skewsymmetric form $\omega(x, y)$ associated with the metric φ, the two sides of the inequality (8.35) are both multiplicative. For the volume this is well known. For the left-hand side, we can write $\omega_{|F} = \omega_1 \oplus \omega_2$, where

$$\begin{aligned}
&\omega_1(f_\alpha, f_\beta) = \omega(f_\alpha, f_\beta) \quad \text{and} \quad \omega_2(f_\alpha, f_\beta) = 0 \quad \text{for } \alpha, \beta = 1, \dots, 2r,\\
&\omega_1(f_\alpha, f_\beta) = 0 \quad \text{and} \quad \omega_2(f_\alpha, f_\beta) = \omega(f_\alpha, f_\beta) \quad \text{for } \alpha, \beta = 2r+1, \dots, 2m.
\end{aligned}$$

Then by exterior algebra

$$\frac{1}{m!}\omega^m = \frac{1}{m!}(\omega_1 + \omega_2)^m = \frac{1}{m!}\binom{m}{r}\omega_1^r\omega_2^{m-r} = \frac{1}{r!}\omega_1^r\frac{1}{(m-r)!}\omega_2^{m-r},$$

so that

$$\frac{1}{m!}\omega^m(f_1, \dots, f_{2m}) = \frac{1}{r!}\omega^r(f_1, \dots, f_{2r})\frac{1}{(m-r)!}\omega^{m-r}(f_{2r+1}, \dots, f_{2m}).$$

Using the scalar product $\alpha(x, y)$, we can write $\omega_{|F}$ in the form $\omega(x, y) = \alpha(A(x), y)$ where $A\colon F \to F$ is a skewsymmetric linear map. Now A has a 2-dimensional invariant subspace $F_0 \subset F$, since every linear map does. Because A is skewsymmetric, the subspace F_1 orthogonal to F_0 is also invariant. Thus $F = F_0 \oplus F_1$, where F_0 and F_1 are orthogonal with respect to both α and ω.

Now note that on passing to another basis, both sides of the inequality (8.35) are multiplied by the absolute value of the determinant of the matrix of the change of basis. Hence we need only prove it for one particular basis. In particular, we can assume that $f_1, f_2 \in F_0$ and $f_3, \dots, f_{2m} \in F_1$. By what we have proved above, it is enough to prove the inequality for F_0 and F_1 separately. The proof thus reduces by induction to the case of a 2-dimensional subspace F. Moreover, we can assume that the vectors f_1 and f_2 are orthogonal with respect to α. Then

$$|\omega(f_1, f_2)| = |\varphi(if_1, f_2)| \le \left(|\varphi(if_1, if_1)| \cdot |\varphi(f_2, f_2)|\right)^{1/2}. \tag{8.36}$$

Since φ is Hermitian, $\varphi(if_1, if_1) = \varphi(f_1, f_1)$ and the right-hand side of (8.36) is $|\operatorname{Vol}(f_1, f_2)|$. The lemma is proved.

We summarise what we have proved.

Theorem 8.7 *In a Kähler complex manifold, the volume of an m-dimensional complex submanifold Y is expressed as $(1/m!)\int_{[Y]}\omega^m$, where ω is the 2-form associated with the Kähler metric. This volume is the same for all homologous complex submanifolds, and realises the minimum volume for real submanifolds in the same homology class.*

Now consider nonsingular projective varieties $X \subset \mathbb{P}^N$. As we saw in Section 4.3, $\mathbb{P}^N$ has a Kähler metric, the Fubini–Study metric, and hence all projective submanifolds are Kähler. It is well known that the $2n$-dimensional homology group $H_{2n}(\mathbb{P}^N, \mathbb{Z})$ is isomorphic to $\mathbb{Z}$, with the class of a projective linear subspace $\mathbb{P}^n \subset \mathbb{P}^N$ as generator. Hence for the cohomology of an n-dimensional complex submanifold we have $[X] = h[\mathbb{P}^n]$ with $h \in \mathbb{Z}$. To determine the value of the coefficient h, we need to consider the intersection number with the cycle $[Z]$ of a projective linear subspace $Z = \mathbb{P}^{N-n}$ of complementary dimension. We have seen that $[X]\cdot[Z] = \deg X$ (7.5). But $[X]\cdot[Z] = h([\mathbb{P}^n]\cdot[Z]) = h$, and therefore $h = \deg X$. On the other hand, by Wirtinger's theorem $\operatorname{Vol} X = (1/n!)\int_{[X]}\omega^n = (h/n!)\int_{[\mathbb{P}^n]}\omega^n$. We use γ_n to denote the absolute value of the constant $\int_{[\mathbb{P}^n]}\omega^n$ (in fact $\gamma_n = \pi^n$). We thus get the following version of Wirtinger's theorem:

$$\deg X = \frac{n!}{\gamma_n}\operatorname{Vol} X.$$

A relation analogous to that just obtained holds for any cycle $\zeta \in H_{2n}(\mathbb{P}^N)$. On the one hand, $\zeta = h[\mathbb{P}^n]$, where $h = \zeta \cdot [\mathbb{P}^{N-n}]$. On the other hand, $\int_\zeta \omega^n = h\int_{[\mathbb{P}^n]}\omega^n = h\gamma_n$, so that $\zeta \cdot [\mathbb{P}^{N-n}] = (1/\gamma_n)\int_\zeta \omega^n$. We see that $(1/\gamma_n)\omega^n$ defines the same cohomology class as $[\mathbb{P}^{N-n}]$. In particular, the cohomology class of the 2-form ω/π is dual to the class of a hyperplane in the sense of the duality between H^2 and H_{2n-2}. All of these relations also hold for the homology class of projective varieties $X \subset \mathbb{P}^N$, since the homology class $\zeta \in H_k(X, \mathbb{Z})$ defines a class $i_*\zeta \in H_k(\mathbb{P}^N, \mathbb{Z})$ where $i\colon X \hookrightarrow \mathbb{P}^N$ is the inclusion map, and the class $\eta \in H^k(\mathbb{P}^N, \mathbb{C})$ defines the class $i^*\eta \in H_k(X, \mathbb{Z})$. The standard formula $i^*\eta \cdot \zeta = \eta \cdot i_*\zeta$ (the so-called *projection formula*) shows that on any projective manifold, the 2-form ω/π defines the cohomology class dual to the class of the hyperplane section.

All the above arguments were based on integrating powers of the 2-form ω of the Kähler form of a Kähler manifold X. We give another very simple but important application of this idea. We have seen that $\int_{[X]}\omega^n = n!\operatorname{Vol} X$, and hence in particular $\int_{[X]} \omega^n \neq 0$. Hence the form ω^n is not homologous to 0. A fortiori ω^m for $m < n$ is not homologous to 0, since $\omega^m = \mathrm{d}\xi$ would imply that $\omega^n = \omega^{n-m} \wedge \mathrm{d}\xi = \mathrm{d}(\omega^{n-m} \wedge \xi)$. Hence we get a generalisation of Proposition of Section 1.3, Chapter 7.

Proposition *For an n-dimensional compact Kähler manifold X,*

$$H^{2m}(X, \mathbb{C}) \neq 0 \quad \text{for } m \leq n.$$

It follows of course by duality that also $H^{2m}(X, \mathbb{C}) \neq 0$.

In particular we see that the Hopf manifold (Example 8.5) does not admit a Kähler metric. We see that, in contrast to Hermitian metrics, by no means every compact complex manifold can be given a Kähler metric.

4.5 Hodge Theory

The most powerful applications of Kähler metrics are related to *Hodge theory*. Although Hodge theory is not easy to construct, it is easy to describe its results.

As noted in (8.5)–(8.6), writing out a differential m-form in the coordinates z_i and $\overline{z}_i$ and separating out terms involving p of the $\mathrm{d}z_i$, and q of the $\mathrm{d}\overline{z}_i$ defines a decomposition

$$\eta = \sum_{p+q=m} \eta^{(p,q)}, \tag{8.37}$$

where the $\eta^{(p,q)}$ are forms of type (p, q).

Can this decomposition be carried over to the cohomology classes defined by differential forms? To do this, in the first instance we need to know that a closed form η has a decomposition (8.37), in which the $\eta^{(p,q)}$ are closed forms; we do not require equality in (8.37), but only that the two sides of (8.37) are cohomologous. That is, we need that any closed m-form η is cohomologous to a sum $\sum_{p+q=m} \eta^{(p,q)}$ with $\mathrm{d}\eta^{(p,q)} = 0$. Secondly we need to know that the decomposition (1) is unique in terms of cohomology classes; that is, if $\eta = \sum_{p+q=m} \eta^{(p,q)}$ with $\mathrm{d}\eta^{(p,q)} = 0$ is an exact differential, then all the $\eta^{(p,q)}$ are exact differentials.

There is of course no reason whatsoever for these properties to hold in general. Hodge theory asserts that they are true for arbitrary compact Kähler manifolds. Note that the assertions themselves express properties of the cohomology of complex manifolds, and do not depend in any way on a metric. A complex manifold may admit many different Kähler metrics (for example, we could embed a projective manifold in many different ways into projective space and take the corresponding Fubini–Study metrics). Any of these will do equally well in the foundation of Hodge theory, but it only plays a role as auxiliary apparatus for proof. Moreover, there exist important cases when there is no Kähler metric on a complex manifold, but the assertions of Hodge theory are nevertheless true. For example, this is the case for the 2-dimensional cohomology of compact complex surfaces (see for example Barth, Peters and Van de Ven [9, Chapter IV, Section 2]).

We recall that every projective manifold is Kähler, so that Hodge theory holds for it.

According to Hodge theory, on a Kähler manifold we have a decomposition of cohomology groups similar to (8.37)

$$H^m(X,\mathbb{C}) = \bigoplus_{p+q=m} H^{p,q}(X), \tag{8.38}$$

where $H^{p,q}(X)$ is the subspace spanned by closed (p,q)-forms. The dimension of $H^m(X,\mathbb{C})$ is the mth Betti number $b_m(X)$ of X. The dimension of $H^{p,q}(X)$ is denoted by $h^{p,q}(X)$. It follows from (8.38) that

$$b_m = \sum_{p+q=m} h^{p,q}(X).$$

In (8.38), the complex conjugation operator obviously takes exact differentials to exact differentials, and hence extends to cohomology groups $H^m(X,\mathbb{C})$. In particular $H^{q,p}(X) = \overline{H^{p,q}(X)}$, so that $h^{q,p} = h^{p,q}$. If m is odd if follows from this that

$$b_m = \sum_{p+q=m} h^{p,q}(X) = 2\sum_{p<m} h^{p,m-p}(X), \tag{8.39}$$

and we see that the odd dimensional Betti number of a compact Kähler manifold are even.

Some of the spaces $H^{p,q}(X)$ have a simpler interpretation. By definition $H^{p,0}(X)$ consists of classes of closed forms of the form

$$\eta = f_{i_1\dots i_p}\mathrm{d}z_{i_1}\wedge\cdots\wedge\mathrm{d}z_{i_p}.$$

The condition $\mathrm{d}\eta = 0$ breaks up into the two conditions $\mathrm{d}'\eta = 0$ and $\mathrm{d}''\eta$, the second of which just means that the $f_{i_1\dots i_p}$ are holomorphic functions. Suppose that such a form η is exact, $\eta = \mathrm{d}\xi$; we prove that then $\eta = 0$. For this we construct the form $\eta\wedge\overline{\eta}\wedge\omega^{n-p}$, where ω is the 2-form associated with the Kähler metric. We prove that if $\eta \neq 0$ then $\eta\wedge\overline{\eta}\wedge\omega^{n-p} = \gamma U\Omega$, where γ is a nonzero constant, U a positive function and Ω the volume element. From this it follows that $\int_{[X]}\eta\wedge\overline{\eta}\wedge\omega^{n-p} \neq 0$, whereas at the same time, if $\eta = \mathrm{d}\xi$ then

$$\eta\wedge\overline{\eta}\wedge\omega^{n-p} = \mathrm{d}\big(\xi\wedge\overline{\eta}\wedge\omega^{n-p}\big) \quad\text{and}\quad \int_{[X]}\eta\wedge\overline{\eta}\wedge\omega^{n-p} = 0.$$

The relation $\eta\wedge\overline{\eta}\wedge\omega^{n-p} = \gamma U\Omega$ is purely local in character, and we can check it in the tangent space at every point. Suppose that in a suitable coordinate system

$\omega = \sum \mathrm{d}z_\alpha \wedge \mathrm{d}\overline{z}_\alpha$, and for brevity write $\eta = \sum c_A \mathrm{d}z_A$, where A runs through multiple indexes $A = (\alpha_1, \dots, \alpha_p)$ and $\mathrm{d}z_A = \mathrm{d}z_{\alpha_1} \wedge \cdots \wedge \mathrm{d}z_{\alpha_p}$. Then

$$\eta \wedge \overline{\eta} \wedge \omega^{n-p} = \left(\frac{i}{2}\right)^{n-p} \left(\sum c_A \mathrm{d}z_A\right) \wedge \left(\sum \overline{c}_A \mathrm{d}\overline{z}_A\right) \wedge \left(\sum \mathrm{d}z_B \wedge \mathrm{d}\overline{z}_B\right),$$

where B runs through multiple indexes $B = (\beta_1, \dots, \beta_{n-p})$. It follows from an obvious calculation in the exterior algebra that

$$\eta \wedge \overline{\eta} \wedge \omega^{n-p} = \gamma \left(\sum |c_A|^2\right) \Omega,$$

where γ is a nonzero constant.

Finally, for any p-form $\eta = \sum f_{\alpha_1 \dots \alpha_p} \mathrm{d}z_{\alpha_1} \wedge \cdots \wedge \mathrm{d}z_{\alpha_p}$ with holomorphic coefficients, consider $\xi = \mathrm{d}\eta$. This form is obviously cohomologous to 0, and is of the kind just considered. By the argument just given, it equals 0. Putting together what we have said, we see that $H^{p,0}(X)$ is the space of holomorphic p-forms, that is, forms $\eta = \sum f_{\alpha_1 \dots \alpha_p} \mathrm{d}z_{\alpha_1} \wedge \cdots \wedge \mathrm{d}z_{\alpha_p}$ where the coefficients $f_{\alpha_1 \dots \alpha_p}$ are holomorphic functions. One can prove that if X is a projective variety, then these forms are the regular rational differential forms on X, that is, in the notation of Section 5.3, Chapter 3, $H^{p,0}(X) = \Omega^p[X]$, and $h^{p,0}(X) = h^p(X)$. For $p = 0$ this is Theorem 8.2.

For example, for $H^1(X, \mathbb{C})$, (8.39) says that $b_1 = 2h^{1,0}$, where $h^{1,0}$ is the dimension of the space of holomorphic 1-forms.

We state one final result which is easy to deduce if you accept Hodge theory on trust: taking the product of forms $\eta \mapsto \eta \wedge \omega$ with the Kähler 2-form ω induces an inclusion

$$H^m(X, \mathbb{C}) \hookrightarrow H^{m+2}(X, \mathbb{C}) \quad \text{for } m + 1 \le n = \dim X,$$

the so-called *hard Lefschetz theorem.* It follows from this that $b_m \le b_{m+2}$ for $m + 1 \le n$. By Poincaré duality, also $b_m \le b_{m-2}$ for $m \ge n + 1$. Thus the odd or even Betti numbers of a Kähler manifold form a monotone sequence, the "Hodge staircase": up to the middle dimension they are monotonically nondecreasing, and from the middle dimension upwards nonincreasing.

4.6 Exercises to Section 4

1 Let L be a complex vector space and $L_{\mathbb{R}}$ the same space viewed as a vector space over $\mathbb{R}$ of twice the dimension. Prove that a symmetric bilinear form $\alpha(x, y)$ on $L_{\mathbb{R}}$ is the real part of a Hermitian form φ on L if and only if $\alpha(ix, iy) = \alpha(x, y)$. Moreover, φ is unique.

2 Prove that every 1-dimensional compact complex manifold is Kähler (assume known that it can be given a Hermitian metric).

3 Calculate the integral over the 2-cycle $[\mathbb{P}^1]$ of the 2-form ω on $\mathbb{P}^1$ associated with the Fubini–Study metric.

4 Prove that if X and Y are Kähler manifolds then so is $X \times Y$ with the natural product metric.

5 Prove that in the inequality (8.35), equality holds if and only if F is a complex subspace of L. Deduce that a $2m$-dimensional real submanifold Y of a Kähler manifold X that minimises the volume in its homology class is a complex manifold.

6 Use Hodge theory to prove that there are no holomorphic differential forms on $\mathbb{P}^n$.

7 Let X be a compact 1-dimensional complex manifold (therefore a Kähler manifold). Multiplication defines a bilinear form $Q(x, y)$ on the cohomology $H^1(X, \mathbb{C}) = H^{1,0}(X) \oplus H^{0,1}(X)$. If φ and ψ are differential forms belonging to cohomology classes x and y, prove that $Q(x, y) = \int_{[X]} \varphi \wedge \psi$. Deduce from this that $Q(\varphi, \overline{\varphi}) > 0$ for $\varphi \in H^{1,0}(X)$.

8 Let L be a 2-dimensional complex vector space with a Hermitian metric that can be written $z_1\overline{z}_1 + z_2\overline{z}_2$ in some coordinate system, and ω the associated skewsymmetric form. Prove that if a $(1, 1)$-form φ on L satisfies $\omega \wedge \varphi = 0$ then $\varphi \wedge \overline{\varphi} = c\Omega$ with $c < 0$, where Ω is the volume element.

9 Let X be a compact Kähler surface and

$$H^2(X, \mathbb{C}) = H^{2,0}(X) \oplus H^{1,1}(X) \oplus H^{0,2}(X)$$

the Hodge decomposition (8.38) of its cohomology; suppose that $Q(x, y)$ is defined for $x, y \in H^2(X, \mathbb{C})$ by $Q(x, y) = \int_{[X]} \varphi \wedge \psi$. Let ω be a Kähler form. Use Exercise 8 to prove that if $\varphi \in H^{1,1}(X)$ satisfies $Q(\varphi, \omega) = 0$ then $Q(\varphi, \overline{\varphi}) < 0$. Deduce that $Q(x, \overline{y})$ defines a Hermitian form on $H^{1,1}(X)$ whose canonical diagonalised form $\sum \lambda_i \xi_i \overline{\xi}_i$ has one positive coefficient λ_i and the remainder negative. The corresponding Hermitian form on $H^2(X, \mathbb{C})$ has $2h^{2,0} + 1$ positive coefficients.

10 Let X be a compact Kähler surface, Y a nonsingular curve on X and $\langle Y \rangle \in H^2(X, \mathbb{C})$ the dual cohomology class of Y. Prove that Y is a real class and $\langle Y \rangle \in H^{1,1}(X)$. [Hint: Check that $Q(\langle Y \rangle, \varphi) = 0$ for $\varphi \in H^{2,0}(X) \oplus H^{0,2}(X)$.]

11 Compare the result of Exercise 9 with the Hodge index theorem (Section 2.4, Chapter 4). What do the results have in common, and how do they differ?

Chapter 9
Uniformisation

1 The Universal Cover

1.1 The Universal Cover of a Complex Manifold

In previous sections of this book we have used the notion of quotient space to construct many important examples of complex manifolds. We now show that the notion leads to a general method of studying complex manifolds.

We start by recalling some simple topological facts (see, for example, Pontryagin [66, §§49–50]). Let X be a path-connected, locally connected and locally simply connected space; later X will be a connected manifold and all these conditions will be satisfied. The *universal cover* of X is a topological space $\widetilde{X}$ having a projection $p\colon \widetilde{X} \to X$ that makes it into an unramified cover (Section 1.2, Chapter 8). Homeomorphisms $g\colon \widetilde{X} \to \widetilde{X}$ satisfying the condition $p \circ g = p$ form a group G isomorphic to the fundamental group $\pi_1(X)$ of X. This group acts freely and discretely on X and

$$X = \widetilde{X}/G. \tag{9.1}$$

Now suppose that X is a complex manifold, and write $\mathcal{O}_X$ for the structure sheaf. The universal cover $\widetilde{X}$ (or indeed any unramified cover) can also be made into a complex manifold in such a way that the projection p is a holomorphic map. For this, consider the presheaf $\widetilde{\mathcal{O}}$ on $\widetilde{X}$ defined by

$$\widetilde{\mathcal{O}}(\widetilde{U}) = \mathcal{O}_X\bigl(p(\widetilde{U})\bigr)$$

for any open set $\widetilde{U} \subset \widetilde{X}$; then also $p(\widetilde{U})$ is open in X, since p is an unramified cover. Write $\mathcal{O}_{\widetilde{X}}$ for the sheafication of $\widetilde{\mathcal{O}}$. Every point $\widetilde{x} \in \widetilde{X}$ has a neighbourhood $\widetilde{U}$ that is mapped homeomorphically by p to $p(\widetilde{U})$. Thus the sheaf $\mathcal{O}_{\widetilde{X}}$ is uniquely determined by its restriction to these opens $\widetilde{U}$. It is easy to see that on them, $\mathcal{O}_{\widetilde{X}}$ is just the sheaf $\mathcal{O}_X$ pulled back by the homeomorphism p.

I.R. Shafarevich, *Basic Algebraic Geometry 2*, DOI 10.1007/978-3-642-38010-5_5,

It follows from what we have said that the pair $\widetilde{X}, \mathcal{O}_{\widetilde{X}}$ defines a complex manifold. Indeed, if $p\colon \widetilde{U} \to p(\widetilde{U})$ is a homeomorphism then the projection p defines an isomorphism of the ringed spaces $\widetilde{U}, \mathcal{O}_{\widetilde{X}|\widetilde{U}}$ and $p(\widetilde{U}), \mathcal{O}_{X|p(\widetilde{U})}$. Hence if $p(\widetilde{U})$ is isomorphic to a domain in $\mathbb{C}^n$ then the same is true of $\widetilde{U}$. It is also obvious that p is holomorphic. Moreover, since the complex structure on $\widetilde{X}$ is determined by the projection p, and the homeomorphisms $g \in G$ do not change this projection, they are automorphisms of the complex manifold $\widetilde{X}$. It follows that (9.1) is an isomorphism of complex manifolds.

Suppose that two manifolds X and X' have a common universal cover $\widetilde{X}$. Then

$$X = \widetilde{X}/G \quad \text{and} \quad X' = \widetilde{X}/G',$$

and there are two unramified covers $p\colon \widetilde{X} \to X$ and $p'\colon \widetilde{X} \to X'$. We determine when X and X' are isomorphic. For this we use the following elementary topological fact, which justifies the term *universal* cover: if $p\colon \widetilde{X} \to X$ is the universal cover and $q\colon X_1 \to X$ is any connected unramified cover then there exists a continuous map $\varphi\colon \widetilde{X} \to X_1$ such that $q \circ \varphi = p$. Suppose that $f\colon X' \to X$ is an isomorphism. Then $q = f \circ p'$ defines an unramified cover $q\colon \widetilde{X} \to X$. Using the result stated above once more, we have thus constructed a continuous map $\varphi\colon \widetilde{X} \to \widetilde{X}$ such that the diagram

$$\begin{array}{ccc} \widetilde{X} & \xrightarrow{\varphi} & \widetilde{X} \\ {\scriptstyle p'}\downarrow & & \downarrow{\scriptstyle p} \\ X' & \xrightarrow[f]{} & X \end{array} \tag{9.2}$$

commutes. It follows that the map φ is holomorphic. Indeed, it follows from the commutativity of the diagram that $p \circ \varphi$ is holomorphic, that is, for functions $u \in \mathcal{O}_{X,x}$ the function $(p \circ \varphi)^*(u) = \varphi^*(p^*(u))$ is holomorphic at points $\widetilde{x} \in (p \circ \varphi)^{-1}(x) = \varphi^{-1}(p^{-1}(x))$. But all functions that are holomorphic in a neighbourhood of a point $\widetilde{x} \in p^{-1}(x)$ are locally of the form $p^*(u)$, from which it follows that φ is holomorphic. Interchanging X and X' in this argument, we see that φ is an automorphism of the complex manifold $\widetilde{X}$.

Now recall that the two groups G and G' consist of all automorphisms of $\widetilde{X}$ for which

$$p\gamma = p \quad \text{for } \gamma \in G \quad \text{and} \quad p'\gamma' = p' \quad \text{for } \gamma' \in G'.$$

Composing the first equality with f and using the commutativity of (9.2) we get that $G' = \varphi G \varphi^{-1}$. We have proved the following result.

Theorem 9.1 *Any connected complex manifold X can be written in the form $X = \widetilde{X}/G$ where $\widetilde{X}$ is a simply connected complex manifold and G is a group of automorphisms of $\widetilde{X}$ acting freely and discretely on it. For any two such representations of the same complex manifold, the groups G and G' are conjugate in the group of all automorphisms of $\widetilde{X}$.*

1.2 Universal Covers of Algebraic Curves

Theorem 9.1 allows us to reduce the study of arbitrary complex manifolds to that of simply connected manifolds and their automorphism groups. Of course, this only moves the problem somewhere else—it all depends how much we know about simply connected complex manifolds and their discrete groups of automorphisms. In general, very little; we will say more about this in Section 4. The exception is provided by 1-dimensional complex manifolds, that we mainly study in what follows.

The classification of connected, simply connected 1-dimensional complex manifolds is very simple. There are just 3 of them:

(1) The projective line $\mathbb{P}^1_{\text{an}}$;
(2) the affine line $\mathbb{C}^1 = \mathbb{A}^1_{\text{an}}$;
(3) the open unit disc $D \subset \mathbb{C}$ defined by $|z| < 1$.

In the theory of analytic functions, (1) and (2) are called the *Riemann sphere* and the *finite complex plane*. This theorem is proved by the same methods as the Riemann existence theorem. The proof is given, for example, in Springer [74, 9–2]. It is easy to see that the three complex manifolds (1)–(3) are not isomorphic: (1) is not isomorphic to the other two because it is compact, and they are not; (3) is not isomorphic to (2) because it admits nonconstant bounded holomorphic functions, and by Liouville's theorem (2) does not.

Thus connected 1-dimensional complex manifolds divide into three classes, depending on whether the universal cover is isomorphic to (1), (2) or (3). The three classes corresponding to (1), (2) and (3) are called *elliptic*, *parabolic* or *hyperbolic* complex manifolds; the same terminology is applied to noncompact 1-dimensional complex manifolds.

To study complex manifolds in these three classes, we need to know what are the groups of automorphisms acting freely and discretely on the universal covers. The answer follows easily from simple facts from the theory of analytic functions of one complex variable.

Proposition *Any automorphism of $\mathbb{P}^1_{\text{an}}$ has a fixed point. An automorphism group G of $\mathbb{C}^1$ acting freely and discretely and with compact quotient $\mathbb{C}^1/G$ consists of translations $z \mapsto z + a$, where a runs through the vectors of some lattice of rank 2 in $\mathbb{C}$. All automorphisms of the unit disc are of the form*

$$z \mapsto \theta \frac{z - \alpha}{1 - \overline{\alpha} z} \quad \text{with } |\theta| = 1 \text{ and } |\alpha| < 1. \tag{9.3}$$

Proof By Theorem 8.5, any automorphism of $\mathbb{P}^1_{\text{an}}$ is of the form g_{an} where g is an automorphism of the algebraic variety $\mathbb{P}^1$, and is hence a fractional linear transformation. Since any fractional linear transformation has a fixed point, this proves the first assertion of the proposition.

An automorphism of $\mathbb{C}^1$ is given by an entire function $f(z)$. If f had an essential singularity at the point ∞ then in any neighbourhood of ∞ it would take values arbitrary close to any given value, by Weierstrass' theorem. This would contradict the assumption that f defines an automorphism. Indeed, if $f(a)=b$ then f already takes all values sufficiently close to b in a neighbourhood of a, and so cannot take them in a neighbourhood of ∞. Therefore f is a polynomial. If it has degree n then it takes every value n times. Hence f defines an automorphism only if $n=1$. In other words, any automorphism of $\mathbb{C}^1$ is of the form

$$f(z)=az+b \quad \text{with } a \neq 0. \tag{9.4}$$

An automorphism belonging to a group G that acts freely does not have fixed points. Hence $a=1$ in (9.4). We see that G must consist of translations $f(z)=z+b$. If we use the group structure on $\mathbb{C}^1$ we can restate our result by saying that G is a subgroup of $\mathbb{C}^1$ and X the quotient group $\mathbb{C}^1/G$. In Section 1.4, Chapter 8, we have already used a simple theorem that determines all discrete subgroups $G\subset\mathbb{C}^1$ with compact quotient (see Pontryagin [66, Exercise 33 of Section 19, p. 110]). In our case it shows that $G=\mathbb{Z}\omega_1+\mathbb{Z}\omega_2$ is a lattice of rank 2, where $\omega_1,\ \omega_2\in\mathbb{C}^1$ are linearly independent over $\mathbb{R}$.

Finally, suppose that D is the open unit disc. A substitution show that transformations of type (9.3) form a group, and that this group acts transitively on D. Thus composing any automorphism with some automorphism of type (9.3) we get an automorphism γ that fixes 0. Hence it is enough to prove than any automorphism fixing 0 is of the form (9.3) with $\alpha=0$. If $\gamma(0)=0$ then

$$\left|\gamma(z)/z\right|\leq 1 \quad \text{for } z\in D,$$

by Schwarz' lemma (Section 2.3, Chapter 8), and since the relation between $\gamma(z)$ and z is symmetric, also $|z/\gamma(z)|\leq 1$. It follows that $\gamma(z)/z$ is constant:

$$\gamma(z)=\theta z \quad \text{with } |\theta|=1.$$

The proposition is proved. □

Thus the classification of manifolds of elliptic type is trivial: they are all isomorphic: indeed, in (9.1) we have $\widetilde{X}=\mathbb{P}^1_{\mathrm{an}}$ and $G=e$, so that $X=\widetilde{X}=\mathbb{P}^1_{\mathrm{an}}$.

In Sections 2–3 we treat compact manifolds of parabolic and hyperbolic type. We prove that for any group G acting freely and discretely and such that $\widetilde{X}/G$ is compact, the quotient manifold is a projective algebraic curve, and we construct an explicit projective embedding of these manifolds. We will thus give a proof of the Riemann existence theorem, starting from the classification theory of simply connected 1-dimensional complex manifolds. We show moreover that compact complex manifolds of parabolic type coincide with algebraic curves of genus 1, that is, elliptic curves, and manifolds of hyperbolic type with curves of genus ≥ 2. (The terminology is clearly a mess—an elliptic curve is a manifold of parabolic type, and the projective line is of elliptic type; but this has been in general use for so long that we do not attempt to correct it.)

1.3 Projective Embedding of Quotient Spaces

In what follows we need to investigate various particular cases of the following general situation. Let $\widetilde{X}$ be a 1-dimensional complex manifold and G a group of automorphisms of $\widetilde{X}$ acting freely and discretely. Assume that the quotient space $X = \widetilde{X}/G$ is compact; then how do we construct an embedding of it into projective space $\mathbb{P}^n$?

We specify such an embedding by $n+1$ functions $f_0, \dots, f_n$ that are holomorphic on the whole of $\widetilde{X}$. Suppose that they are not simultaneously 0 at any point $\widetilde{x} \in \widetilde{X}$. Then

$$\widetilde{f}\colon \widetilde{X} \to \mathbb{P}^n \quad \text{defined by } \widetilde{f}(\widetilde{x}) = \bigl(f_0(\widetilde{x}) : \dots : f_n(\widetilde{x})\bigr) \tag{9.5}$$

is a holomorphic map.

In order that $\widetilde{f}$ induces a map of X to $\mathbb{P}^n$, we could require the functions f_i to be invariant under all $g \in G$. But then they would be holomorphic functions on X, and would thus be constant by Theorem 8.2. However, this condition can be weakened, by requiring only that for every $g \in G$ there exists a function φ_g on $\widetilde{X}$ such that

$$g^*(f_i) = \varphi_g f_i \quad \text{for } i = 0, \dots, n. \tag{9.6}$$

It follows already from this that $\widetilde{f} \circ g = \widetilde{f}$ for every $g \in G$, and hence $\widetilde{f}$ factors as $\widetilde{f} = f \circ \pi$ where π is the projection $\widetilde{X} \to X$ and f is some holomorphic map $X \to \mathbb{P}^n$. We will say that $\widetilde{f}$ defines the map $f\colon X \to \mathbb{P}^n$. Because the functions f_i are holomorphic and not simultaneously 0 on $\widetilde{X}$, it follows that each function φ_g is also holomorphic and does not vanish anywhere on $\widetilde{X}$.

We determine when such a system of functions defines an isomorphic embedding $f\colon X \to \mathbb{P}^n$.

Proposition *Let $\widetilde{X}$ be a 1-dimensional complex manifold, G a group of automorphisms of $\widetilde{X}$ acting freely and discretely, and $f_0, \dots, f_n$ holomorphic functions on $\widetilde{X}$ satisfying* (9.6), *where the φ_g are holomorphic functions with no zeros on $\widetilde{X}$.*

Suppose that the following conditions hold:

$$\operatorname{rank}\begin{pmatrix} f_0(x_1) & \dots & f_n(x_1) \\ f_0(x_2) & \dots & f_n(x_2) \end{pmatrix} = 2 \tag{A}$$

for any pair of points $x_1, x_2 \in \widetilde{X}$ such that $x_1 \neq g(x_2)$ for all $g \in G$ and

$$\operatorname{rank}\begin{pmatrix} f_0(x) & \dots & f_n(x) \\ f_0'(x) & \dots & f_n'(x) \end{pmatrix} = 2 \tag{B}$$

for all $x \in X$. Here $f'(x)$ denotes the derivative of f as a function of a local parameter at x; condition (B) *does not depend on the choice of this parameter. Then the map* (9.5) *defines an isomorphic embedding of $X = \widetilde{X}/G$ into $\mathbb{P}^n$.*

Proof The proof boils down to a simple verification. (A) guarantees that all the functions f_i are not simultaneously 0 at any point $x \in \widetilde{X}$, so that (9.5) really defines a point of projective space. Equation (9.6) shows that f defines a map

$$f: X \to \mathbb{P}^n$$

that is holomorphic because of the preceding remark. Then (A) guarantees that f is one-to-one.

Suppose that for some $x_0 \in \widetilde{X}$ we have $f_0(x_0) \neq 0$. The image $x \in X$ has a neighbourhood U in which f is given by the equations

$$y_i = g_i(x) = f_i(x)/f_0(x) \quad \text{for } i = 1, \dots, n,$$

where $y_1, \dots, y_n$ are coordinates in the affine piece $\mathbb{A}^n$ which U maps to. Now (B) implies that $g_i'(x_0) \neq 0$ for some $i > 0$; we assume that $i = 1$, so that $g_1'(x_0) \neq 0$. Because of this, the local parameter z at x_0 can be expressed as an analytic function of $y_1 = g_1(z)$:

$$z = h(y_1).$$

We see that $f(X)$ is defined in a neighbourhood of $f(x)$ by the analytic equations

$$y_i - g_i\big(h(y_1)\big) = 0 \quad \text{for } i = 2, \dots, n,$$

where moreover the functions $u_1 = y_1$, $u_i = y_i - g_i(h(y_1))$ for $i = 2, \dots, n$ form a system of local coordinates in a neighbourhood of $f(x)$ in $\mathbb{P}^n$. This proves that $f(X)$ is a complex submanifold of $\mathbb{P}^n$.

Finally the inverse map of f is given in a neighbourhood of $f(x)$ by the function $z = h(y_1)$ (recall that z can be viewed as a local coordinate on X). Hence f is an isomorphic embedding. This proves the proposition. □

1.4 Exercises to Section 1

1 Prove that the universal cover of an n-dimensional Abelian variety over $\mathbb{C}$ is isomorphic to $\mathbb{C}^n$.

2 Prove that two nonsingular projective surfaces that are birational have isomorphic fundamental groups.

3 Let X be a compact complex manifold. Prove that there exists only a finite number of nonisomorphic manifolds Y having a finite holomorphic map $f: Y \to X$ such that Y is an unramified cover of X of given finite degree m.

4 Determine the fundamental group and the universal cover $\widetilde{X}$ of the manifold $X = \mathbb{P}^1(\mathbb{C}) \setminus \{0, \infty\}$, and find a representation $X = \widetilde{X}/G$ where G is a discrete group of automorphisms of $\widetilde{X}$.

5 The same as Exercise 4 for $X = D \setminus 0$, where $D = \{z||z| < 1\}$.

6 Prove that the universal cover of $X = \mathbb{P}^1(\mathbb{C}) \setminus \{\alpha, \beta, \gamma\}$, where $\alpha, \beta, \gamma \in \mathbb{P}^1(\mathbb{C})$ are 3 distinct points, is isomorphic to the disc D. [Hint: Use the classification of simply connected 1-dimensional complex manifolds given at the start of Section 1.2.]

7 Use the result of Exercise 6 to deduce the theorem of Picard that an entire function f which does not take two values α and β (with $\alpha \neq \beta$) is constant. [Hint: Interpret f as a map $\mathbb{C}^1 \to \mathbb{P}^1(\mathbb{C}) \setminus \{\alpha, \beta, \infty\}$.]

2 Curves of Parabolic Type

2.1 Theta Functions

It follows from Proposition of Section 1.2, that any compact complex curve of parabolic type is a 1-dimensional torus, that is, it is of the form $\mathbb{C}^1/\Omega$ where Ω is a lattice of rank 2. By Theorem 9.1, two lattices Ω and Ω' have isomorphic quotient spaces $\mathbb{C}^1/\Omega$ and $\mathbb{C}^1/\Omega'$ if and only if the groups of translation corresponding to them are conjugate under some automorphism f of the complex manifold $\mathbb{C}^1$. This is obviously equivalent to $\Omega' = f(\Omega)$. Since f can be written as $f(z) = az + b$, it follows that Ω and Ω' are similar lattices.

Our aim now is to show that every 1-dimensional torus X is of the form Y_{an}, where Y is a projective curve. For this we use the method described in Section 1.3. First of all, since we can replace the lattice by a similar lattice without changing the torus $\mathbb{C}^1/\Omega$, we will assume that Ω has a basis $1, \tau$ where $\operatorname{Im}\tau > 0$. We attempt to embed the torus $\mathbb{C}^1/\Omega$ into $\mathbb{P}^n$ by means of functions $f_0, \dots, f_n$ satisfying the following special form of the relations (9.6):

$$\left.\begin{aligned} f_i(z+1) &= f_i(z) \\ f_i(z+\tau) &= e^{-2\pi ikz} f_i(z) \end{aligned}\right\} \quad \text{for } i = 0, \dots, n, \tag{9.9}$$

where k is a positive integer. Formally speaking, we do not need to justify the choice we have made, provided that we can prove that for some k we can find linearly independent functions satisfying (9.9) and the assumptions of Proposition, Section 1.3. However, it can be shown that in fact functions defining any embedding of X into $\mathbb{P}^n$ can be reduced to this form. The point is that we do not change the map if we multiply all the functions $f_i(z)$ by $e^{u(z)}$, where $u(z)$ is an entire function. Using this, it is not hard to show that the relations (9.6) can always be put in the special form (9.9).

Definition An entire function satisfying the conditions (9.9) is called a *theta function* of weight k.

All theta functions of weight k obviously form a vector space over $\mathbb{C}$, which we denote by $\mathcal{L}_k$.

Theorem 9.2 $\dim \mathcal{L}_k = k$.

Proof Let $f(z)$ be any of the functions f_i. The first of the conditions (9.9) shows that $f(z) = \varphi(t)$, where φ is a holomorphic function on $\mathbb{C}^1 \setminus 0$, and $t = e^{2\pi i z}$. Indeed, $\varphi(t) = f((\log t)/2\pi i)$ is a well-defined analytic function on $\mathbb{C}^1 \setminus 0$. Suppose that

$$\varphi(t) = \sum_{m=-\infty}^{\infty} c_m t^m$$

is the Laurent series expansion of this function. Setting $e^{2\pi i \tau} = \lambda$ transforms the second of the conditions (9.9) into

$$\varphi(\lambda t) = \sum_{m \in \mathbb{Z}} c_m \lambda^m t^m = \sum_{m \in \mathbb{Z}} c_m t^{m-k} = \sum_{m \in \mathbb{Z}} c_{m+k} t^m,$$

or

$$c_{m+k} = c_m \lambda^m \quad \text{for } m \in \mathbb{Z}. \tag{9.10}$$

We set

$$m = kr + a \quad \text{with } 0 \le a < k. \tag{9.11}$$

Then (9.10) implies

$$c_m = c_a \lambda^{ra + k\frac{r(r-1)}{2}}.$$

Thus the function φ is uniquely determined by the numbers $c_0, \dots, c_{k-1}$, and it follows that $\dim \mathcal{L}_k \le k$.

To complete the proof of the theorem, it is enough to prove that the series corresponding to any sequence of numbers satisfying (9.10) converges. We can restrict ourselves to one arithmetic progression (9.11). We get a series

$$c_a t^a \sum_{r \in \mathbb{Z}} u^r \mu^{\frac{r(r-1)}{2}}, \quad \text{where } u = t^k \lambda^a \text{ and } \mu = \lambda^k.$$

Since by assumption $\operatorname{Im} \tau > 0$ and $k > 0$, it follows that $|\mu| < 1$. Thus the series

$$\sum |u|^r |\mu|^{\frac{r(r-1)}{2}}$$

obviously converges. The theorem is proved. □

Remark It follows from the theorem that up to a factor there exists a unique theta function of weight 1. If we set $c_0 = 1$ in (9.10) then this function is uniquely determined, and we denote it by $\theta(z)$.

2.2 Projective Embedding

Now we can prove the main result of this section.

Theorem 9.3 *For any $k \geq 3$, the theta functions of weight k define an isomorphic embedding of $X = \mathbb{C}^1/\Omega$ into $\mathbb{P}^{k-1}$.*

Proof We give the proof for $k = 3$; the general case is entirely similar. We use the following obvious remark: if $f(z)$ is a theta function of weight k and $a_1, \dots, a_m$ are complex numbers such that $a_1 + \cdots + a_m = 0$ then the function

$$g(z) = \prod_{i=1}^{m} f(z + a_i)$$

is a theta function of weight mk. In particular, for any a and b the function

$$f(z) = \theta(z + a)\theta(z + b)\theta(z - a - b)$$

is a theta function of weight 3; here θ is the function defined at the end of Section 2.1.

We have to prove that 3 linearly independent theta functions of weight 3 satisfy the conditions (A) and (B) of Proposition, Section 1.3. If condition (A) is not satisfied by the 3 basis elements of $\mathcal{L}_3$ then there exist two points z', z'' such that $z' - z'' \notin \Omega$ and numbers α and β, not both 0, such that $\alpha f(z') = \beta f(z'')$ for every $f \in \mathcal{L}_3$. In particular

$$\alpha\theta(z' + u)\theta(z' + a)\theta(z' - u - a) = \beta\theta(z'' + u)\theta(z'' + a)\theta(z'' - u - a)$$

for any a and u. We set $z' + u = z$, $z'' - z' = \zeta$, viewing z as the variable, and the other quantities as fixed. We saw that

$$\alpha\theta(z)\theta(z' + a)\theta(2z' - a - z) = \beta\theta(z + \zeta)\theta(z'' + a)\theta(z' + z'' - a - z),$$

that is,

$$\frac{\theta(z)}{\theta(z + \zeta)} = \frac{\beta}{\alpha} \times \frac{\theta(z'' + a)}{\theta(z' + a)} \times \frac{\theta(z' + z'' - a - z)}{\theta(2z' - a - z)} = \text{const.} \frac{\theta(z' + z'' - a - z)}{\theta(2z' - a - z)}.$$

We choose a such that the functions $\theta(z)$ and $\theta(z' + z'' - a - z)$ have no common zeros. Then the functions $\theta(z + \zeta)$ and $\theta(2z' - a - z)$ have the same property. Therefore $\theta(z)/\theta(z + \zeta)$ has no zeros or poles, and it follows that $\theta(z + \zeta) = e^{g(z)}\theta(z)$ where g is an entire function. From the definition of $\theta(z)$ it follows that

$$g(z + 1) = g(z) + 2\pi i k, \tag{9.12}$$

$$g(z + \tau) = g(z) - 2\pi i \zeta + 2\pi i l, \tag{9.13}$$

with $k, l \in \mathbb{Z}$. Thus $g'(z)$ is a function with periods 1 and τ. Hence it is bounded on $\mathbb{C}^1$, and since it is entire, it is constant. We see that $g(z) = \alpha z + \beta$, and it follows from (9.12) that $\alpha = 2\pi ik$, and from (9.13) that

$$2\pi ik\tau = -2\pi i\zeta - 2\pi il, \quad \text{and} \quad \zeta = l - k\tau \in \Omega.$$

This contradiction proves condition (A).

Condition (B) is proved in a similar way. Namely, if it does not hold then there exists $z_0 \in \mathbb{C}^1$ such that $(f'/f)(z_0) = 0$ for all $f \in \mathcal{L}_3$. In particular we can take $f = \theta(z+u)\theta(z+a)\theta(z-u-a)$. We get that

$$\frac{\theta'(z_0+u)}{\theta(z_0+u)} + \frac{\theta'(z_0+a)}{\theta(z_0+a)} + \frac{\theta'(z_0-u-a)}{\theta(z_0-u-a)} = 0. \tag{9.14}$$

We again think of u as the variable and choose a so that the functions $\theta(z_0+u)$ and $\theta(z_0-u-a)$ have no common zeros. Then equality (9.14) is only possible if all 3 factors on the left-hand side are entire functions of u. This in turn is only possible if $\theta(z)$ has no zeros, that is $\theta(z) = e^{g(z)}$, with $g(z)$ an entire function. This expression leads at once to a contradiction with the definition of the function θ. Indeed, from its definition it follows that

$$g(z+1) = g(z) + 2\pi im, \tag{9.15}$$

$$g(z+\tau) = g(z) - 2\pi i\zeta - 2\pi ik + 2\pi il. \tag{9.16}$$

From this, as before, we get that $g''(z)$ is constant. Hence $g(z) = \alpha z^2 + \beta z + \gamma$, and then we see from (9.15) that $\alpha = 0$ and from (9.16) that $k = 0$, contradicting the assumption $k > 0$. The theorem is proved. □

Remark The proof that a higher dimensional complex torus is projective if its period matrix satisfies the Frobenius relations (as mentioned in Remark 8.1) is a similar argument, but more complicated.

2.3 Elliptic Functions, Elliptic Curves and Elliptic Integrals

Now that we have constructed a map $f: X \to \mathbb{P}^n$, it is interesting to study it in more detail. We have seen that $Y = f(X)$ is a nonsingular algebraic curve. Adding points on the torus defines a group structure also on Y. Moreover, the addition map defines a holomorphic map $\mu: Y \times Y \to Y$ of the corresponding complex manifolds. By Theorem 8.5 it follows that μ is a morphism. Thus Y is a 1-dimensional Abelian variety. We saw in Section 6.3, Chapter 3, that in this case the canonical class of Y is equal to 0, and hence the genus equals 1. Thus we have proved that compact manifolds of parabolic type are nonsingular projective curves of genus 1, that is, elliptic curves, and only them.

Note that the zeros of theta functions on X make sense. Although the value of $\theta(z)$ changes under the substitution $z \mapsto z+a$ with $a \in \Omega$, if $\theta(z)=0$ then also $\theta(z+a)=0$. The usual definition allows us to talk of the divisor of a theta function on X.

Theta functions of weight 3 give an isomorphic embedding of X into $\mathbb{P}^2$. In this case $Y = f(X) \subset \mathbb{P}^2$ is a nonsingular plane curve of genus 1. It follows from the formula for the genus of a plane curve (Section 2.3, Chapter 4) that Y has degree 3. In particular any theta function of weight 3 defines a divisor on X, and the map $f\colon X \to \mathbb{P}^2$ associates with it a divisor of Y, the intersection of Y with a line of $\mathbb{P}^2$, which has degree 3. Applying this remark to the function θ^3, where θ is the theta function of weight 1, implies that the divisor of θ on X consists of a single point with multiplicity 1. If x_0 is this point, then $\theta(z-a+x_0)$ has divisor a. This remark throws new light on the role of theta functions: if we allow theta functions (which are of course not meromorphic functions, nor indeed functions on X), then every divisor is principal.

The embedding f we have constructed defines an isomorphism of fields $f^*\colon \mathbb{C}(Y) \to \mathcal{M}(X)$. On the other hand, $\mathcal{M}$ can be described as the field of meromorphic functions on $\mathbb{C}^1$ having the two periods 1 and τ. Functions of this type are called *elliptic functions*. Thus the field $\mathbb{C}(Y)$ is isomorphic to the field of elliptic functions. In particular if

$$F(x,y)=0$$

is the equation of an affine model of Y then there exists a parametrisation

$$x=\varphi(z), \qquad y=\psi(z)$$

of it by elliptic functions. A parametrisation of this type is called a *uniformisation* of Y. This establishes the relation between elliptic functions and elliptic curves: elliptic functions uniformise elliptic curves.

Suppose given an elliptic curve Y. How do we find the lattice Ω corresponding to Y, for which $Y = \mathbb{C}^1/\Omega$? Let ω be a regular differential form on Y; it is uniquely determined up to a constant factor, since Y has genus 1. If $f\colon C^1 \to Y$ is the holomorphic map that we are looking for, then $f^*(\omega)$ is a holomorphic differential form on $\mathbb{C}^1$, which must moreover be invariant under translations by vectors of the lattice Ω. This means that $f^*(\omega) = u(z)\mathrm{d}z$, where $u(z)$ is an entire function, invariant under translations in Ω. Therefore $u(z)$ is constant, and normalising the arbitrary choice of ω, we can assume that

$$f^*(\omega)=\mathrm{d}z.$$

Suppose that $z_0 \in \Omega$, that is, $f(z_0)=f(0)$, and let s be a path joining 0 and z_0 in $\mathbb{C}^1$. Then

$$z_0 = \int_0^{z_0} \mathrm{d}z = \int_s f^*(\omega) = \int_{f(s)} \omega. \tag{9.17}$$

The path $f(s)$ is closed in $Y(\mathbb{C})$, that is, it defines an element of $H_1(Y(\mathbb{C}),\mathbb{Z})$. In this way we obviously obtain all elements of this group. Equation (9.17) shows that the lattice Ω consists of all complex numbers

$$\int_\sigma \omega \quad \text{with } \sigma \in H_1\big(Y(\mathbb{C}),\mathbb{Z}\big).$$

In particular, it has a basis consisting of the two numbers

$$\int_{\sigma_1} \omega, \int_{\sigma_2} \omega,$$

where σ_1, σ_2 is a basis of $H_1(Y(\mathbb{C}),\mathbb{Z})$. For example, if Y is the curve given by $v^2 = u^3 + Au + B$ then we can set $\omega = du/v$, and the basis of Ω consists of the two numbers

$$\int_{\sigma_1} \frac{du}{\sqrt{u^3+Au+B}}, \quad \int_{\sigma_2} \frac{du}{\sqrt{u^3+Au+B}}.$$

The integrals $\int \omega$ are called *elliptic integrals*, and the numbers $\int_\sigma \omega$ for $\sigma \in H_1(Y(\mathbb{C}),\mathbb{Z})$ their *periods*. Thus the lattice Ω that defines the torus $\mathbb{C}^1/\Omega$ isomorphic to an elliptic curve Y consists of the periods of the elliptic integral associated with this curve.

In conclusion we remark that uniformisation of elliptic curves gives us another point of view on the fundamental fact that we discussed in Section 7.1, Chapter 3 (see also Exercise 8 of Section 3.6, Chapter 3), that not all curves of genus 1 are isomorphic to one another. We can even form some impression of the structure of the set of equivalence classes of elliptic curves up to isomorphism. For this, we represent every elliptic curve as $\mathbb{C}^1/\Omega$, and, replacing Ω if necessary by an equivalent lattice, choose a basis $1, \tau$ with $\operatorname{Im}\tau > 0$. It is easy to see that any two such bases $1, \tau$ and $1, \tau'$ define similar lattices if and only if

$$\tau' = \frac{a\tau+b}{c\tau+d}, \quad \text{for some } a, b, c, d \in \mathbb{Z} \text{ with } ad-bc=1. \tag{9.18}$$

The set of all transformations (9.18) form a group G, called the *modular group*. Write H for the upper half-plane $\operatorname{Im}\tau > 0$. Since elliptic curves are isomorphic if and only if the corresponding lattices are similar, the set of isomorphism classes of elliptic curves is in one-to-one correspondence with points of the quotient space H/G.

It can be shown that G acts on H discretely, although not freely (that is, it has fixed points). Nevertheless, the quotient space H/G is a 1-dimensional complex manifold. Moreover, it is isomorphic to $\mathbb{C}$. The function $j\colon H/G \to \mathbb{C}$ realising this isomorphism establishes a one-to-one correspondence between the set of isomorphism classes of elliptic curves and the complex numbers. An algebraic description of j can be extracted from Exercise 8 of Section 3.6, Chapter 3.

2.4 Exercises to Section 2

1 Prove that if an elliptic curve X is defined by an equation with real coefficients then it is isomorphic to $\mathbb{C}^1/\Omega$, where the lattice Ω is either rectangular or rhombic (that is, generated by two vectors of the same length). In the first case $X(\mathbb{R})$ consists of one oval, and in the second, of two ovals.

2 Prove that if a real elliptic curve X has one oval then this oval is not homologous to 0 in $X(\mathbb{C})$.

3 Prove that if a real elliptic curve X has two ovals T_1 and T_2 then for suitable orientations of T_1 and T_2 they are homologous in $X(\mathbb{C})$.

4 Prove that for given periods 1 and τ, all the theta functions of weights $0, 1, \dots$ form a ring, and that this ring is generated by theta functions of weight ≤ 3.

5 Let f be an elliptic function with period lattice Ω. Prove that the number of zeros and poles of f not equivalent under Ω are equal.

6 In the notation of Exercise 5, let $\alpha_1, \dots, \alpha_m$ and $\beta_1, \dots, \beta_m$ be the zeros, respectively the poles, of f not equivalent under Ω. Prove that $\alpha_1 + \dots + \alpha_m - \beta_1 - \dots - \beta_m \in \Omega$. Prove that any numbers $\alpha_1, \dots, \alpha_m$ and $\beta_1, \dots, \beta_m$ satisfying this condition is the set of zeros and poles of some elliptic function.

7 Let $X = \mathbb{C}^1/\Omega$ and $X' = \mathbb{C}^1/\Omega'$ be two elliptic curves. Prove that the group $\operatorname{Hom}(X, X')$ of homomorphisms of algebraic groups $X \to X'$ is isomorphic to the group of complex numbers $\alpha \in \mathbb{C}$ such that $\alpha\Omega \subset \Omega'$.

8 Prove that the regular forms ω on X and ω' on X' can be chosen so that the number $\alpha \in \mathbb{C}$ in Exercise 7 corresponding to a homomorphism $f \in \operatorname{Hom}(X, X')$ is determined by $f^*(\omega') = \alpha\omega$.

9 Prove that for an elliptic curve X defined over $\mathbb{C}$, the ring $\operatorname{End} X = \operatorname{Hom}(X, X)$ is isomorphic either to $\mathbb{Z}$ or to $\mathbb{Z} + \mathbb{Z}\gamma$, where γ satisfies an equation $\gamma^2 + a\gamma + b = 0$ with $a, b \in \mathbb{Z}$ having no real roots. In the second case $X = \mathbb{C}^1/\Omega$, where the lattice Ω is similar to an ideal of the ring of numbers $\mathbb{Z} + \mathbb{Z}\gamma$.

3 Curves of Hyperbolic Type

3.1 Poincaré Series

Consider the open unit disc D and a group G of automorphisms of D; we assume that G acts freely and discretely on D with compact quotient space $X = D/G$. We are going to construct an embedding of X into projective space. As in the previous section, this embedding is constructed by studying holomorphic functions f on D that satisfy the condition

$$g^*(f) = \varphi_g f \quad \text{for } g \in G, \tag{9.19}$$

where φ_g are nowhere vanishing holomorphic functions on D. It follows at once from (9.19) that

$$\varphi_{g_1 g_2} = g_2^*(\varphi_{g_1})\varphi_{g_2} \quad \text{for } g_1, g_2 \in G. \tag{9.20}$$

For an automorphism g, we set

$$J_g = \frac{\mathrm{d}g}{\mathrm{d}z}.$$

The chain rule for differentiating a function of a function shows that equality (9.20) hold for $\varphi_g = J_g$, and hence also for $\varphi_g = J_g^k$ for any positive integer k.

It can be shown that any solution of (9.20) can be reduced using certain trivial transformations to the form $\varphi_g = J_g^k$. In what follows we only consider this case.

Definition A holomorphic function f in D satisfying the relation

$$g^*(f) = J_g^k f \quad \text{for } g \in G,$$

is an *automorphic form* of weight k with respect to G.

Our immediate aim is to construct automorphic forms. For this, we take an arbitrary holomorphic function h that is bounded on D, and consider the series

$$\sum_{g \in G} J_g^k g^*(h), \tag{9.21}$$

A function of this type is called a *Poincaré series*. If it defines an analytic function, then a formal verification shows that this function is an automorphic form. Thus it remains to prove the following result.

Proposition *If $k \geq 2$ then a Poincaré series converges absolutely and uniformly on any compact set $K \subset D$.*

We use the following simple properties of analytic functions.

Lemma *If a function f is analytic in the disc $|z| \leq r$ then*

$$|f(0)|^2 \leq \frac{1}{\pi r^2} \int_{|z| \leq r} |f(z)|^2 \mathrm{d}x \wedge \mathrm{d}y,$$

where $z = x + iy$.

Proof For any ρ with $0 < \rho < r$ we have

$$f(0)^2 = \frac{1}{2\pi i} \int_0^{2\pi} f\left(\rho e^{i\varphi}\right)^2 d\varphi.$$

Multiplying this equality by $\rho d\rho$ and integrating with respect to ρ from 0 to r gives

$$\left(\frac{r^2}{2}\right) f(0)^2 = \frac{1}{2\pi i} \int_0^r \int_0^{2\pi} f\left(\rho e^{i\varphi}\right)^2 \rho d\rho \wedge d\varphi = \frac{1}{2\pi i} \int_{|z|\le r} f(z)^2 dx \wedge dy.$$

Hence

$$\left|f(0)\right|^2 = \frac{1}{\pi r^2} \left| \int_{|z|\le r} f(z)^2 dx \wedge dy \right| \le \frac{1}{\pi r^2} \int_{|z|\le r} \left|f(z)\right|^2 dx \wedge dy.$$

The lemma is proved. □

Proof of the Proposition Since h is bounded on D, it is enough to prove the convergence of the series $\sum_{g\in G} |J_g|^k$, or even just of the series

$$\sum_{g\in G} |J_g|^2. \tag{9.22}$$

The proof uses the fact that $|J_g|^2$ is the Jacobian determinant of the map g. Write $s(U)$ for the area of a domain U defined by the Euclidean metric of the plane $\mathbb{C}^1$ containing D. Then

$$s\big(g(U)\big) = \int_{g(U)} dx \wedge dy = \int_U |J_g|^2 dx \wedge dy. \tag{9.23}$$

The convergence of (9.22) follows at once from this remark. Indeed, let U be a disc with centre z_0, and with small enough radius that $g(U) \cap U = \emptyset$ for all $g \in G$ with $g \ne e$. According to the Lemma and the remark just made,

$$\begin{aligned} \sum_{g\in G} \left|J_g(z_0)\right|^2 &\le \frac{1}{\pi r^2} \sum_{g\in G} \int_U \left|J_g(z)\right|^2 dx \wedge dy \\ &= \frac{1}{\pi r^2} \sum_{g\in G} s\big(g(U)\big) \le \frac{s(D)}{\pi r^2} = \frac{1}{r^2}, \end{aligned} \tag{9.24}$$

which proves convergence.

To prove uniform convergence, we note that if K_1, K_2 are any two compact sets then there are at most a finite number of $g \in G$ such that $g(K_1) \cap K_2 \ne \emptyset$. Indeed, by the definition of a free and discrete group action in Section 1.2, Chapter 8, any two points x, y have neighbourhoods U and V such that $g(U) \cap V = \emptyset$ for all but possibly one $g \in G$. We take an arbitrary point $x \in K_1$, and for any point $y \in K_2$, choose neighbourhoods $U_y \ni x$ and $V_y \ni y$ such that $g(U_y) \cap V_y = \emptyset$ for all but possibly one $g \in G$. It follows at once from the compactness of K_2 that there exists a neighbourhood $U \ni x$ such that $g(U) \cap K_2 = \emptyset$ for all but finitely many $g \in G$. The assertion we need now follows at once from the compactness of K_1.

Now choose $r > 0$ sufficiently small so that the disc of radius r with centre in any point of K is contained in a compact set $K' \subset D$. For any $\varepsilon > 0$, let $C \subset D$ be a disc sufficiently close to D such that $K' \subset C$ and $s(D \setminus C) < \varepsilon$. Write q for the number of elements $g \in G$ such that $g(K') \cap K' \neq \emptyset$. For all but finitely many $g \in G$, we have $g(K') \subset D \setminus C$. Write $\sum'$ for the sum taken over these g; then, as in the course of deducing (9.23), we get that

$$\sum{}' \left|J_g(z)\right|^2 \leq \frac{1}{\pi r^2} \sum{}' s\big(g(U)\big) \leq \frac{q}{\pi r^2} s(D \setminus C) < \frac{q\varepsilon}{\pi r^2},$$

which implies that (9.22) converges uniformly. The proposition is proved. □

Remark Denote by M the multiplicative group of nowhere vanishing holomorphic functions on D. It is a G-module under the action $f \mapsto g^*(f)$. The condition (9.20) is exactly the definition of a 1-cocycle of G with values in M. In the construction of an automorphic form by means of a Poincaré series one can recognise the idea of the proof of the so-called Hilbert Theorem 90 in homological algebra.

3.2 Projective Embedding

We can now proceed to the main result.

Theorem *Let G be a group of automorphisms of D acting freely and discretely, and with compact quotient $X = D/G$. Then there exists a finite number of automorphic forms of the same weight k that define an isomorphic embedding of X into $\mathbb{P}^n$.*

The point is, of course, to check that there exist automorphic forms that satisfy conditions (A) and (B) of Proposition, Section 1.3. We first arrange that these are satisfied locally.

Lemma *For any two points $z_1, z_2 \in D$ such that $z_2 \neq g(z_1)$ for all $g \in G$, there exist automorphic forms f_0 and f_1 satisfying* (A) *of Proposition, Section* 1.3 *for these points. For any point $z_0 \in D$ there exist automorphic forms f_0 and f_1 satisfying* (B) *of Proposition, Section* 1.3 *at z_0. In either case we can assume that*

$$f_0(z) \neq 0 \quad \text{and} \quad f_1(z) \neq 0 \quad \text{for } z = z_1, z_2 \text{ or } z_0.$$

Proof We look for automorphic forms f_i satisfying condition (A) for z_1, z_2 as Poincaré series

$$f_i = \sum g^*(h_i) J_g^k \quad \text{for } i = 0, 1. \tag{9.25}$$

Because the Poincaré series converges, it follows that $|J_g(z_1)| < 1$ and $|J_g(z_2)| < 1$ for all but finitely many $g \in G$. Let $g_0 = e$ and suppose that $g_1, \dots, g_N$ are these exceptional elements. Then as $k \to \infty$

$$\sum_{g \neq g_0, \dots, g_N} |J_g(z)|^k \to 0 \quad \text{for } z = z_1, z_2 \text{ or } z_0. \tag{9.26}$$

Now choose functions h_0 and h_1 satisfying the following conditions:

$$h_i\big(g_m(z_1)\big) = h_i\big(g_m(z_2)\big) = 0 \quad \text{for } i = 0, 1 \text{ and } m = 1, \dots, N;$$
$$h_i(z_1) \neq 0, \quad h_i(z_2) \neq 0 \quad \text{for } i = 0, 1;$$
$$h_0(z_1)h_1(z_2) - h_0(z_2)h_1(z_1) \neq 0.$$

These can be found for example among polynomials.

Then for $i = 0, 1$, we have

$$f_i(z_1) = h_i(z_1) + \sum_{g \neq g_0, \dots, g_N} h_i\big(g(z_1)\big)J_g(z_1)^k = h_i(z_1) + u_i^{(k)}(z_1),$$

where $u_i^{(k)}(z_1) \to 0$ as $k \to \infty$ by (9.26). The same holds for z_2. It follows from this that if k is sufficiently large, then $f_i(z_1) \neq 0$, $f_i(z_2) \neq 0$ for $i = 0, 1$, and

$$f_0(z_1)f_1(z_2) - f_0(z_2)f_1(z_1) \neq 0.$$

Now we construct functions that satisfy condition (B) of Proposition, Section 1.3. We again look for them in the form (9.25). Let $g_0 = e$ and suppose that $g_1, \dots, g_N$ are the elements $g \in G$ such that $|J_g(z_0)| \geq 1$. Choose h_0 and h_1 such that

$$h_i\big(g_m(z_0)\big) = h_i'\big(g_m(z_0)\big) = 0 \quad \text{for } i = 0, 1 \text{ and } m = 1, \dots, N;$$
$$h_i(z_0) \neq 0 \quad \text{for } i = 0, 1;$$
$$h_0(z_0)h_1'(z_0) - h_1(z_0)h_0'(z_0) \neq 0.$$

As before, we have

$$f_i(z_0) = h_i(z_0) + u_i^{(k)}(z_0) \quad \text{and} \quad f_i'(z_0) = h_i'(z_0) + v_i^{(k)}(z_0)$$

for $i = 0, 1$, where $u_i^{(k)}(z_0)$, $v_i^{(k)}(z_0) \to 0$ as $k \to \infty$. Hence

$$f_0(z_0)f_1'(z_0) - f_1(z_0)f_0'(z_0) \neq 0$$

for k sufficiently large. This proves the lemma. □

Proof of the Theorem We note first that if functions f_0 and f_1 satisfy (B) of Proposition, Section 1.3 at a point z_0, then they also satisfy (A) for all points z_1, z_2 with $z_1 \neq z_2$ in a sufficiently small neighbourhood of z_0. Indeed, the function

$$F(z_1, z_2) = \big(f_1(z_1)f_0(z_2) - f_1(z_2)f_0(z_1)\big)/(z_1 - z_2)$$

$$= f_0(z_1)\left(\frac{f_1(z_1) - f_1(z_2)}{z_1 - z_2}\right) - f_1(z_1)\left(\frac{f_0(z_1) - f_0(z_2)}{z_1 - z_2}\right)$$

is analytic, and

$$F(z, z) = f_0(z) f_1'(z) - f_1(z) f_0'(z).$$

Hence $F(z_0, z_0) \neq 0$, and therefore $F(z_1, z_2) \neq 0$ for points z_1 and z_2 sufficiently close to z_0, which gives our assertion.

Obviously, if conditions (A) or (B) of Proposition, Section 1.3 are satisfied for some functions and points z_1, z_2 or z_0, then they are also satisfied for sufficiently close points. Using the lemma, we choose a finite cover of the compact manifold $X = D/G$ by open sets U_i for $i = 1, \dots, N_0$ such that in each U_i, condition (B) is satisfied by functions f_{0i} and f_{1i}. By the remark that we made above, there exists a neighbourhood U of the diagonal in $X \times X$ such that at every point of this set some pair of functions f_{0i}, f_{1i} satisfies (A). Since the set $X \times X \setminus U$ is compact, we can extend the set of functions to a finite set $\{f_{0i}, f_{1i}\}$ for $i = 1, \dots, N$ such that some pair of functions f_{0i}, f_{1i} also satisfies (A) for every point of $X \times X$.

Suppose that f_{0i} and f_{1i} have weight m_i, and set $M = \prod m_1$ and $l_i = M/m_i$. Consider the system of functions consisting of all the products of the form

$$f_{0i}^{2l_i}, \quad f_{1i}^{2l_i}, \quad f_{0i}^{l_i} f_{1i}^{l_i} \quad \text{and} \quad f_{0i}^{l_i-1} f_{1i}^{l_i+1} \quad \text{for } i = 1, \dots, N.$$

These are obviously all automorphic forms of weight $2M$. Let us prove that they satisfy (A) and (B) of Proposition, Section 1.3. Indeed, if f_{0i} and f_{1i} satisfy (A) at points z_1, z_2 then the following minor is nonzero:

$$\begin{aligned}
&\bigl(f_{0i}^{l_i} f_{1i}^{l_i}\bigr)(z_1)\bigl(f_{0i}^{l_i-1} f_{1i}^{l_i+1}\bigr)(z_2) - \bigl(f_{0i}^{l_i} f_{1i}^{l_i}\bigr)(z_2)\bigl(f_{0i}^{l_i-1} f_{1i}^{l_i+1}\bigr)(z_1) \\
&\quad = f_{0i}^{l_i-1}(z_1) f_{0i}^{l_i-1}(z_2) f_{1i}^{l_i}(z_1) f_{1i}^{l_i}(z_2)\bigl(f_{0i}(z_1) f_{1i}(z_2) - f_{0i}(z_2) f_{1i}(z_1)\bigr).
\end{aligned}$$

The verification of (B) is similar. The theorem is proved. □

Remark The proof of the theorem makes very little use of specific properties of the open unit disc D. Even the assumption that X is 1-dimensional plays no essential role. The proof carries over almost without change to the case when D is any bounded domain in $\mathbb{C}^n$ and G is a group of automorphism of D acting freely and discretely such that the quotient $X = D/G$ is compact. One needs only take J_g to be the Jacobian of the transformation $g \in G$ in the definition of automorphic form.

3.3 Algebraic Curves and Automorphic Functions

In Sections 1–2 we proved that algebraic curves of elliptic and parabolic types are exactly the curves of genus 0 and 1. Therefore curves of parabolic type are the curves of genus $g \geq 2$. The theorem shows that these curves are the compact manifolds of the form D/G, where D is the open unit ball and G a group of automorphisms of D acting freely and discretely.

We now give the algebraic description of the embedding of the curve $X = D/G$ into projective space defined by automorphic forms. Let $f(z)$ be an automorphic form of weight k. The expression $\eta = f(z)(\mathrm{d}z)^k$ defines a holomorphic differential form of weight k on D. The definition of a holomorphic differential form of weight k is given in Exercise 2 of Section 2.4, Chapter 8, and that of a differential form of weight k in Exercise 7 of Section 8.1, Chapter 3. It follows from the definition of automorphic form that η is invariant under automorphisms of G. In fact

$$g^*(\eta) = g^*(f)\big(\mathrm{d}g(z)\big)^k = J_g^k f J_g^{-k}(\mathrm{d}z)^k = \eta.$$

Therefore $\eta = \pi^*(\omega)$, where π is the projection $D \to D/G$ and ω a holomorphic differential form of weight k on D/G. Finally if $\varphi\colon D/G \to X$ is an isomorphism with an algebraic curve then $\omega' = (\varphi^{-1})^*(\omega)$ is a holomorphic differential form of weight k on X. It follows from this that ω' is a rational differential form on X. To see this, it is enough to take its ratio with any rational differential form of the same weight k on X; by Theorem 8.4, this ratio will be a rational function on X. Lemma of Section 3.1, Chapter 8 shows that ω' is a regular differential form of weight k on X. It is easy to show that, conversely, every regular differential form of weight k on X is obtained in this way.

We see that the space of automorphic forms of weight k is isomorphic to the space of regular differential forms of weight k on the algebraic curve X. Thus the map to projective space defined by all automorphic forms of weight $k \geq 2$ coincides with the map corresponding to the divisor class kK_X. In Section 7.1, Chapter 3, we deduced from the Riemann–Roch theorem that this map is an embedding for $k \geq 3$. Hence the same is true for the map defined by automorphic forms of weight k. Furthermore, we obtain an interesting analytic application of the Riemann–Roch theorem: the space of automorphic forms of weight k is finite dimensional, and by the Riemann–Roch theorem, its has dimension

$$l(kK_X) = (2k-1)(g_X - 1),$$

As in Section 2.3, it follows from Theorem 9.2 that the field $\mathbb{C}(X)$ is isomorphic to the field of meromorphic functions on D invariant under G. Such functions are called *automorphic functions*. Thus every curve of genus $g \geq 2$ is uniformised by automorphic functions.

Let us compare the picture we have obtained with that in the parabolic case. In either case, the description of curves reduces to the description of certain discrete groups. In the parabolic case, the discrete groups are extremely simple: they are lattices in $\mathbb{C}$. What happens in the hyperbolic case?

Poincaré discovered a general method of constructing groups of automorphisms of the unit disc that act freely and discretely. His method is based on the fact that one can define a metric on D such that the orientation-preserving isometries coincide with analytic automorphisms of D; moreover, with this metric, D is isomorphic as a metric space to the Lobachevsky plane. In this isomorphism, the lines of

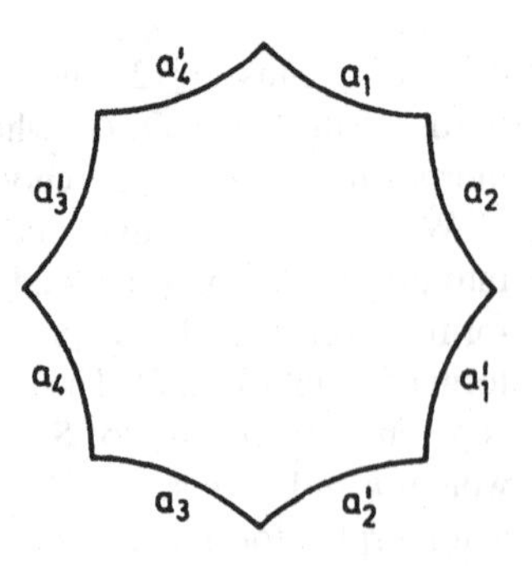

Fig. 33 A fundamental polygon (the case $g = 2$)

Lobachevsky geometry correspond to arcs of circles in D orthogonal to the unit circle, the boundary of D. We do not require the definition of this metric. Note only that in it, the magnitude of an angle is equal to the magnitude of the angle between circles (that is, between the tangent lines at their point of intersection) in the Euclidean metric of the plane $\mathbb{C}^1$ of a complex variable containing D.

Poincaré showed that any group G of automorphisms of D which acts freely and discretely on D such that the quotient D/G is compact is defined by some polygon in the geometry just described. This polygon plays the same role in the hyperbolic case as the fundamental parallelogram of the lattice Ω in the parabolic case, and is called the *fundamental polygon* of G. If the algebraic curve $X = D/G$ has genus g then the fundamental polygon has $4g$ sides. We choose a direction of circumnavigation of the polygon (that is, clockwise or anticlockwise), and denote its sides in the chosen cyclic order by $a_1, a_2, a_1', a_2', a_3, a_4, a_3', a_4', \dots, a_{2g-1}, a_{2g}, a_{2g-1}', a_{2g}'$; the sides are marked with the direction of circumnavigation. Figure 33 shows the case $g = 2$.

Then the following relations hold:

(1) the sides a_i and a_i' are congruent;
(2) the sum of the $4g$ internal angles of the polygon equals 2π.

The group G is defined by its fundamental polygon in the following way: for $i = 1, \dots, 2g$, write g_i for the motion that takes the side a_i into a_i', reversing the direction of circumnavigation but preserving the orientation. The motions g_i generate G.

Conversely, given a polygon Φ satisfying conditions (1) and (2), the group G generated by the transformations g_i acts freely and discretely on D, with fundamental polygon Φ. Geometrically, this means that if F is the interior of Φ then applying the generators g_i first to F, then to the domains $g_j(F)$, and so on, we tile the whole of D with polygonal domains that intersect only along the sides of the boundary.

It would be natural to try to use this picture to describe the set of isomorphism classes of curves of genus $g \geq 2$, by analogy with the way this was done at the end of Section 2.3 for $g = 1$. However, the situation here is much more complicated. The complex space corresponding to the problem, or even the algebraic variety (for the definitions, see Ahlfors [3] and Bers [10] the analytic case and Mumford and Fogarty [64] in the algebraic case) can be defined precisely. It is called the *moduli space* of curves of genus g. The detailed study of its properties is the subject of much current activity, but we do not have space to discuss the results here, and for the most part they are not definitive. Of the problems that can be easily stated, the following is one of the most interesting: is the moduli space rational, or at least unirational? It is only known that the moduli space is rational for $g = 2$. The question of unirationality seems to be easier: it can be proved easily for small genus ($g = 3, 4, 5$). It is proved that $\mathcal{M}_g$ is unirational for $g \leq 13$, and is not unirational if $g \geq 23$.

3.4 Exercises to Section 3

1 Prove that already the automorphic forms of weight 3 define a projective embedding of a compact hyperbolic curve D/G.

2 Prove that for a fixed group of automorphisms G of the open disc D with compact quotient D/G, the equation $\sum_{g\in G} g^*(h) J_g^k = 0$ has infinitely many linearly independent solutions with h a bounded holomorphic function on D.

3 Prove that the genus g of a curve D/G and the area S of the fundamental domain Φ in the sense of Lobachevsky geometry are related by $g - 1 = S/4\pi$. [Hint: Use the theorem of Lobachevsky geometry that the sum of angles of a triangle is 2π minus its area; and the relation between the Euler characteristic and the genus.]

4 Uniformising Higher Dimensional Varieties

4.1 Complete Intersections are Simply Connected

Almost nothing is known about the universal covers and fundamental groups of complex manifolds of dimension ≥ 2. We give a few simple examples and remarks, with the aim of throwing some light on the nature of the problems that arise.

The main new phenomenon we come up against is the following. Among the nonsingular complete algebraic curves, only one is simply connected, namely the projective line; therefore, passing to the universal cover almost always reduces the study of a curve to that of another manifold, which one hopes is easier to study, and which turns out to be so. For complex manifolds of dimension ≥ 2 this is just not the case: very many of them are simply connected, so that passing to the universal cover gives nothing new. To be slightly more precise about what "very many" means, we discuss a very wide class of complex manifolds of dimension ≥ 2, containing in particular all the nonsingular projective hypersurfaces, and prove that they are all simply connected.

Definition We say that a projective variety $X \subset \mathbb{P}^N$ of codimension n is a *complete intersection* if it the intersection of n hypersurfaces that meet transversally at each point of intersection.

By definition, complete intersections are nonsingular algebraic varieties. From now on, we consider them over the complex number field. We will prove that if $\dim X \geq 2$ then the topological manifold $X(\mathbb{C})$ is simply connected. This is a consequence of a general result that we will use several times in what follows.

Proposition *If V is an n-dimensional projective variety over the complex number field and $W \subset V$ a hyperplane section of V such that $V \setminus W$ is nonsingular then the embedding $W(\mathbb{C}) \hookrightarrow V(\mathbb{C})$ induces isomorphisms of homotopy groups*

$$\pi_r\big(W(\mathbb{C})\big) \cong \pi_r\big(V(\mathbb{C})\big) \quad \textit{for } r < n-1.$$

The proposition is a simple application of Morse theory (see for example Milnor [58]). One uses [58, Theorem 7.4] and the exact homotopy sequence of the pair $(V(\mathbb{C}), W(\mathbb{C}))$.

We prove that complete intersections are simply connected by induction on their codimension in projective space. At the first step of the induction we must use the simple connectedness of $\mathbb{P}^n(\mathbb{C})$.

Let $X \subset \mathbb{P}^N$ be an intersection of n transversal hypersurfaces $E_1, \ldots, E_n$ of degrees $m_1, \ldots, m_n$; by reordering, we can assume that $m_1 \geq \cdots \geq m_n$. Since $E_1, \ldots, E_n$ intersect transversally along X and have no common zeros outside X, it is an easy exercise in Bertini's theorem to see that, after replacing the E_i if necessary by more general forms of the same degree generating the ideal of X, the intersection $E_1 \cap \cdots \cap E_i$ is a nonsingular complete intersection for each $i = 1, \ldots, n$.

In particular, $Y = E_1 \cap \cdots \cap E_{n-1}$ is a complete intersection. Then $Y(\mathbb{C})$ is simply connected by induction on n. Now consider the Veronese embedding

$$v_m \colon \mathbb{P}^N \hookrightarrow \mathbb{P}^M, \quad \text{where } m = m_n = \deg E_n \text{ and } M = \binom{N+m}{N} - 1$$

(see Example 1.28 of Section 4.4, Chapter 1). Let $V = v_m(Y)$ and $W = v_m(X)$. Obviously W is the hyperplane section of V by the hyperplane corresponding to E_n. Since $V(\mathbb{C})$ is homeomorphic to $Y(\mathbb{C})$, it is simply connected. We can apply the proposition, and deduce that

$$\pi_1\big(W(\mathbb{C})\big) = \pi_1\big(V(\mathbb{C})\big) = 0 \quad \text{if } \dim V > 2,$$

that is, $\dim X \geq 2$. Since $X(\mathbb{C})$ is homeomorphic to $W(\mathbb{C})$, it is simply connected.

4.2 Example of Manifold with π_1 a Given Finite Group

Despite what was said in Section 4.1, there exist many nonsimply connected algebraic varieties of any given dimension. We now illustrate this phenomenon, which is in a certain sense the opposite of that discussed in Section 4.1. Namely, we prove that for any finite group Γ and any integer $n \geq 2$ there exists an n-dimensional complete algebraic variety whose fundamental group is isomorphic to Γ.

We construct the example first in the case that Γ is the symmetric group on m elements $\Gamma = \mathfrak{S}_m$. For this, consider the product of m copies of s-dimensional projective spaces, $\Pi = \mathbb{P}^s \times \cdots \times \mathbb{P}^s$. Points $x \in \Pi$ are denoted by $x = (x_1, \ldots, x_m)$ with $x_i \in \mathbb{P}^s$. The group $\mathfrak{S}_m$ acts on Π by permuting the m points:

$$g(x_1, \ldots, x_m) = (x_{i_1}, \ldots, x_{i_m}), \quad \text{where } g = \begin{pmatrix} 1 & \ldots & m \\ i_1 & \ldots & i_m \end{pmatrix} \in \mathfrak{S}_m.$$

The basic step in the construction of the example is the construction of the quotient space $\Pi' = \Pi/\mathfrak{S}_m$. By definition, Π' is a normal variety with a finite morphism $\varphi\colon \Pi \to \Pi'$ such that $\varphi(x) = \varphi(x')$ if and only if $x' = g(x)$ for some $g \in \mathfrak{S}_m$.

Write $x_j = (x_{0j}, \ldots, x_{sj})$ to denote the homogeneous coordinates in the jth copy of $\mathbb{P}^s$. We introduce $s + 1$ auxiliary variables $t_0, \ldots, t_s$ and consider the form

$$F(x, t) = \prod_{j=1}^{m} L_j(x_j, t), \quad \text{where } L_j = \sum_{i=0}^{s} x_{ij} t_i. \tag{9.27}$$

Write $T^1, \ldots, T^N$ for all the monomials of degree m in $t_0, \ldots, t_s$. Then

$$F(x, t) = \sum_{\alpha=1}^{N} F_\alpha(x) T^\alpha,$$

where the $F_\alpha(x)$ are forms in the variables x_{ij} that are linear in each of the m sets of variables $x_{0j}, \ldots, x_{sj}$. Consider the rational map

$$\varphi(x) = \bigl(F_1(x) : \cdots : F_N(x)\bigr)$$

defined by these forms. This map is regular: if all $F_1(x) = \cdots = F_N(x) = 0$ for some $x \in \Pi$ then $F(x, t) \equiv 0$ as a polynomial in t, and this means that $L_j(x_j, t) \equiv 0$ for some j, that is, all the coordinates of the jth point x_j are zero.

There is a simple relation between the map φ and the embedding of Π as a closed subvariety $\overline{\Pi}$ of some projective space constructed in Section 5.1, Chapter 1. Namely, it is easy to check that φ is a projection of $\overline{\Pi}$, and moreover, the assumptions of Theorem 1.16 of Section 5.3, Chapter 1, hold for this projection, so that $\varphi\colon \overline{\Pi} \to \Pi' \subset \mathbb{P}^N$ is finite onto its image $\Pi' = \varphi(\overline{\Pi})$.

The action of $g \in \mathfrak{S}_m$ on Π interchanges the factors in (9.27), from which it follows that $\varphi \circ g = \varphi$; that is, if $x = g(y)$ then $\varphi(x) = \varphi(y)$. Conversely if $\varphi(x) = \varphi(y)$ then $x = g(y)$ for some $g \in \mathfrak{S}_m$. Indeed, if $\varphi(x) = \varphi(y)$ then $F(x, t) = cF(y, t)$ with $c \neq 0$, and by unique factorisation of polynomials, it follows that the points $y_1, \ldots, y_m$ are obtained by permuting the points $x_1, \ldots, x_m$. Therefore

$$\varphi^{-1}(x') = \bigl\{g(x) \mid g \in \mathfrak{S}_m\bigr\} \quad \text{for } x' \in \Pi'. \tag{9.28}$$

We now prove that Π' is normal. For this note that the polynomials $F_\alpha(x_1, \dots, x_m)$ do not change on interchanging the points $x_1, \dots, x_m$. The converse is also true: any polynomial in the homogeneous coordinates of the m points $x_1, \dots, x_m$ that is homogeneous in each of the m groups of variables and invariant under any permutation of the points is a form in the polynomials F_α. This is an analogue of the main theorem on symmetric functions, and is also proved in a completely analogous way. The proof can be found in old algebra textbooks, for example Bôcher [12, Theorem 1 of Section 89, Chapter XIX].

Let H be a form in the homogeneous coordinates of a point of Π', $Y \subset \Pi'$ the affine open set defined by $H \neq 0$, and $X = \varphi^{-1}(Y)$. Then X is also affine and defined by the condition $\varphi^*(H) \neq 0$. We check that the ring $\varphi^*(k[Y])$ is precisely the ring of elements of $k[X]$ invariant under $\mathfrak{S}_m$, that is

$$\varphi^*\bigl(k[Y]\bigr) = k[X]^{\mathfrak{S}_m}. \tag{9.29}$$

Indeed, the function $f \in k[X]$ is of the form

$$f = \frac{H_1}{(\varphi^*(H))^k},$$

where H_1 is a form of degree equal to that of $(\varphi^*(H))^k$. If f is invariant under $\mathfrak{S}_m$ then the same holds for H_1 (the form $\varphi^*(H)$ is obviously invariant). Hence from the generalised version just given of the theorem on symmetric functions it follows that H_1 is a form in the polynomials F_α, and this means that $f \in \varphi^*(k[Y])$.

Thus the affine variety Y is the quotient of X under G, that is, $Y = X/G$. From Example of Section 5.1, Chapter 2, it follows that Y is normal. Since the open sets Y corresponding to different forms H cover Π', it follows that Π' is normal.

From (9.29), the analogous equality $\varphi^*(k(\Pi')) = k(\Pi)^{\mathfrak{S}_m}$ for the field of fractions follows easily. From the elementary set-up of Galois theory it now follows that $k(\Pi') \subset k(\Pi)$ is a Galois extension with Galois group $\mathfrak{S}_m$. In particular

$$\deg \varphi = m!. \tag{9.30}$$

We write $\Delta \subset \Pi$ for the closed set consisting of all points $(x_1, \dots, x_m)$ such that $x_i = x_j$ for some $i \neq j$, and set $\Delta' = \varphi(\Delta) \subset \Pi'$ and $W = \Pi \setminus \Delta$, $W' = \Pi' \setminus \Delta'$. If $x' \in W'$ then by (9.28), $\varphi^{-1}(x')$ consists of $m!$ distinct points. Comparing this with (9.30) we see that $\varphi\colon W \to W'$ is an unramified cover. It follows from Example of Section 2.1, Chapter 2, that W' is nonsingular.

We have constructed two nonsingular varieties W and W' and an unramified cover $\varphi\colon W \to W'$ with automorphism group $\mathfrak{S}_m$. However, this is not what we need, since both of our varieties are incomplete. To overcome this defect, we intersect Π' with a projective linear subspace $L \subset \mathbb{P}^N$ such that L is disjoint from Δ, and the variety $Y = L \cap \Pi'$ is nonsingular. Such a subspace exists, and can be defined by d linearly independent linear equations, provided that

$$d > \dim \Delta'. \tag{9.31}$$

The set of points $(x_1, \dots, x_m) \subset \Pi$ for which $x_p = x_q$ has codimension s in Π, and hence

$$\operatorname{codim}(\Delta \subset \Pi) = \operatorname{codim}(\Delta' \subset \Pi') = s,$$

so that (9.31) takes the form $d > (m-1)s$. We can choose the linear subspace such that the dimension of Y is given by the theorem on dimensions of intersection

$$\dim Y = \dim \Pi' - d = ms - d.$$

It is obvious that taking s sufficiently large we can arrange that

$$\dim Y = ms - d = n \quad \text{with } d > s.$$

For this it is enough that $s > n$.

Since $Y \cap \Delta' = \emptyset$, that is, $Y \subset W'$, it follows that $X = \varphi^{-1}(Y)$ is an unramified cover of Y with automorphism group $\mathfrak{S}_m$.

The preceding arguments were all purely algebraic. Suppose now that all the varieties are defined over the field of complex numbers. As we saw in Section 6.3, Chapter 2, the map $\varphi\colon X(\mathbb{C}) \to Y(\mathbb{C})$ is an unramified cover. Applying Proposition of Section 4.1, with $r = 0$ shows that $X(\mathbb{C})$ is connected. Indeed, X is obtained from Π by intersecting with hypersurfaces, that we can view as intersecting with hyperplanes under the Veronese embedding of Π to projective space. The same proposition can be applied when $r = 1$. Since $\Pi(\mathbb{C})$ is simply connected, the proposition implies also that $X(\mathbb{C})$ is simply connected.

We see that $X(\mathbb{C})$ is the universal cover of $Y(\mathbb{C})$, and $\pi_1(Y(\mathbb{C})) = \mathfrak{S}_m$. Note that $Y(\mathbb{C})$ is projective by construction.

Starting from an unramified cover $\varphi\colon X \to Y$ with group $\mathfrak{S}_m$, we can easily get a cover with an arbitrary finite group Γ. For this, suppose that $\Gamma \subset \mathfrak{S}_m$. We have seen that the field extension $\mathbb{C}(Y) \subset \mathbb{C}(X)$ is Galois with group $\mathfrak{S}_m$; by Galois theory, the subgroup Γ corresponds to an intermediate subfield K such that $\mathbb{C}(Y) \subset K \subset \mathbb{C}(X)$ and $K \subset \mathbb{C}(X)$ is Galois with group Γ. Let $\overline{Y}$ be the normalisation of Y in K. By general properties of normalisation, we have morphisms

$$X \xrightarrow{\overline{\varphi}} \overline{Y} \xrightarrow{\psi} Y, \quad \text{with } \psi \circ \overline{\varphi} = \varphi.$$

It follows from general properties of finite morphisms that $\overline{\varphi}$ and ψ are finite. We now prove that $\overline{Y}$ is nonsingular, and $\overline{\varphi}$ is unramified. Indeed, since $\deg \varphi = \deg \overline{\varphi} \deg \psi$, and the number of inverse images $\varphi^{-1}(y)$ of a closed point $y \in Y$ is equal to $\deg \varphi$, it follows that for every $y \in Y$ and $\overline{y} \in \overline{Y}$, the number of inverse images $\psi^{-1}(y)$ and $\overline{\varphi}^{-1}(\overline{y})$ equals $\deg \psi$ and $\deg \overline{\varphi}$ respectively. Thus $\overline{\varphi}$ and ψ are unramified, and since Y is nonsingular, Theorem 2.30 of Section 6.3, Chapter 2, implies that $\overline{Y}$ is nonsingular. We see that $X(\mathbb{C})$ is the universal cover of $\overline{Y}(\mathbb{C})$, and $\pi_1(\overline{Y}(\mathbb{C})) = \Gamma$. This completes the construction of the example. Note that from the theorem that the normalisation of a projective variety is projective (which we have not proved) it follows that the example $\overline{Y}$ just constructed is projective.

4.3 Remarks

Concerning the examples constructed in Section 2, we should add that one can construct many examples of projective varieties with infinite fundamental groups. Thus, for an n-dimensional Abelian variety X, the complex manifold $X(\mathbb{C})$ is homeomorphic to the $2n$-dimensional torus by the results of Section 1.3, Chapter 8, so that $\pi_1(X(\mathbb{C})) = \mathbb{Z}^{2n}$. If $n \geq 3$ then by Proposition of Section 4.1, a nonsingular hyperplane section Y of X has the same fundamental group; its universal cover is a subvariety of $\mathbb{C}^n$ about which nothing seems to be known.

In these examples we meet two types of construction of fundamental groups and universal covers of algebraic varieties. Type I is when the fundamental group $\pi_1(X)$ is finite. In this case one can show that the universal cover $\widetilde{X}$ is a complete algebraic variety, and if X is projective then so is $\widetilde{X}$. The only 1-dimensional variety that can be represented in this form is $\mathbb{P}^1$, the unique complex curve of elliptic type.

Type II is hard to define precisely at present, beyond the fact that the fundamental group is infinite. In this case, the universal cover is a "very big" complex manifold, very far from being a projective or complete algebraic variety. In case of dimension 1, this is the class of curves of parabolic or hyperbolic type. In case of dimension ≥ 2, it includes Abelian varieties, and (by the remark at the end of Section 3.2) manifolds of the form D/G, where D is a bounded domain in $\mathbb{C}^n$ and G a group of automorphisms of D acting freely and discretely and such that D/G is compact. It also includes hyperplane sections of these manifolds.

To attempt to characterise the second type of manifold more precisely, we give the definition of two types of complex space that play a basic role in the general theory of complex spaces and manifolds.

We say that a complex space X is *holomorphically convex* if for any sequence of points $x_n \in X$ not having an accumulation point in X, there exists a holomorphic function f on X such that $|f(x_n)| \to \infty$ as $n \to \infty$. Every compact space is holomorphically convex, trivially. Another example is given by the spaces X_{an}, where X is an affine algebraic variety: if $X \subset \mathbb{A}^n$ then the required function f can already be found among the coordinate functions on X.

A holomorphically convex complex space X is said to be *holomorphically complete* or a *Stein space* if the holomorphic functions on X separate points, that is, for any two distinct points $x', x'' \in X$, there exists a function f, holomorphic on the whole of X, such that $f(x') \neq f(x'')$. It follows from Theorem of Section 2.2, Chapter 8, that a compact complex variety is holomorphically complete only if it consists of a single point. This also holds for compact complex spaces. Complex spaces of the form X_{an} with X an affine algebraic variety are obviously holomorphically convex. Generally speaking holomorphically complete spaces play a role in the theory of complex spaces analogous to that of affine varieties in algebraic geometry. For example, they are the "opposites" of compact spaces, in the same way that affine varieties are the opposites of projective or complete varieties.

Now we can give a more precise description of the two types of examples of universal covers of algebraic manifolds: in Type I the universal cover is compact, and in Type II holomorphically complete. It is natural to hope that the general case is in some sense a "mixture" of these two extreme types. There is a fundamental result in the theory of complex spaces that can be viewed as making the term "mixture" more precise. This is the so-called Remmert reduction theorem, which asserts that any holomorphically convex normal complex space X has a proper map $f: X \to Y$ to a holomorphically complete space Y (a map f is *proper* if the inverse image of any compact set is compact; in particular, its fibres are compact).

In the light of this result it is an interesting question to know whether the universal cover of a complete algebraic variety is holomorphically convex. (We could be more cautious and restrict ourselves to projective varieties.)

All compact complex manifolds are obviously holomorphically convex. A typical example of a manifold that is not holomorphically convex is $\mathbb{C}^2 \setminus 0$. Indeed, one can see that a function holomorphic on $\mathbb{C}^2 \setminus 0$ is also holomorphic at 0; (in the same way that a rational function on a nonsingular algebraic variety fails to be regular at points of a whole divisor.) Hence if $x_n \to 0$ then any function f that is holomorphic on $\mathbb{C}^2 \setminus 0$ satisfies $f(x_n) \to f(0)$. The manifold $\mathbb{C}^2 \setminus 0$ is indeed the universal cover of a nonalgebraic complex manifold, for example the Hopf surface (Example 8.2). Kodaira proved that a compact complex manifold with universal cover $\mathbb{C}^2 \setminus 0$ is not algebraic.

In Section 3.2 we observed that if a bounded domain $D \subset \mathbb{C}^n$ is the universal cover of a compact manifold then this manifold is algebraic. On the other hand, it can be proved that any such domain D is holomorphically convex. These examples speak in favour of our conjecture.[4]

4.4 Exercises to Section 4

1 Let $C \subset \mathbb{P}^2$ be a nonsingular projective plane curve given by an equation $F(x_0, x_1, x_2) = 0$ of degree n and $V \subset \mathbb{P}^3$ the surface given by $F(x_0, x_1, x_2) = x_3^n$; let $f: V \to \mathbb{P}^2$ be the projection from the point $(0:0:0:1)$. Prove that $f: V \setminus f^{-1}C \to \mathbb{P}^2 \setminus C$ is an unramified cover, and that $V \setminus f^{-1}C$ is the universal cover of $\mathbb{P}^2 \setminus C$. Deduce that $\pi_1(\mathbb{P}^2 \setminus C) \cong \mathbb{Z}/n$.

2 Prove that $(\mathbb{P}^1)^m / \mathfrak{S}_m = \mathbb{P}^m$.

3 Let X be a nonsingular projective curve, and $G = \{1, g\}$ the group of order 2 where $g: X \times X \to X \times X$ is given by $g(x, x') = (x', x)$. Prove that the ringed space $Y = (X \times X)/G$ is a complex space and even a complex manifold (notwithstanding the fixed points of g). Prove that $\pi_1(Y) \cong H_1(X)$.

[4] J. Kollár [50, 51] has recently introduced a number of formal algebraic analogues of this conjecture, and has proved them in some cases, and discovered many applications of these ideas to complex varieties.

4 Find the mistake in the following "proof" of the Jacobian conjecture with $k = \mathbb{C}$ (compare Section 2.3, Chapter 1). Let $\varphi\colon \mathbb{A}^2 \to \mathbb{A}^2$ be a regular map with constant nonzero Jacobian. Then $U = \mathbb{A}^2(\mathbb{C}) \setminus \varphi(\mathbb{A}^2(\mathbb{C}))$ is a finite set of points: for if a curve $f = 0$ intersected $\varphi(\mathbb{A}^2(\mathbb{C}))$ in a finite number of points, the polynomial $\varphi^*(f)$ would only have a finite number of zeros on $\mathbb{A}^2$. Then from "general position" arguments it follows easily that U is simply connected. However, $\varphi\colon \mathbb{A}^2(\mathbb{C}) \to U$ is an unramified cover, and since $\mathbb{A}^2(\mathbb{C})$ is connected, it must be an isomorphism. It is easy to prove that $\mathbb{A}^2 \setminus \{x_1, \dots, x_r\}$ is not an affine variety. Therefore $\varphi(\mathbb{A}^2) = \mathbb{A}^2$, and φ is an automorphism.

Historical Sketch

This sketch makes no pretence at a systematic treatment of the history of algebraic geometry. It aims only to describe in very broad terms how the ideas and notions discussed in the book came to be created. Because of this, when discussing the research of this or that mathematician we often omit to mention important works, sometimes even his or her most important works, if they do not bear on the contents of our book.

We try to state results in language as close as possible to that of their authors, only occasionally using modern notation and terminology. In cases when the interpretation is not obvious, we give a discussion from the point of view of notions and results of our book; parenthetical sections of this nature are printed in italics.

1 Elliptic Integrals

Naturally enough, algebraic geometry arose first as the theory of algebraic curves. Properties of algebraic curves that are specific to algebraic geometry only arise when we go beyond the context of rational curves. We thus leave to one side the theory of curves of degree 2*, which are all rational. The next case in order of difficulty, and therefore the first nontrivial examples, are curves of genus* 1*, that is, elliptic curves, and in particular nonsingular cubics. And, historically, the first stage in the development of the theory of algebraic curves consisted of working out its basic notions and ideas in the example of elliptic curves.*

Thus it might seem that these ideas developed in the same sequence in which they are now treated (as for example in Section 1*, Chapter* 1*). However, in one respect this is not at all the case. For the web of ideas and results that we now call the theory of elliptic curves arose as a branch of analysis, and not of geometry: as the theory of integrals of rational functions on an elliptic curve. It was these integrals that were first given the name elliptic (because they turned up in connection with calculating the arc length of an ellipse), which spread subsequently to functions and to curves.*

I.R. Shafarevich, *Basic Algebraic Geometry 2*, DOI 10.1007/978-3-642-38010-5,

Elliptic integrals arose as objects of study already in the 17th century, as examples of integrals that cannot be expressed in terms of elementary functions, thus leading to new transcendental functions.

At the very end of the 17th century, first Jakob and then Johann Bernoulli came across a new interesting property of these integrals (see Bernoulli [103, Vol. I, p. 252]). In their study they considered the integrals that express the arc length of certain curves. They discovered certain transformations of one curve into another that preserve the arc length of the curves, although the arcs themselves are not congruent. Clearly from the point of view of analysis this leads to a transformation of one integral into another. In some cases transformations from an integral into itself occur. In the first half of the 18th century many examples of such transformations we discovered by Fagnano.

In its general form the problem was stated and solved by Euler. The first results in this direction were communicated in a letter to Goldbach in 1752. His investigations of elliptic integrals were published from 1756 to 1781 (see Euler [116, Section II, Chapter VI], [117]).

Euler considers an arbitrary polynomial $f(x)$ of degree 4 and poses the problem of the possible relations between x and y under which

$$\frac{\mathrm{d}x}{\sqrt{f(x)}} = \frac{\mathrm{d}y}{\sqrt{f(y)}}. \tag{A.1}$$

He treats this equality as a differential equation relating x and y. The required relation is the general integral of this differential equation. He finds this relation, which turns out to be algebraic of degree 2 in both x and y. Its coefficients depend on the coefficients of the polynomial $f(x)$ and on an independent parameter c.

Euler also states this result in another way, as saying that the sum of two definite integrals is equal to a third:

$$\int_0^\alpha \frac{\mathrm{d}x}{\sqrt{f(x)}} + \int_0^\beta \frac{\mathrm{d}x}{\sqrt{f(x)}} = \int_0^\gamma \frac{\mathrm{d}x}{\sqrt{f(x)}}, \tag{A.2}$$

where γ can be expressed as a rational function of α and β. Moreover, Euler gives an argument why such a relation cannot hold if $f(x)$ is a polynomial of degree >4.

For arbitrary elliptic integrals of the form $\int r(x)\mathrm{d}x/\sqrt{f(x)}$, Euler proves a relation generalising (A.2):

$$\int_0^\alpha \frac{r(x)\mathrm{d}x}{\sqrt{f(x)}} + \int_0^\beta \frac{r(x)\mathrm{d}x}{\sqrt{f(x)}} - \int_0^\gamma \frac{r(x)\mathrm{d}x}{\sqrt{f(x)}} = \int_0^\delta V(y)\mathrm{d}y, \tag{A.3}$$

where γ is the same rational function as in (A.2), and δ and V are also rational functions.

The reason for the existence of the integral of (A.1), *and for all the particular cases discovered by Bernoulli and Fagnano is the group law on the elliptic curve with equation* $s^2 = f(t)$, *and the fact that the everywhere regular differential form*

$\mathrm{d}t/s$ is invariant under translations by elements of this group. The relations holding between x and y in (A.1) *discovered by Euler can be written in the form*

$$(x, \sqrt{f(x)}) \oplus (c, \sqrt{f(c)}) = (y, \sqrt{f(y)}),$$

where $\oplus$ denotes the addition of points on the elliptic curve $s^2 = f(t)$. Thus these results contain at once the group law on an elliptic curve and the existence of an invariant differential form on it.

The relation (A.2) *is also an immediate consequence of the invariance of the form $\varphi = \mathrm{d}x/\sqrt{f(x)}$. In it*

$$(\gamma, \sqrt{f(\gamma)}) = (\alpha, \sqrt{f(\alpha)}) \oplus (\beta, \sqrt{f(\beta)}),$$

and

$$\int_0^\alpha \varphi + \int_0^\beta \varphi = \int_0^\alpha \varphi + \int_\alpha^\gamma t_g^* \varphi = \int_0^\alpha \varphi + \int_\alpha^\gamma \varphi = \int_0^\gamma \varphi,$$

where t_g is the translation by the point $g = (\alpha, \sqrt{f(\alpha)})$. Notice that here we write equalities between integrals in a formal way, without indicating the path of integration. In essence this is an equality "up to a constant of integrations", that is, an equality between the corresponding differential forms. This is also how Euler understood them.

Finally, the significance of the relation (A.3) *will become clear later, in connection with Abel's theorem* (*see Section* 3 *below*).

2 Elliptic Functions

Following Euler, the theory of elliptic integrals was developed mainly by Legendre. His researches, starting in 1786 are collected in his three-tome work *Traité des fonctions elliptiques et des intégrales Eulériennes* [134]. Legendre used elliptic function to mean what we now call elliptic integral. The modern terminology became firmly rooted after Jacobi. In the first paragraph of the *premier supplement*, published in 1828, Legendre writes as follows (Tome III, p. 1):

> Après m'être occupé pendant un grand nombre d'années de la théorie des fonctions elliptiques, dont l'immortel Euler avait posé les fondemens, j'ai cru devoir rassembler les résultats de ce long travail dans un Traité qui a été rendu public au mois de janvier 1827. Jusque là les géomètres n'avaient pris presque aucune part à ce genre de recherches; mais à peine mon traité avait-il vu le jour, à peine son titre pouvait-il être connu des savans étrangers, que j'appris avec autant d'étonnement que de satisfaction, que deux jeunes géomètres, MM. *Jacobi* (C.-G.-J.) de Koenigsberg et *Abel* de Christiania, avaient reussi, par leurs travauxs particuliers, à perfectionner considérablement la théorie des fonctions elliptiques dans ses points les plus élevés.

Abel's work on elliptic functions appeared in 1827–1829. His starting point is the elliptic integral (see Abel [101, Vol. I, Nos. XVI and XXIV])

$$\theta = \int_0^\lambda \frac{\mathrm{d}x}{\sqrt{(1-c^2x^2)(1-e^2x^2)}}, \qquad \text{with } c, e \in \mathbb{C},$$

which he views as a function $\theta(\lambda)$ of the upper limit; he introduces the inverse function $\lambda(\theta)$ and the function $\Delta(\theta) = \sqrt{(1-c^2\lambda^2)(1-e^2\lambda^2)}$. From the properties of elliptic integrals known at the time (in essence, from Euler's relations 1, (A.2)) he deduces that the functions $\lambda(\theta \pm \theta')$ and $\Delta(\theta \pm \theta')$ can be simply expressed as rational functions of $\lambda(\theta)$, $\lambda(\theta')$, $\Delta(\theta)$, $\Delta(\theta')$. Abel proves that both of these functions are periodic in the complex domain, with two periods 2ω and $2\widetilde{\omega}$:

$$\omega = \frac{1}{2}\int_0^{1/c} \frac{\mathrm{d}x}{\sqrt{(1-c^2x^2)(1-e^2x^2)}},$$

$$\widetilde{\omega} = \frac{1}{2}\int_0^{1/e} \frac{\mathrm{d}x}{\sqrt{(1-c^2x^2)(1-e^2x^2)}}.$$

He finds an infinite product expansion for the functions $\lambda(\theta)$ and $\Delta(\theta)$ as a product taken over all their zeros.

As a direct generalisation of a problem which Euler had worked on, Abel poses the following question (see Abel [101, Vol. I, No. XIX]): catalogue all cases in which the differential equation

$$\frac{\mathrm{d}y}{\sqrt{(1-c_1^2y^2)(1-e_1^2y^2)}} = \pm a \frac{\mathrm{d}x}{\sqrt{(1-c^2x^2)(1-e^2x^2)}}. \tag{A.4}$$

can be satisfied by taking y to be an algebraic function of x, either rational or irrational.

This question became know as the transformation problem for elliptic functions. Abel proved that if (A.1) can be satisfied with y an algebraic function, then it can also be done with a rational function. He proved that if $c_1 = c$ and $e_1 = e$ then a must be a rational number, or a number of the form $\mu' + \sqrt{-\mu}$ where μ, μ' are rational and $\mu > 0$. In the general case, he proved that the periods ω_1, ω_1' of the integral on the left-hand side of (A.1), after multiplying by a common factor, can be expressed as an integral linear combination of the periods ω, ω' of the integral on the right-hand side.

A little after Abel, but independently of him, Jacobi [125, Vol. I, Nos. 3, 4] also considered the inverse function to an elliptic integral, proved that it has two independent periods, and obtained a series of results in the transformation problem. Reworking as series the infinite product expansions of elliptic functions found by Abel, Jacobi arrived at the notion of theta functions (these appeared earlier in 1822 in Fourier's book on the heat equation[5]), and found a multitude of applications of them, not only in the theory of elliptic functions, but in number theory and mechanics.

Finally, after the publication of Gauss' posthumous papers, and especially his diaries, it became clear that he had already to a greater or lesser extent mastered some of these ideas long before the work of Abel and Jacobi.

The first part of Abel's result requires practically no comment. The map $x = \lambda(\theta)$, $y = \Delta(\theta)$ *defines a uniformisation of the elliptic curve* $y^2 = (1 - c^2x^2)(1 - e^2x^2)$ *by elliptic functions. Under this map* $f\colon \mathbb{C}^1 \to X$ *the regular differential form* $\varphi = \mathrm{d}x/y$ *pulls back to a regular differential form on* $\mathbb{C}^1$ *invariant under translations by vectors of the lattice* $2\omega\mathbb{Z} + 2\widetilde{\omega}\mathbb{Z}$; *such a form is a constant factor times* $\mathrm{d}\theta$, *and we can assume that* $\mathrm{d}\theta = f^*(\mathrm{d}x/y)$, *that is,* $\theta = \int \mathrm{d}x/y$.

Integrating the differential equation (A.4) *has the following geometric meaning. Let* X *and* X_1 *be the elliptic curves given by*

$$u^2 = (1 - c^2x^2)(1 - e^2x^2) \quad \text{and} \quad v^2 = (1 - c_1^2y^2)(1 - e_1^2y^2).$$

The question concerns the study of curves $C \subset X \times X_1$, *(which correspond to algebraic relations between* x *and* y*). Since an elliptic curve is its own Picard variety (see Section* 4.4, *Chapter* 3), C *defines a morphism* $f\colon X \to X_1$. *This explains why the problem reduces to the case* y *a rational function of* x. *By Theorem* 3.13 *of Section* 3.3, *Chapter* 3, *we can assume that* $f\colon X \to X_1$ *is a homomorphism of algebraic groups. Thus Abel studied the group* $\mathrm{Hom}(X, X_1)$, *and when* $X = X_1$, *the ring* $\mathrm{End}\, X$. *The homomorphism* $f \in \mathrm{Hom}(X, X_1)$ *defines a linear map of* 1*-dimensional vector spaces* $f^*\colon \Omega^1[X_1] \to \Omega^1[X]$, *which is determined by one number: this is the factor* $\pm a$ *in* (A.1). *See also Exercises* 7–9 *of Section* 2.4, *Chapter* 9.

3 Abelian Integrals

The step from elliptic curves to the study of arbitrary algebraic curves took place still within the context of analysis. Abel showed that the basic properties of elliptic integrals can be generalised to integrals of arbitrary algebraic functions. These integrals subsequently became known as Abelian integrals.

In 1826 Abel wrote a paper (see [101, Vol. I, No. XII]) which marks the birth of the general theory of algebraic curves. In it, he considers the algebraic function y defined by

$$\chi(x, y) = 0. \tag{A.5}$$

[5]The reference, given by Krazer [150, p. 5], is probably to Fourier [118, Chapter IV, §§ 238–246]. The chapter, part of Fourier's 1807 prize essay, treats the heat flow on the circle; compare Grattan-Guiness [149, p. 254]. Theta functions are certainly among the general trigonometric series treated in the chapter, but I doubt whether they occur specifically in Fourier. In any case, Grattan-Guiness seems to suggest that Euler and Daniel Bernoulli considered related series more than a hundred years earlier in connection with the equation of the plucked string.

He considers a second equation

$$\theta(x, y) = 0, \tag{A.6}$$

where $\theta(x, y)$ is a polynomial depending on x and y, and additionally depending linearly on a number α of extra parameters $a, a', \ldots$. As these parameters vary, there may be some common solutions of (A.5) and (A.6) that do not change. Let $(x_1, y_1), \ldots, (x_\mu, y_\mu)$ be the variable solutions, and $f(x, y)$ an arbitrary rational function. Abel proves that

$$\int_0^{x_1} f(x, y)\mathrm{d}x + \cdots + \int_0^{x_\mu} f(x, y)\mathrm{d}x = \int V(g)\mathrm{d}g, \tag{A.7}$$

where $V(t)$ and $g(x, y)$ are rational functions that also depend on the parameters $a, a', \ldots$. Abel interpreted this result as saying that the left-hand side of (A.7) is an elementary function.

Using the arbitrary choice of the parameters $a, a', \ldots,$ Abel shows that the sum of any number of integrals $\int_0^{x_i} f(x, y)\mathrm{d}x$ can be expressed in terms of $\mu - \alpha$ of them, and a summand of the same type as the right-hand side of (A.7). He establishes that the number $\mu - \alpha$ depends only on (A.5). For example, for the equation $y^2 + p(x)$ where p is a polynomial of degree $2m$, we have $\mu - \alpha = m - 1$.

Next, Abel studies the functions f for which the right-hand side of (A.7) does not depend on the parameters $a, a', \ldots$. Writing f in the form

$$\frac{f_1(x, y)}{f_2(x, y)\chi'_y}, \quad \text{where } \chi'_y = \frac{\partial \chi}{\partial y},$$

he proves that $f_2 = 1$, and that f_1 satisfies a series of restrictions which imply that there are at most a finite number γ of linearly independent functions f satisfying the current assumption. Abel proved that $\gamma \geq \mu - \alpha$, and that $\gamma = \mu - \alpha$ if, for example, the curve $\chi(x, y)$ has no singular points (in subsequent terminology).

Considering the solutions $(x_1, y_1), \ldots, (x_\mu, y_\mu)$ *of the system of* (A.5) *and* (A.6) *brings us at once to the modern notion of linear equivalence of divisors. Namely, let* X *be the curve with* (A.5) *and* D_λ *the divisor cut out on* X *by the form* θ_λ *(in homogeneous coordinates), where* $\lambda = (a, a', \ldots)$ *is the system of parameters. By assumption* $D_\lambda = \overline{D}_\lambda + D_0$ *where* D_0 *does not depend on* λ*. Hence all the divisors* $\overline{D}_\lambda = (x_1, y_1) + \cdots + (x_\mu, y_\mu)$ *are linearly equivalent. The problem Abel considered reduces to the study of sums* $\int_{\alpha_1}^{\beta_1} \varphi + \cdots + \int_{\alpha_\mu}^{\beta_\mu} \varphi$*, where* φ *is a differential form on* X*, and* α_i *and* β_i *are points of* X *such that* $\alpha_1 + \cdots + \alpha_\mu \sim \beta_1 + \cdots + \beta_\mu$*. We give a sketch proof of Abel's theorem, which is close in spirit to the original proof. We can assume that* $\alpha_1 + \cdots + \alpha_\mu - \beta_1 - \cdots - \beta_\mu = \operatorname{div} g$*, that is* $g \in \mathbb{C}(X)$*, with*

$$\alpha_1 + \cdots + \alpha_\mu = (\operatorname{div} g)_0 \quad \text{and} \quad \beta_1 + \cdots + \beta_\mu = (\operatorname{div} g)_\infty.$$

Consider the morphism $g\colon X \to \mathbb{P}^1$ *and the corresponding field extension* $\mathbb{C}(g) \subset \mathbb{C}(X)$*. We assume for simplicity that this extension is Galois (the general case reduces easily to this), with group* G*. The automorphisms* $\sigma \in G$ *act on* $\mathbb{C}(X)$ *and* *on X, and take the points $\alpha_1, \dots, \alpha_\mu$ to one another, since $\{\alpha_1, \dots, \alpha_\mu\} = g^{-1}(0)$. Hence $\{\alpha_1, \dots, \alpha_\mu\} = \{\sigma(\alpha) \mid \sigma \in G\}$, where α is one of the points α_i. Similarly, $\{\beta_1, \dots, \beta_\mu\} = \{\sigma(\beta) \mid \sigma \in G\}$. Writing φ as $u\mathrm{d}g$ we see that*

$$\sum_{i=1}^{\mu} \int_{\alpha_i}^{\beta_i} \varphi = \sum_{\sigma \in G} \int_{\sigma(\alpha)}^{\sigma(\beta)} u\mathrm{d}g = \int_{\alpha}^{\beta} \Big(\sum_{\sigma \in G} \sigma(u)\mathrm{d}g \Big). \tag{A.8}$$

The function $v = \sum_{\sigma \in G} \sigma(u)$ is contained in $\mathbb{C}(g)$, and Abel's theorem follows from this.

We conclude that any sum of integrals $\sum_i \int_0^{x_i} f(x, y)\mathrm{d}x$ can be expressed as $\sum_{j=1}^{k} \int_0^{x'_j} f(x, y)\mathrm{d}x + \int V(g)\mathrm{d}g$, in terms of a sum of k integrals, provided that we have a linear equivalence of the type:

$$\sum_i (\alpha_i - O) \sim \sum_{j=1}^{k} (\alpha'_j - O), \tag{A.9}$$

where $\alpha_i = (x_i, y_i)$, $\alpha'_j = (x'_j, y'_j)$, and $O \in X$ is a point with $x = 0$. From the Riemann–Roch theorem, it follows at once that if $k = g$ then a linear equivalence (A.9) *always holds (for any points α_i, and certain corresponding α'_j) (see Exercise* 14 *of Section* 8.1, *Chapter* 3). *Thus the constant $\mu - \alpha$ introduced by Abel is the genus of the curve X.*

If the form $\varphi \in \Omega^1[X]$ then also $v\mathrm{d}g \in \Omega^1[\mathbb{P}^1]$, where $v = \sum_{\sigma \in G} \sigma(u)$ in (A.8). *Since $\Omega^1[\mathbb{P}^1] = 0$, the term on the right-hand side of* (A.7) *vanishes in this case. It follows that $\gamma \geq g$. In natural cases the two numbers are equal.*

We see that this work of Abel contains the notions of genus of an algebraic curve and linear equivalence of divisors, and gives a criterion for linear equivalence in terms of integrals. In this final respect it leads to the theory of the Jacobian variety of an algebraic curve (see Section 5 *below).*

4 Riemann Surfaces

In his 1851 dissertation [142, No. I], Riemann applied an entirely new principle of studying functions of a complex variable. He proposed that such a function is defined not on the plane of a complex variable, but on a certain surface that covers this plane in a many sheeted way. The real and imaginary parts of such a function satisfy the Laplace equation. A function is uniquely determined by this property if we know the points at which it is infinite, together with the nature of its singularities at these points, and the curve one needs to cut along to make it single valued, together with the nature of its many valuedness on crossing over these curves. Riemann also develops a method of constructing a function from data of this kind, based on a variational principle which he called the "Dirichlet principle".

In the first part of his paper on the theory of Abelian functions which appeared in 1857 [142, No. VI], he applied these ideas to the theory of algebraic functions and their integrals. The paper starts with a study of the properties of the corresponding surfaces, which relate, in Riemann's words, to *analysis situs* (that is, in modern terminology, topology). After making an even number $2p$ of cuts, the surface becomes a simply connected domain. By means of ideas taken from *analysis situs* he proves that $p = 1 - n + w/2$, where n is the number of sheets, and w the number of ramification points (counted with appropriate multiplicities) of the surface as a cover over the plane of one complex variable.

One considers functions, in general many valued on the surface, but that become single valued in the domain obtained after the making the cuts, and on crossing over the cuts their values change by constants called moduli of periodicity. The Dirichlet principle provides a method of constructing functions of this kind. In particular, there exist p linearly independent functions that are everywhere finite, the "integrals of the first kind". In a similar way one constructs functions that are infinite at given points. To pick out those that are single valued on the surface, one has to set to zero their moduli of periodicity. It follows from this that there are at least $m - p + 1$ linearly independent single valued functions that are infinite only at m given points.

Riemann proves that all the functions that are single valued on a given surface are rational functions of two of them, s and z, which are connected by a relation $F(s, z) = 0$. He says that two such relations belong to one "class" if they can be rationally transformed into one another. In this case the corresponding surfaces have the same value of the number p. But the converse is not true. By studying the possible position of the ramification of points of a surface, Riemann proves that for $p > 1$, the set of classes depends on $3p - 3$ parameters that he calls "moduli".

The surfaces introduced by Riemann correspond closely to the modern notion of 1*-dimensional complex manifold; these are the sets on which analytic functions are defined. Riemann poses and solves the problem of the relation between this notion and that of algebraic curve* (*the corresponding result is nowadays called the Riemann existence theorem*).

This circle of ideas of Riemann did not by any means become immediately clear. Klein's lectures [129] *played an important role in explaining them. Klein stresses that a Riemann surface is not a priori related to an algebraic curve or algebraic function. A definition of a Riemann surface differing only in terminology from the definition of* 1*-dimensional complex manifold in current use* (*for example, in this book*) *was given by H. Weyl* [148].

Riemann's work initiated the study of the topology of algebraic curves. It made clear the topological significance of the number $p = \dim \Omega^1[X]$: *it is equal to one half of the dimension of the* 1*-dimensional homology of the topological space* $X(\mathbb{C})$. *Riemann used analysis to prove the inequality* $l(D) \geq \deg D - p + 1$. *The Riemann–Roch equality was proved by Roch, his student. Finally, in this work, the function field* $k(X)$ *first appears as an object of primary importance associated with a curve* X, *together with the notion of birational equivalence.*

5 The Inversion of Abelian Integrals

Already Abel posed the problem of inverting the integrals of arbitrary algebraic functions. He discovered in particular that the inverse of a hyperelliptic integral associated with $\sqrt{\psi(x)}$ is a function with periods equal to one half of the value of this integral taken between two roots of ψ (see [101, Vol. II, No. VII]).

Jacobi observed that, except when the integral is actually elliptic, the inverse should be a function of one complex variable which has more that 2 periods, which is impossible for a reasonable function. In the case of a polynomial $\psi(x)$ of degree 5 or 6, Jacobi proposed to consider the pair of functions

$$u = \int_0^x \frac{dt}{\sqrt{\psi(t)}} + \int_0^y \frac{dt}{\sqrt{\psi(t)}}, \qquad v = \int_0^x \frac{tdt}{\sqrt{\psi(t)}} + \int_0^y \frac{tdt}{\sqrt{\psi(t)}}.$$

He proposes to express $x + y$ and xy as analytic functions of the two variables u and v, and conjectures that such an expression is possible in terms of a generalisation of theta functions (see Jacobi [125, Vol. II, Nos. 2 and 4]). This conjecture was confirmed in a paper of Göpel [120], published in 1847.

The relation between theta functions and the inversion problem in the general case is the subject of the second part of Riemann's paper [142, No. VI] on Abelian functions. He considers a series in p variables

$$\theta(v) = \sum_m e^{F(m)+2(m,v)}, \tag{A.10}$$

where $v = (v_1, \ldots, v_p) \in \mathbb{C}^p$; here the sum takes place over all integer valued p-vectors $m = (m_1, \ldots, m_p) \in \mathbb{Z}^p$, with $(m, v) = \sum m_i v_i$ and $F(m) = \sum_{j,k} \alpha_{jk} m_j m_k$ for some $\alpha_{jk} = \alpha_{kj}$. This series converges for all values of v if the real part of the quadratic from F is negative definite. The basic property of the function θ is the functional equation

$$\theta(v + \pi i r) = \theta(v) \quad \text{and} \quad \theta(v + \alpha_j) = e^{L_j(v)}\theta(v), \tag{A.11}$$

where r is any integer valued p-vector, and α_j the jth column of the matrix (α_{jk}); here the $L_j(v)$ are linear functions.

Riemann proves that the cuts $a_1, \ldots, a_p, b_1, \ldots, b_p$ needed to make the surface he introduced simply connected, and the basis $u_1, \ldots, u_p$ of the differentials that are finite everywhere on the surface, can be chosen in such a way that the integrals of u_j along a_k are $= 0$ for $j \neq k$ and $= \pi i$ if $j = k$, and the integrals of v_j along b_k form a symmetric matrix (α_{jk}) satisfying the conditions required to make the right-hand side of (A.10) converge. He considers the function θ corresponding to these coefficients α_{jk} and the function $\theta(u - e)$, where $u = (u_1, \ldots, u_p)$ (the u_i are the differentials that are finite everywhere on the surface) and $e \in \mathbb{C}^p$ is an arbitrary vector.

Riemann proves that the function $\theta(u - e)$ either has p zeros $\eta_1, \ldots, \eta_p$ on the surface, or is identically 0. For suitable choice of the lower limit of integration in the integrals u_i, in the first case

$$e \equiv u(\eta_1) + \cdots + u(\eta_p) \quad \text{modulo periods,} \tag{A.12}$$

where the congruence is considered modulo integral linear combinations of the periods of the integrals u_i. The points $\eta_1, \ldots, \eta_p$ are uniquely determined by this. In the second case there exist points $\eta_1, \ldots, \eta_{p-2}$ such that

$$e \equiv -\big(u(\eta_1) + \cdots + u(\eta_{p-2})\big). \tag{A.13}$$

Already Riemann knew that the periods of an arbitrary $2n$-periodic function of n variables satisfy relations analogous to those required for the convergence of the series (A.10) defining theta functions. These relations between periods were written out explicitly by Frobenius [119], who proved that they are necessary and sufficient conditions for the existence of nontrivial functions satisfying the functional equation (A.11). It follows that these relations are necessary and sufficient conditions for the existence of a meromorphic function with $2n$ given periods that cannot be reduced by a linear change of coordinates to a function of fewer variables. One need only apply the theorem that any meromorphic function with $2n$ periods can be represented as a quotient of entire functions satisfying the functional equation of a theta function. This theorem, stated by Weierstrass, was proved by Poincaré [141]. In 1921 Lefschetz [131] proved that if the Frobenius relations are satisfied then theta functions define an embedding of the manifold $\mathbb{C}^n/\Omega$ into projective space, where Ω is the lattice corresponding to the given period matrix.

The problem of inverting Abelian integrals relates to questions that we only touched on in passing in the book, often without proof. The subject under discussion is the construction of the Jacobian variety of an algebraic curve and the properties of arbitrary Abelian varieties (Sections 4.3–4.4, *Chapter* 3 *and Section* 1.3, *Chapter* 8).

If $O \in X$ is a fixed point then $\{f(x) = x - O \mid x \in X\}$ is obviously an algebraic family of divisors of degree 0 *on X, parametrised by X itself. By definition of the Jacobian $J(X)$ of X (recall from Section* 4.4, *Chapter* 3 *that for a curve X, the Picard variety is called the Jacobian), there exists a morphism $\varphi\colon X \to J(X)$ which is an embedding if X has genus $p \neq 0$. It can be proved that $\varphi^*\colon \Omega^1[J(X)] \to \Omega^1[X]$ is an isomorphism. Hence in the representation*

$$J(X) = \mathbb{C}^p/\Omega, \tag{A.14}$$

$\Omega \subset \mathbb{C}^p$ is the lattice of rank $2p$ consisting of the periods of the p linearly independent differential forms $\omega \in \Omega^1[X]$. This analytic representation of the Jacobian is Riemann's starting point, and he goes on to develop the algebraic method of studying it.

If D_0 is an arbitrary effective divisor of degree p, then $\{g(x_1, \ldots, x_p) = x_1 + \cdots + x_p - D_0 \mid x_1, \ldots x_p \in X\}$ is a family of divisors of degree 0 *on X whose parameter space we can take to be symmetric product $X^p/\mathfrak{S}_p$, that is, the quotient space of the product of p copies of X by the symmetric group acting by permuting the factors. By definition of the Jacobian, there exists a morphism $\psi\colon X^p/\mathfrak{S}_p \to J(X)$. It follows easily from the Riemann–Roch theorem that this is onto, and is a one-to-one* *correspondence on an open set of $J(X)$. The map ψ is not one-to-one, by definition, at points $(x_1, \dots, x_p)$ such that $l(x_1 + \cdots + x_p) > 1$. It follows from the Riemann–Roch theorem that this is equivalent to $l(K - x_1 - \cdots - x_p) > 0$, that is,* (since $\deg K = 2g - 2$) *to the condition that $x_1 + \cdots + x_p \sim K - y_1 - \cdots - y_{p-2}$ for certain points $y_1, \dots, y_{p-2} \in X$. This final relation coincides with* (A.13), *up to adding the summand K, and thus up to a translation by a point of $J(X)$.*

The Frobenius relations are the condition for a complex torus $\mathbb{C}^p/\Omega$ to be projective. They are written out in (8.14) *and* (8.15).

6 The Geometry of Algebraic Curves

So far, we have seen how the notions and results that now form the basis of the theory of algebraic curves were created under the influence and in the context of the analytic theory of algebraic functions and their integrals. Independently of this direction a purely algebraic theory of curves was developing. For example, in a book appearing in 1839, Plücker [139] found the formulas relating the degree and class of a plane curve and its number of double points (see Exercise 2 of Section 4.5, Chapter 4). He proved there that a plane curve of degree 3 has 9 inflexion points (compare Example 3.4 of Section 3.4, Chapter 3). But studies of this kind played a secondary role in the math of that period, and did not relate to the deeper ideas.

It was only in the period following Riemann that the geometry of algebraic curves occupied a central place in the math of the time, on a level with the theory of Abelian integrals and Abelian functions. This change of viewpoint is connected especially with the name of Clebsch. While for Riemann the basic object was a function, for Clebsch it is an algebraic curve. One could say that Riemann considered a finite morphism $f: X \to \mathbb{P}^1$, whereas Clebsch considered the algebraic curve X itself. The book Clebsch and Gordan [112] derives a formula for the number p of linearly independent integrals of the first kind (that is, the genus of X), expressing it in terms of the degree of the curve and the number of singular points (see Exercise 2 of Section 4.5, Chapter 4). They also prove that if $p = 0$ then the curve has a rational parametrisation, and if $p = 1$ then it can be transformed into a plane curve of degree 3.

A mistake of Riemann turned out to be exceptionally profitable for the development of the algebraic geometric aspect of the theory of algebraic curves. In the proof of his existence theorems, he considered it obvious that a certain variational problem could be solved, the "Dirichlet principle". Soon after this, Weierstrass showed that not every variational problem has a solution. Hence for a certain period, Riemann's results remained without rigorous foundation. One of the gains from this was the appearance of algebraic proofs of these theorems; their statement was in essence algebraic. These investigations undertaken by Clebsch (see Clebsch and Gordan [112]) facilitated to a considerable extent the recognition of the essentially algebraic geometric nature of the results of Abel and Riemann, from under its mantle of analysis.

The direction initiated by Clebsch reached its zenith in the work of his student Max Noether. The circle of ideas of Noether is especially clearly delineated in his joint work with Brill [105]. The task this sets itself is to develop geometry on an algebraic curve contained in the projective plane as a body of results that are invariant under mutually single valued transformations (that is, birational transformations). The basic notion is that of a group of points of a curve (distinct or with coincidences). One considers systems of groups of points cut out on the original curve by linear systems of curves (that is, having equations that form a linear space). It can happen that all the groups of some system have some group G in common, that is, they consist of G plus groups of another system G'. The system G' obtained in this way is called a linear system. If the dimension of the linear (projective) space of equations of the curves cutting out a linear system is equal to q, and the group G' consists of Q points, then the system is denoted by $g_Q^{(q)}$. Two groups of one system are said to be coresidual. This obviously corresponds to the modern notion of linear equivalence of effective divisors, and if a group G is contained in a linear system $g_Q^{(q)}$ then in modern notation $\deg G = Q$ and $l(G) \geq q + 1$ (recall that $l(G)$ is the dimension of a vector space, and q that of the corresponding projective space).

Every group of points G defines a biggest possible linear system $g_Q^{(q)}$ containing all groups coresidual to G. The numbers q and Q are related by the Riemann–Roch theorem, which is proved purely algebraically.

Of course, the statement of the Riemann–Roch theorem assumes a definition of an analogue of the canonical class. This can be given without appealing to the notion of differential form, but the relation with this notion is very easy to establish. Namely, if a curve of degree n has equation $F = 0$ and is smooth, then a differential form $\omega = \Omega^1[X]$ can be written as

$$\omega = \frac{\varphi \mathrm{d}x}{F'_y}, \tag{A.15}$$

where φ is a homogeneous polynomial of degree $n - 3$ (Section 6.4, Chapter 3). It can be proved that if a curve has only the simplest possible singularities then the expression (A.15) remains valid if we require that φ should be 0 at all the singular points. Such polynomials are said to be adjoint; adjoint polynomials of degree $n - 3$ define the linear system that is the analogue of the canonical class. Brill and Noether consider the map of a curve to $(p - 1)$-dimensional projective space defined by adjoint forms of degree $n - 3$. Its image is called a normal curve (in the case of nonhyperelliptic curves). It is proved that a single valued (that is, birational) correspondence between curves leads to a projective transformation of the normal curves.

Noether [135] applies these ideas to the investigation of space curves. In modern language one can say that this paper studies irreducible components of the Chow variety of curves in $\mathbb{P}^3$.

7 Higher Dimensional Geometry

By the middle of the 19th century, a large number of special properties of algebraic varieties of dimension >1, mainly surfaces, had been discovered. For example, surfaces of degree 3 were studied in detail, and in particular Salmon and Cayley proved in 1849 that every cubic surface with no singular points contains 27 distinct lines. However, for a long time these ideas were not unified by any general principles, and were not connected to the deep ideas worked out up to this time in the theory of algebraic curves.

The decisive step in this direction seems to have been made by Clebsch. In 1868 he published a short note [111] in which he considers (in modern terminology) algebraic surfaces from the point of view of birational equivalence. He considers double integrals on a surface that are everywhere finite, and observes that the maximal number of linearly independent integrals is invariant under birational equivalence.

These ideas were developed in the two-part work of Noether [136]. As is clear already from the title, he considers algebraic varieties of any number of dimensions. However, most of his results relate to surfaces. This is typical for all of the subsequent period of algebraic geometry. Although very many results are actually true for varieties of arbitrary dimension, they are stated and proved only for surfaces.

In the first part, Noether considers "differential expressions" on varieties of any dimension, and it is interesting that the integral sign only appears once. Thus the algebraic nature of the notion of differential form already becomes formally obvious here. Noether only considers differential forms of the highest order. He proves that they form a finite dimensional space, whose dimension is invariant under single valued (that is, birational) transformations.

In the second part he considers curves and surfaces; the final section only contains some interesting remarks on 3-dimensional varieties. Noether describes the canonical class (in modern terminology) using adjoint surfaces, by analogy with the way this was done before for curves. He formulates the question of the surfaces cutting out (again in modern terminology) the canonical class of a curve C lying on a surface V; he calls the curves on V that these cut out the adjoint curves of C, and he gives an explicit description of them, which leads him to the formula for the genus on a curve on a surface. This formula is essentially identical with that of (4.28) of Section 2.3, Chapter 4; however, the insight that the adjoint curve is of the form $K+C$ was only achieved 20 years later in papers of Enriques.

In the same paper Noether studies the notion of exceptional curve contracted to a point under a birational map.

The ideas of Clebsch and Noether found their most brilliant development not in Germany but in Italy. The Italian school of algebraic geometry had an enormous influence on the development of the subject. Without doubt many of the ideas of this school have not to this day been fully understood and developed. The founders of the Italian school of geometry were Cremona, C. Segre and Bertini. Its most significant representatives were Castelnuovo, Enriques and Severi. The papers of Castelnuovo began appearing in the late 1880s. Enriques was a student (and a relative) of Castelnuovo; his papers appear from the early 1890s. Severi started work about 10 years after Enriques and Castelnuovo.

One of the fundamental achievements of the Italian school is the classification of algebraic surfaces. The first result here can be considered to be the paper of Bertini [104], in which he gives a classification of involutive transformations of the plane. In modern terminology, the subject is the classification of elements of order 2 of the group of birational self-maps of the plane, up to conjugacy in this group. The classification turns out to be very simple, and in particular it follows easily from it that a quotient of the plane by a group of order 2 is a rational surface. In other words, if X is a unirational surface and the map $f\colon \mathbb{P}^2 \to X$ has degree 2 then X is rational.

The general case of the Lüroth problem for algebraic surfaces was solved (positively) by Castelnuovo [107]. After this he posed the question of characterising rational surfaces by numerical invariants, and solved this in [108]. The classification of surfaces, discussed briefly in Section 6.7, Chapter 3, was obtained by Enriques in a series of papers, which were finished already in the 1910s (see Enriques [115]).

In connection with the Lüroth problem for 3-folds, Fano studied certain types of 3-folds, and proposed proofs of their irrationality. Enriques had proved that many of these are unirational. This would have given counterexamples to the Lüroth problem, but many unclear points were found in Fano's proofs. Certain intermediate propositions turned out to be false. The problem was definitively settled only while the final pages of the first edition of this book were being written. V.A. Iskovskikh and Yu.I. Manin showed that the basic idea of Fano can be salvaged. They proved the irrationality of smooth hypersurfaces of degree 4 in $\mathbb{P}^4$; B. Segre had previously proved that some of these are unirational. At the same time Clemens and Griffiths found a new analytic method of proof of the irrationality of certain varieties. They proved the irrationality of smooth hypersurfaces of degree 3 in $\mathbb{P}^4$ (see Exercise 13 of Section 8.1, Chapter 3). These results are of course only the first steps on the path to a classification of unirational varieties.

The basic tool of the Italian school was the study of families of curves on surfaces, both linear families and algebraic families (which they called continuous families). This led to the notion of linear and algebraic equivalence. The relation between these two notions was first studied by Castelnuovo [109]. He discovered the connection of this question with the important invariant of a surface called its irregularity. We do not treat here the definition of irregularity used by Castelnuovo, which is closely related to the ideas of sheaf cohomology. Another interpretation of this notion is given by (A.16) below.

Castelnuovo [109] proved that if not every continuous system of curves is contained in a linear system (that is, if algebraic equivalence is not equal to linear equivalence) then the irregularity of the surface is nonzero. Enriques [114] proves the converse. He shows moreover that every sufficiently general curve (in a precisely defined sense) on a surface of irregularity q is contained in a continuous family that is algebraically complete (that is, maximal) and fibred in linear families all having the same dimension, where the base of the fibration is a variety of dimension q. Castelnuovo proved [110] that the q-dimensional base of the fibration constructed by Enriques has a group law, defined by addition of linear systems (that is, divisor classes), under which this base is an Abelian variety, and is hence uniformised by

Abelian functions (with $2q$ periods). This Abelian variety is determined by the surface, and does not depend on the curve from which we started. It is called the Picard variety of the surface.

The irregularity turned out to be related to the theory of differential 1-forms on surfaces, the foundations of which were laid by Picard [137]; in this paper he proves that the space of everywhere regular 1-forms is finite dimensional. In 1905 Severi and Castelnuovo proved that this dimension is equal to the irregularity; in our notation

$$q = h^1 = \dim \Omega^1[X]. \tag{A.16}$$

Severi [145] studied the group of algebraic equivalence classes, and proved that it is finitely generated. His proof is based on the relation between the notion of algebraic equivalence and the theory of differential 1-forms. Namely, algebraic equivalence $n_1 C_1 + \cdots + n_r C_r \approx 0$ holds if and only if there exists a differential 1-form whose set of "logarithmic singularities" equals the curves $C_1, \ldots, C_r$ taken with multiplicities $n_1, \ldots, n_r$. (A 1-form ω has logarithmic singularity of multiplicity n along a curve C if locally $\omega = n df/f$, where f is a local equation of C.) Picard had already proved that the equivalence relation defined in this way in terms of differential 1-forms defines a group of classes with a finite number of generators (see Picard and Simart [138]).

8 The Analytic Theory of Complex Manifolds

Although a substantial proportion of the ideas of algebraic geometry arose in analytic form, their algebraic significance eventually became clear. We now proceed to notions and results which are (at least from a modern point of view) related to analysis in an essential way.

At the beginning of the 1880s there appeared the papers of Klein and Poincaré on the problem of uniformising algebraic curves by automorphic functions. The aim, by analogy with the way that curves of genus 1 can be uniformised by elliptic functions, is to uniformise any curve by the functions we now call automorphic functions (the term was proposed by Klein after various different terms were used). Klein's starting point was the theory of modular functions (see Klein [128, No. 84]). The field of modular functions is isomorphic to the field of rational functions, but one can consider functions invariant under various subgroups of the modular group, and thus get more complicated fields. In particular, Klein considered automorphic functions with respect to the group consisting of all transformations $z \mapsto (az + b)/(cz + d)$, where $a, b, c, d \in \mathbb{Z}$, $ad - bc = 1$ and He proved that these functions uniformise the curve of genus 3 given by $x^3 y + y^3 z + z^3 x = 0$. One can deform the fundamental polygon of this group and obtain new groups that uniformise new curves of genus 3.

A similar train of thought underlies the papers Klein [128, Nos. 101–103] and Poincaré [140, Nos. 92, 108, 169]; here Poincaré used what are now called Poincaré series in the construction of automorphic functions. They both guessed correctly that any algebraic curve admits a uniformisation by a corresponding group and made significant progress in the direction of proving this result. However, a complete proof was not obtained at that time. The proof was obtained by Poincaré, and independently by Koebe, only in 1907. An important part was played in this by Poincaré's study of the fundamental group and universal cover.

The topology of algebraic curves is very simple and was completely studied by Riemann. To study the topology of algebraic surfaces Picard developed a method based on studying the fibres of a morphism $f\colon X \to \mathbb{P}^1$. The question here is how the topology of the fibres $f^{-1}(a)$ changes as the point $a \in \mathbb{P}^1$ moves, and in particular when the fibre becomes singular. By this method he proved, for example, that smooth surfaces in $\mathbb{P}^3$ are simply connected (see Picard and Simart [138, Vol. I]). Using this method Lefschetz [132, 133] obtained many deep results in the topology of algebraic surfaces, and of higher dimensional varieties.

The study of global properties of complex manifolds began relatively late (see Hopf [124] and Weil [147]). This subject began to develop very actively in the 1950s in connection with the creation and application by Cartan and Serre of the theory of coherent analytic sheaves (see Cartan [106] and Serre [143]). We omit the definition of this notion, which is an exact analogue of that of coherent algebraic sheaf (but we should stress that the analytic definition was introduced earlier than the algebraic). One of the basic results of this theory was the proof that the cohomology groups of a coherent analytic sheaf (in particular its group of sections) on a compact manifold are finite dimensional. In this connection Cartan gave the definition of a complex manifold based on the idea of sheaf, and proposed the idea that the definition of various types of manifolds and varieties is related to specifying sheaves of rings on them.

9 Algebraic Varieties over Arbitrary Fields and Schemes

Formally speaking the study of varieties over an arbitrary field started only in the 20th century, but the foundations for this were laid earlier. An important part was played here by two papers published in 1882 in the same volume of Crelle's journal. Kronecker [130] studies questions that would nowadays be part of the theory of rings of finite type with no zerodivisors and of characteristic 0. In particular, he constructs a theory of divisors for integrally closed rings.

Dedekind and Weber study the theory of algebraic curves in [113]. Their aim is to give a purely algebraic treatment of a substantial part of this theory. The authors stress that they nowhere use the notion of continuity, and their results remain valid if the complex number field is replaced by the field of all algebraic numbers.

The essential significance of the article of Dedekind and Weber is that in it the main object of study is the function field of an algebraic curve. Concrete affine models are only used as a technical means; the authors moreover use the term "invariant" to indicate notions and results that do not depend on the choice of the model. The whole treatment in this paper parallels to a significant extent the theory of algebraic number fields. They stress in particular the analogy between the prime ideals of an algebraic number field and points of the Riemann surface of an algebraic function field. We can say that in either case we are dealing with the maximal spectrum of a 1-dimensional scheme.

Interest in algebraic geometry over "nonclassical" fields arose first in connection with the theory of congruences, that can be interpreted as equations over a finite field. In his paper to the 1908 international congress, Poincaré says that the methods of the theory of algebraic curves can be applied to the study of congruences in two variables.

The groundwork for the systematic construction of algebraic geometry was laid by the general development of the theory of rings and fields in the 1910s and 1920s.

In 1924 E. Artin published a paper [102, No. 1], in which he studied quadratic extensions of the rational function field in one variable over a finite ground field k, basing himself on the analogy with quadratic extensions of the rational number field. Particularly essential for the subsequent development of algebraic geometry was his introduction of the zeta function of such a field and his formulation of the analogue of the Riemann hypothesis for zeta functions. Let us introduce (as Artin did not do) the hyperelliptic curve X defined over the finite field k, so that our field is of the form $k(X)$. Then the Riemann hypothesis gives the best possible estimate for the number N of points $x \in X$ defined over a given finite extension field $k \subset K$ (that is, such that $k(x) \subset K$), much as the Riemann hypothesis for the rational number field gives the best possible bound for the asymptotic distribution of prime numbers. More precisely, the Riemann hypothesis is equivalent to the inequality $|N - (q + 1)| \leq 2g\sqrt{q}$, where q is the number of elements of K and g is the genus of X.

It immediately became clear that the Riemann hypothesis could be stated for any algebraic curve over a finite field, and the attempt to prove it led Hasse and his students in the 1930s to construct the theory of algebraic curves over an arbitrary field. The Riemann hypothesis itself was proved for elliptic curves by Hasse himself [122] (compare Example 3.5 of Section 3.4, Chapter 3).

Properly speaking, this theory discussed not the curves themselves, but the corresponding function fields, and geometric terminology is nowhere used. One can get to know this style from Hasse's book [123] (see the sections on function fields). That such a birationally invariant theory of algebraic curves is possible is related to the uniqueness of the nonsingular projective model of an algebraic curve. Therefore there are substantial difficulties in applying this approach to the higher dimensional case.

On the other hand, in a series of article published in Mathematische Annalen under the general title "Zur algebraischen Geometrie" from the late 1920s to the late 1930s, van der Waerden made progress in constructing algebraic geometry over an arbitrary field. In particular he constructed an intersection theory (or as we would say nowadays, he defined a ring of cycle classes) on a nonsingular projective variety.

In 1940 Weil succeeded in proving the Riemann hypothesis for an arbitrary algebraic curve over a finite field. He found two different methods of proof. One is based on the theory of correspondences on a curve X, that is, divisors on the surface $X \times X$

(compare Exercise 10 of Section 2.7, Chapter 4), and the second on considering its Jacobian variety. Thus higher dimensional varieties are invoked in either case. In this connection, Weil's book [146] contains a construction of algebraic geometry over an arbitrary field: the theory of divisors, cycles, intersection theory. Here "abstract" varieties (not necessarily quasiprojective) are defined for the first time by the process of glueing affine pieces (as in Section 3.2, Chapter 5).

The definition of a variety based on the notion of sheaf is contained in Serre's paper [144], where he also constructs the theory of algebraic coherent sheaves; the model for this was the recently created theory of analytic coherent sheaves (compare 8 above).

Generalisations of the notion of algebraic variety along similar lines to the definition of scheme introduced subsequently were proposed in the early 1950s. It seems that the first and at the time very systematic development of these ideas is due to Kähler [126, 127]. The idea of a scheme is due to Grothendieck, along with most of the results in general scheme theory. The first systematic exposition of these ideas is contained in Grothendieck's paper to the 1958 international congress [121].

References

1. Abhyankar, S.S.: Local Analytic Geometry. Academic Press, New York (1964); MR **31**–173
2. Abraham, R., Robbin, J.: Transversal Mappings and Flows. Benjamin, New York (1967)
3. Ahlfors, L.: The complex analytic structure of the space of closed Riemann surfaces. In: Analytic Functions, pp. 45–66. Princeton University Press, Princeton (1960)
4. Aleksandrov, P.S., Efimov, V.A.: Combinatorial Topology, Vol. 1. Graylock, Rochester (1956)
5. Altman, A.B., Kleiman, S.L.: Compactifying the Picard scheme, I. Adv. Math. **35**, 50–112 (1980). MR **81**f:14025a
6. Altman, A.B., Kleiman, S.L.: Compactifying the Picard scheme, II. Amer. J. Math. **101**, 10–41 (1979); MR **81**f:14025b
7. Artin, M., Mumford, D.: Some elementary examples of unirational varieties which are not rational. Proc. Lond. Math. Soc. **25**, 75–95 (1972); MR **48** #299
8. Atiyah, M.F., Macdonald, I.G.: Introduction to Commutative Algebra. Addison-Wesley, Reading (1969); MR **39**–4129
9. Barth, W., Peters, C., Van de Ven, A.D.M.: Compact Complex Surfaces. Springer, Berlin (1984)
10. Bers, L.: Spaces of Riemann surfaces. In: Proc. Int. Congr. Math., pp. 349–361. Edinburgh (1958)
11. Birkar, C., Cascini, P., Hacon, C., McKernan, J.: Existence of minimal models for varieties of log general type. J. Am. Math. Soc. **23**, 405–468 (2010)
12. Bôcher, M.: Introduction to Higher Algebra. Dover, New York (1964)
13. Bogomolov, F.A.: Brauer groups of quotient varieties. Izv. Akad. Nauk SSSR, Ser. Mat. **51**, 485–516 (1987). English translation: Math. USSR, Izv. **30**, 455–485 (1988)
14. Bombieri, E., Husemoller, D.: Classification and embeddings of surfaces. In: Proc. Symp. in Pure Math., vol. 29, pp. 329–420. AMS, Providence (1975); MR 58 #22085
15. Borevich, Z.I., Shafarevich, I.R.: Number Theory, 2 edn. Nauka, Moscow (1985). English translation: Academic Press, New York (1966)
16. Bourbaki, N.: Élements de Mathématiques, Topologie générale. Hermann, Paris. English translation: General Topology, I–II, Addison-Wesley, Reading (1966); reprint, Springer, Berlin (1989)
17. Bourbaki, N.: Élements de Mathématiques, Algèbre commutative. Masson, Paris (1983–1985). English translation: Addison-Wesley, Reading (1972)
18. Bourbaki, N.: Élements de Mathématiques, Algèbre. Hermann, Paris (1962). Chap. 2 (Algèbre linéaire)
19. Bourbaki, N.: Élements de Mathématiques, Groupes et algèbre de Lie. Hermann, Paris (1960–1975). (Chapter I: 1960, Chapters IV–VI: 1968, Chapters II–III: 1972, Chapters VII–

I.R. Shafarevich, *Basic Algebraic Geometry 2*, DOI 10.1007/978-3-642-38010-5,

VIII: 1975) and Masson, Paris (Chapter IX: 1982); English translation of Chapters 1–3: Lie groups and Lie algebras, Springer, Berlin (1989)
20. Cartan, H.: Théorie élémentaire des fonctions analytiques d'une ou plusieurs variables complexes. Hermann, Paris (1961). English translation: Elementary theory of analytic functions of one or several complex variables, Hermann, Paris (1963), and Addison Wesley, Reading, Palo Alto, London (1963); MR **26** #5138
21. Cartier, P.: Équivalence linéaire des ideaux de polynomes. In: Séminaire Bourbaki 1964–1965, Éxposé 283. Benjamin, New York (1966)
22. Chern, S.S.: Complex Manifolds Without Potential Theory. Van Nostrand, Princeton (1967); MR **37** #940
23. Clemens, C.H., Griffiths, P.A.: The intermediate Jacobian of the cubic threefold. Ann. Math. (2) **95**, 281–356 (1972); MR **46** #1796
24. de la Harpe, P., Siegfried, P.: Singularités de Klein, Enseign. Math. (2) **25**, 207–256 (1979); MR **82**e:32010
25. de Rham, G.: Variétés différentiables. Formes, courants, formes harmoniques. Hermann, Paris (1965). English translation: Differentiable Manifolds, Springer, Berlin (1984); MR **16**–957
26. Esnault, H.: Classification des variétés de dimension 3 et plus. In: Séminaire Bourbaki 1980–1981, Éxposé 586. Lecture Notes in Math., vol. 901 (1981)
27. Fleming, W.: Functions of Several Variables. Springer, Berlin (1965)
28. Forster, O.: Riemannsche Flächen. Springer, Berlin (1977). English translation: Lectures on Riemann Surfaces, Springer (1981); MR **56** #5867
29. Fulton, W.: Intersection Theory. Springer, Berlin (1983)
30. Fulton, W.: Algebraic Curves. Benjamin, New York (1969)
31. Gizatullin, M.H.: Defining relations for the Cremona group of the plane. Izv. Akad. Nauk SSSR, Ser. Mat. **46**, 909–970 (1982). English translation: Math. USSR, Izv. **21**, 211–268 (1983)
32. Goursat, É.: Cours d'Analyse Mathématique, 3 vols. Gauthier-Villar, Paris (1902). English translation: A Course in Mathematical Analysis, 3 vols. Dover, New York (1959–1964); MR **21** #4889
33. Griffiths, P.A., Harris, J.: Principles of Algebraic Geometry. Wiley, New York (1978)
34. Grothendieck, A.: Cohomologie locale des faisceaux cohérents et théorèmes de Lefschetz locaux et globaux (SGA 2). North-Holland, Amsterdam (1968)
35. Grothendieck, A.: Technique de descente et théorèmes d'existence en géométrie algébrique, IV, Séminaire Bourbaki t. 13 Éxposé 221, May 1961. V, Séminaire Bourbaki t. 14 Éxposé 232, Feb 1962. V, Séminaire Bourbaki t. 14 Éxposé 236, May 1962. Reprinted in Fondements de la géométrie algébrique (extraits du Séminaire Bourbaki 1957–1962), Secrétariat mathématique, Paris (1962); MR **26** #3566.
36. Gunning, G., Rossi, H.: Analytic Functions of Several Complex Variables. Prentice Hall International, Englewood Cliffs (1965); MR **31** #4927
37. Hartshorne, R.: Algebraic Geometry. Springer, Berlin (1977)
38. Hilbert, D.: Mathematical Problems (Lecture delivered before the International Congress of Mathematicians at Paris in 1900), Göttinger Nachrichten, pp. 253–297 (1900); English translation reprinted in Proc. of Symposia in Pure Math., vol. 28, pp. 1–34. AMS, Providence (1976)
39. Hironaka, H.: On the equivalence of singularities. I. In: Schilling, O.F.G. (ed.) Arithmetic Algebraic Geometry, Proc. Conf., Purdue Univ., 1963, pp. 153–200. Harper and Rowe, New York (1965); MR **34** #1317
40. Humphreys, J.E.: Linear Algebraic Groups. Springer, Berlin (1975)
41. Husemoller, D.: Fibre Bundles, McGraw-Hill, New York (1966); 2nd edn., Springer, Berlin (1975)
42. Iskovskikh, V.A.: A simple proof of a theorem of a theorem of Gizatullin. Tr. Mat. Inst. Steklova **183**, 111–116 (1990). Translated in Proc. Steklov Inst. Math. Issue **4**, 127–133 (1991)

43. Iskovskikh, V.A., Manin, Yu.A.: Three-dimensional quartics and counterexamples to the Lüroth problem. Math. USSR Sb. **86**(128), 140–166 (1971). English translation: Math. USSR Sb. **15**, 141–166 (1971); MR **45** #266
44. Kähler, E.: Über die Verzweigung einer algebraischen Funktion zweier Veränderlichen in der Umgebung einer singuläre Stelle. Math. Z. **30**, 188–204 (1929)
45. Kawamata, Y.: Minimal models and the Kodaira dimension of algebraic fibre spaces. J. Reine Angew. Math. **363**, 1–46 (1985); MR **87**a:14013
46. Kawamata, Y., Matsuda, K., Matsuki, K.: Introduction to the minimal model problem. In: Oda, T. (ed.) Proc. Sympos. Algebraic Geometry, Sendai, 1985. Adv. Stud. Pure Math., vol. 10, pp. 283–360. Kinokuniya, Tokyo (1987)
47. Kleiman, S., Laksov, D.: Schubert calculus. Am. Math. Mon. **79**, 1061–1082 (1972); MR **48** #2152
48. Knutson, D.: Algebraic spaces, Lect. Notes Math. **203** (1971); MR **46** #1791 (1971)
49. Koblitz, N.: p-Adic Numbers, p-Adic Analysis and Zeta Functions. Springer, Berlin (1977); MR **57** #5964
50. Kollár, J.: Shafarevich maps and the plurigenera of algebraic varieties. Invent. Math. **113**, 177–215 (1993)
51. Kollár, J.: Shafarevich Maps and Automorphic, M.B. Porter Lectures. Princeton University Press, Princeton (1995); MR1341589
52. Kostrikin, A.I., Manin, Yu.I.: Linear Algebra and Geometry. Moscow University Publications, Moscow (1980). English translation: Gordon and Breach, New York (1989)
53. Kurosh, A.G.: The Theory of Groups. Gos. Izdat. Teor.-Tekh. Lit., Moscow (1944). English translation: Vols. I, II, Chelsea, New York (1955, 1956). Zbl. 64, 251.
54. Lang, S.: Algebra, 2nd edn. Addison-Wesley, Menlo Park (1984)
55. Lang, S.: Introduction to Algebraic Geometry. Wiley-Interscience, New York (1958)
56. Lang, S.: Introduction to the Theory of Differentiable Manifolds. Wiley-Interscience, New York (1962); MR **27** #5192
57. Matsumura, H.: Commutative Ring Theory. Cambridge University Press, Cambridge (1986)
58. Milnor, J.: Morse Theory. Princeton University Press, Princeton (1963); MR **29** #634
59. Milnor, J.: Singular Points of Complex Hypersurfaces. Princeton University Press, Princeton (1968); MR **39** #969
60. Mumford, D.: Algebraic Geometry, I. Complex Projective Varieties. Springer, Berlin (1976)
61. Mumford, D.: Introduction to Algebraic Geometry, Harvard Notes 1976. Reissued as the Red Book of Varieties and Schemes, Lecture Notes in Math., vol. 1358 (1988)
62. Mumford, D.: Lectures on Curves on a Algebraic Surface. Princeton University Press, Princeton (1966); MR **35** #187
63. Mumford, D.: Picard groups of moduli problems. In: Arithmetical Algebraic Geometry, pp. 33–81. Harper and Rowe, New York (1965); MR **34** #1327
64. Mumford, D., Fogarty, J.: Geometric Invariant Theory, 2nd edn. Springer, Berlin (1982)
65. Pham, F.: Introduction à l'étude topologique des singularités de Landau. Mém. Sci. Math., Gauthier-Villar, Paris (1967); MR **37** #4837
66. Pontryagin, L.S.: Continuous Groups, Gos. Izdat. Teor.-Tekh. Lit, Moscow (1954). English translation: Topological Groups (Vol. 2 of Selected Works), Gordon and Breach, New York (1986)
67. Saltman, D.J.: Noether's problem over an algebraically closed field. Invent. Math. **77**, 71–84 (1984)
68. Seifert, G., Threlfall, V.: Lehrbuch der Topologie. Chelsea, New York (1934). English translation: Academic Press, New York (1980)
69. Shafarevich, I.R., et al.: Algebraic Surfaces. Proceedings of the Steklov Inst., vol. 75. Nauka, Moscow (1965). English translation: AMS, Providence (1967); MR **32** #7557
70. Shokurov, V.V.: Numerical geometry of algebraic varieties. In: Proc. Int. Congress Math., vol. 1, Berkeley, 1986, pp. 672–681. AMS, Providence (1988)
71. Siegel, C.L.: Automorphic Functions and Abelian Integrals. Wiley-Interscience, New York (1971)

72. Siegel, C.L.: Abelian Functions and Modular Functions of Several Variables. Wiley-Interscience, New York (1973)
73. Siu, Y.-T.: A general non-vanishing theorem and an analytic proof of the finite generation of the canonical ring. arXiv:math/0610740
74. Springer, G.: Introduction to Riemann Surfaces, 2nd edn. Chelsea, New York (1981)
75. Springer, T.: Invariant Theory. Springer, Berlin (1977)
76. van der Waerden, B.L.: Moderne Algebra, Bd. 1, 2, Springer, Berlin (1930, 1931); I: Jrb, 56, 138. II: Zbl. 2, 8. English translation: Algebra, Vols. I, II, Ungar, New York (1970)
77. Walker, R.J.: Algebraic Curves. Springer, Berlin (1978)
78. Wallace, A.: Differential Topology: First Steps. Benjamin, New York (1968)
79. Weil, A.: Introduction à l'étude des variétés kählériennes. Publ. Inst. Math. Univ. Nancago. Hermann, Paris (1958); MR **22** #1921
80. Wilson, P.M.H.: Towards a birational classification of algebraic varieties. Bull. Lond. Math. Soc. **19**, 1–48 (1987)
81. Zariski, O., Samuel, P.: Commutative Algebra, 2 vols. Springer, Berlin (1975)

References for the Historical Sketch

101. Abel, N.H.: Œuvres complètes, I, II. Christiania (1881); Reprinted Johnson reprint corp., New York (1965)
102. Artin, E.: Collected Papers. Addison-Wesley, Reading (1965)
103. Bernoulli, J.: Opera Omnia, 4 vols. Bosquet, Lausannae et Genevae (1742); Reprinted Georg Olms Verlagsbuchhandlung, Hildesheim (1968)
104. Bertini, E.: Ricerche sulle trasformazioni univoche involutorie nel piano. Ann. Mat. Pura Appl. (2) **8**, 244–286 (1877)
105. Brill, A., Noether, M.: Über die algebraischen Funktionen und ihre Anwendung in der Geometrie. Math. Ann. **7** (1873)
106. Cartan, H.: Variétés analytiques complexes et cohomologie. In: Coll. sur les fonctions de plusieurs variables, Bruxelles, 1953, pp. 41–55. George Thone, Liège and Masson Paris (1953); Collected Works, Vol. II, pp. 669–683; MR **16** 235
107. Castelnuovo, G.: Sulla razionalità delle involuzioni piane. Rend. R. Accad. Lincei (V) **2** (1893); Also Math. Ann. **44** (1894); Memorie Scelte No. 20
108. Castelnuovo, G.: Sulle superficie di genere zero. Mem. Soc. Ital. Sci. (III) **10** (1896); Memorie Scelte No. 21
109. Castelnuovo, G.: Alcune proprietà fondamentali dei sistemi lineari di curve tracciati sopra una superficie algebrica. Annali di Mat. (II) **25** (1897); Memorie Scelte No. 23
110. Castelnuovo, G.: Sugli integrali semplici appartenenti ad una superficie irregolare. Rend. R. Accad. Lincei (V) **14** (1905); Memorie Scelte No. 26
111. Clebsch, A.: Sur les surfaces algébriques. C. R. Acad. Sci. Paris **67**, 1238–1239 (1868)
112. Clebsch, A., Gordan, P.: Theorie der abelschen Funktionen. Teubner, Leipzig (1866)
113. Dedekind, R., Weber, H.: Theorie der algebraischen Funktionen einer Veränderlichen. J. Reine Angew. Math. **92**, 181–290 (1882); Dedekind's Werke, Vol. I, pp. 238–349
114. Enriques, F.: Sulla proprietà caratteristica delle superficie algebriche irregolari. Rend. Accad. Bologna **9** (1904)
115. Enriques, F.: Le Superficie algebriche. Zanichelli, Bologna (1949)
116. Euler, L.: Integral Calculus, Vol. I, Chapter VI
117. Euler, L.: Opera Omnia. Ser. I, Vol. XXI, pp. 91–118
118. Fourier, J.-B.-J.: Théorie analytique de la chaleur. Didot, Paris (1822); Second edition Gauthiers Villars, Paris (1888); English translation: The Analytic Theory of Heat, Dover, New York (1955)

119. Frobenius, F.G.: Über die Grundlagen der theorie der Jakobischen Functionen. J. Reine Angew. Math. **97**, 16–48, 188–223 (1884); Gesammelte Abhandlungen, Vol. II, **31 32**, pp. 172–240
120. Göpel: Theoriae transcendentium Abelianarum primi ordinis adumbrato levis. J. Reine Angew. Math. **35** (1847)
121. Grothendieck, A.: The cohomology theory of abstract algebraic varieties. In: Proc. Int. Congr. Math., Edinburgh, 1958, pp. 103–118. Cambridge University Press, Cambridge (1960)
122. Hasse, H.: Zur Theorie der abstrakten elliptischen Funktionenkörper I, II, III. J. Reine Angew. Math. **175**, 55–62, 69–88, 193–208 (1936); Mathematische Abhandlungen, Vol. II, **47 49**, pp. 223–266
123. Hasse, H.: Zahlentheorie. Akademie-Verlag, Berlin (1949); English translation, Number Theory, Springer, Heidelberg (1980)
124. Hopf, H.: Studies and essays presented to R. Courant. In: Zur Topologie der Komplexen Mannigfaltigkeiten, pp. 167–185. Interscience, New York (1948)
125. Jacobi, C.G.J.: Gesammelte Werke, 8 vols. Berlin (1881–1894); Reprint Chelsea, New York (1969)
126. Kähler, E.: Algebra und Differentialrechnung, Bericht über die Math. Tagung. Berlin, 1953. Deutscher Verlag der Wissenschaften, Berlin (1953); MR **21** #4155
127. Kähler, E.: Geometria arithmetica. Ann. Mat. Pura Appl. (4) **45** (1958); MR **21** #4155
128. Klein, F.: Gesammelte mathematische Abhandlungen, Vol. III. Springer, Berlin (1923)
129. Klein, F.:. Riemannsche Flächen, Autographed Lecture Notes, 2 vols. Berlin (1891–1892); Reprinted (1906)
130. Kronecker, L.: Grundzüge einer arithmetischen theorie der allgemeinen algebraischen Grössen. J. Reine Angew. Math. **92**, 1–122 (1882); Mathematische Werke, Vol. II, pp. 237–387
131. Lefschetz, S.: On certain numerical invariants of algebraic varieties with application to Abelian varieties. Trans. Am. Math. Soc., **22**, 327–482 (1921); Selected Papers, Chelsea, New York (1971), pp. 41–196
132. Lefschetz, S.: L'analysis situs et la géométrie algébrique. Gauthier-Villars, Paris (1924); Selected Papers, pp. 283–439
133. Lefschetz, S.: Géométrie sur les surfaces et les variétés algébriques. Mém. Sci. Math., vol. 40. Gauthiers-Villars, Paris (1929)
134. Legendre, A.M.: Traité des fonctions elliptiques et des intégrales eulériennes, 3 vols. Huzard Courcier, Paris (1825–1828)
135. Noether, M.: Zur Grundlegung der Theorie der algebraischen Raumcurven. J. Reine Angew. Math. **93**, 271–318 (1882)
136. Noether, M.: Zur Theorie des eindeutigen Entsprechens algebraischer Gebilde von beliebig vielen Dimensionen. Mat. Ann. **2** (1870) and **8** (1875)
137. Picard, Ch.-E.: Sur les intégrales de différentielles totales algébrique et sur une classe de surfaces algébriques. C. R. Acad. Sci. Paris **99**, 1147–1149 (1884); Œuvres, Vol. III
138. Picard, E., Simart, G.: Théorie des fonctions algébriques de deux variables inépendantes, 2 vols. Gauthier-Villars, Paris (1897 and 1906)
139. Plücker, J.: Theorie der algebraischen Curven gegründet auf eine neue Behandlungsweise der analytischen Geometrie. Bonn (1839)
140. Poincaré, H.: Œuvres, Vol. II, Paris (1916); Second printing Gauthiers-Villars, Paris (1952)
141. Poincaré, H.: Sur les propriétés du potentiel et sur les fonctions Abéliennes. Acta Math. **22**, 89–178 (1899); Œuvres, Tome IV, pp. 162–243
142. Riemann, B.: Gesammelte Werke. Reprint Dover, New York (1953)
143. Serre, J.-P.: Quelques problèmes globaux relatifs aux variétés de Stein. In: Coll. sur les fonctions de plusieurs variables, Bruxelles, 1953, pp. 57–68. George Thone, Liège and Masson, Paris (1953); **MR** 16 235
144. Serre, J.-P.: Faisceaux algébriques cohérents. Ann. Math. (2) **61**, 197–278 (1955); MR **16** 953

145. Severi, F.: La base minima pour la totalité des courbes tracées sur une surface algébrique. Ann. Sci. Éc. Norm. Super. (3), **25**, 449–468 (1908); Opere matematiche (Acc. Naz. dei Lincei, Roma 1971), Vol. 1, 462–477
146. Weil, A.: Foundations of Algebraic Geometry. Amer. Math. Soc., New York (1946); MR **9** 303
147. Weil, A.: Sur la théorie des formes différentielles attachées à une variété analytique complexe. Comment. Math. Helv. **20**, 110–116 (1947); MR **9** 65
148. Weyl, H.: Die Idee der Riemannschen Fläche. Teubner, Berlin (1913); 3rd edn. Teubner, Stuttgart (1955); MR **16** p. 1096; English translation: The Concept of a Riemann Surface, Addison-Wesley, Reading (1955)
149. Grattan-Guinness, I.: Joseph Fourier 1768–1830. MIT Press, Cambridge (1972)
150. Krazer, A.: Lehrbuch der Thetafunktionen. Teubner, Leipzig (1903)

Index[6]

[6]Italic page numbers such as *245* refer to Volume 1.

I.R. Shafarevich, *Basic Algebraic Geometry 2*, DOI 10.1007/978-3-642-38010-5,

S

Printed in the United States
By Bookmasters